MANAGEMENT OF HAZARDOUS MATERIALS AND WASTES: Treatment, Minimization and Environmental Impacts

The Pennsylvania Academy of Science Publications

Books and Journal

Editor: Shyamal K. Majumdar
Professor of Biology
Lafayette College
Easton, Pennsylvania 18042

1. *Energy, Environment, and the Economy,* 1981. ISBN: 0-9606670-0-8. Editor: Shyamal K. Majumdar.

2. *Pennsylvania Coal: Resources, Technology and Utilization,* 1983. ISBN: 0-9606670-1-6. Editors: Shyamal K. Majumdar and E. Willard Miller.

3. *Hazardous and Toxic Wastes: Technology, Management and Health Effects,* 1984. ISBN: 0-9606670-2-4. Editors: Shyamal K. Majumdar and E. Willard Miller.

4. *Solid and Liquid Wastes: Management, Methods and Socioeconomic Considerations,* 1984. ISBN: 0-9606670-3-2. Editors: Shyamal K. Majumdar and E. Willard Miller.

5. *Management of Radioactive Materials and Wastes: Issues and Progress,* 1985. ISBN: 0-9606670-4-0. Editors: Shyamal K. Majumdar and E. Willard Miller.

6. *Endangered and Threatened Species Programs in Pennsylvania and Other States: Causes, Issues and Management,* 1986. ISBN: 0-9606670-5-9. Editors: Shyamal K. Majumdar, Fred J. Brenner, and Ann F. Rhoads.

7. *Environmental Consequences of Energy Production: Problems and Prospects,* 1987. ISBN:0-9606670-6-7. Editors: Shyamal K. Majumdar, Fred J. Brenner and E. Willard Miller.

8. *Contaminant Problems and Management of Living Chesapeake Bay Resources,* 1987. ISBN: 0-9606670-7-5. Editors: Shyamal K. Majumdar, Lenwood W. Hall, Jr. and Herbert M. Austin.

9. *Ecology and Restoration of The Delaware River Basin,* 1988. ISBN: 0-9606670-8-3. Editors: Shyamal K. Majumdar, E. Willard Miller and Louis E. Sage.

10. *Management of Hazardous Materials and Wastes: Treatment, Minimization and Environmental Impacts,* 1989. ISBN 0-9606670-9-1. Editors: Shyamal K. Majumdar, E. Willard Miller and Robert F. Schmalz.

MANAGEMENT OF HAZARDOUS MATERIALS AND WASTES: Treatment, Minimization and Environmental Impacts

EDITED BY
SHYAMAL K. MAJUMDAR, Ph.D.
Professor of Biology
Lafayette College
Easton, Pennsylvania 18042

E. WILLARD MILLER, Ph.D.
Professor of Geography and Associate Dean
for Resident Instruction (Emeritus)
The Pennsylvania State University
University Park, Pennsylvania 16802

ROBERT F. SCHMALZ, Ph.D.
Professor of Geology
The Pennsylvania State University
University Park, Pennsylvania 16802

Founded on April 18, 1924

A Publication of
The Pennsylvania Academy of Science

Library of Congress Cataloging in Publication Data

Management of Hazardous Materials and Wastes: Treatment, Minimization and Environmental Impacts

Bibliography
Index
Appendix

Library of Congress Catalog Card No.: 89-60201

ISBN-0-9606670-9-1

Printed in the United States of America by

Typehouse of Easton
Phillipsburg, New Jersey 08865

PREFACE

Humans have always thought that their environment was so vast that pollution could make only the most gradual changes. The traditional environmental problems consisted of visible threats, often emanating from a single source, like a plant or toxic dump. The remedies were usually simple. Today the problems are far more complex. The sources of hazardous and toxic wastes are many and diverse and the damage may not only be immediate but also have long-term effects.

The problem of the disposal of the various types of wastes created by an industrial society has reached a critical stage. Most waste comes from industry, agriculture, sewage plants and incinerators. Much of the toxic waste is now sent to monitored dumps, but a vast amount is deposited on the landscape with few controls. This pollution brutally tests the resilience of the earth's natural systems. The modern industrial wastes not only devastate the natural environment but have the potential to affect the health and welfare of millions of human beings.

In recent decades, in addition to the traditional sources of hazardous and toxic wastes, the problem of the contamination of the environment from the use of radioactive materials has arisen. The disposal of the radioactive wastes has become a technical and social problem of tremendous magnitude. To illustrate, no permanent place has yet been found for the final disposal of spent radioactive fuel elements from nuclear power plants. The seriousness of this problem will become increasingly evident as the waste material continues to accumulate, and as additional waste is produced by decommissioning of superannuated nuclear power plants.

This volume on Management of Hazardous Materials and Wastes is divided into six parts. The first part considers the generation, treatment and technology of hazardous substances and covers such wide-ranging aspects as the types of hazardous and toxic wastes produced to the management and treatment of the wastes including landfill, incinerator and biotechnology processes. The second part considers waste minimization and reduction methods. Reduction of the bulk of hazardous and toxic materials is of fundamental importance in future plannings. Part three covers regulations, management and transportation. Because of the potential impact on public health, a vast number of regulations and agencies have been created to deal with the problem of transport and disposal of toxic and hazardous materials. The fourth part considers the environmental and health effects. Not only can the health of humans be affected, but the environment can be polluted and virtually destroyed for long periods. Environmental education is needed to inform the population of the potential dangers from these contaminants. Sociological impacts are presented in the six chapters that make up part five. It is now recognized that the public has a responsibility to maintain a healthy environment. Finally, part six presents case studies. These studies treat the problems created by accidents such as the nuclear explosion at the Chernobyl nuclear plant in the Soviet Union, the chemical accident in Bhopal, India, and the Ashland oil spill in Pittsburgh, Pennsylvania.

The Pennsylvania Academy of Science has published a number of volumes in recent years on different types of waste materials. In 1984 two volumes were published on *Hazardous and Toxic Wastes: Technology, Management and Health Effects* and *Solid and Liquid Wastes: Management, Methods and Socioeconomic Considerations,* and in 1985 the volume appeared on *Management of Radioactive Materials and Wastes: Issues and Progress.* The present volume is not a revision of a previous volume but is entirely new. The science and technology of management of hazardous and toxic materials is changing so rapidly that material is soon outdated. This volume presents the current state-of-the-art.

This volume will be of interest to a wide audience. Experts have prepared the chapters, each presenting a concise statement of the field at the present time. The chapters provide a wide perspective so that a specialist in one field can be informed of developments in an allied area. Because the management of hazardous and toxic wastes is a universal problem, this volume will also be of interest to individuals who want to be informed about some of the most critical environmental problems of the day.

We express our deep appreciation for the cooperation and dedication of the contributors, who recognize the importance of solving critical environmental problems. Many other individuals, in addition to the authors, made contributions and we are appreciative of those persons. Thanks are extended to Drs. Robert S. Chase, Head, Department of Biology, Lafayette College, George C. Shoffstall, Executive Secretary of the Academy, Kurt C. Schreiber, President of the Academy, and Leonard M. Rosenfeld, Recording Secretary of the Academy for their encouragement and help in the preparation of this book. Gratitude is extended to Lafayette College and The Pennsylvania State University for providing facilities to the editors for their work. The editors extend heartfelt thanks to their wives for their encouragement and help in the preparation of this book.

Shyamal K. Majumdar
E. Willard Miller
Robert F. Schmalz

Editors, March 1989

Management of Hazardous Materials and Wastes: Treatment, Minimization and Environmental Impacts.

Table of Contents

Part Three: Regulations, Management and Transportation

Part Four: Environmental and Health Effects

Part Five: Social Impacts

COMMONWEALTH OF PENNSYLVANIA
LIEUTENANT GOVERNOR'S OFFICE
HARRISBURG, PA 17120-0002
717-787-3300

Mark S. Singel
Lieutenant Governor

FOREWORD

Dear Friends:

As we work in partnership to increase economic development opportunities and create jobs in Pennsylvania, we also must ensure a safe and healthy environment for all.

We have made significant strides recently in reclaiming our environmental heritage with the enactment of landmark programs to recycle solid wastes, to clean up drinking water through the $2.5 billion PENNVEST environmental infrastructure program, and to clean up toxic waste sites across the Commonwealth.

Still, environmental concerns remain among our top priorities. Intrinsic to those concerns is the question of proper handling, transportation, and disposal of hazardous materials, as well as an effective, efficient response to accidents involving such materials.

Pennsylvania is one of the most populous, industrialized states in this nation. Hazardous materials are in continuous transit across our 1,500 mile interstate network, the 14,912 miles of rail service within the state and 45,000 miles of improved roads which service our communities.

The potential for accidents involving hazardous materials both in transit and in the workplace is a real and serious threat to virtually every community in this Commonwealth.

In partnership with businesses, local governments, and emergency services, it is our joint responsibility to be prepared to respond promptly and properly in those critical situations.

The Pennsylvania Emergency Response Commission was established in April of 1987 to address the requirements of the 1986 federal Superfund Amendments and Reauthorization Act (SARA). Through a comprehensive and cooperative effort involving the state departments of Environmental Resources, Labor and Industry and Health, as well as county and local governments, we have begun the task of developing an emergency response capacity which will significantly improve our ability to ensure the safety of all Pennsylvanians.

While the challenge of SARA, Title III addresses 406 of the most dangerous chemicals in the marketplace, much of what is learned and developed will be applicable across the board in preventing and responding to hazardous materials accidents.

Through a collection of articles representing the science, industry, and government communities, this book explores in depth the complex issues and initiatives involved in the management of hazardous materials and wastes.

The Pennsylvania Academy of Science is to be commended for its efforts in compiling this comprehensive volume. We welcome their continuing assistance in promoting greater awareness of these critical environmental concerns as we continue to address the serious question of hazardous materials and emergency preparedness in Pennsylvania.

Sincerely,

Mark S. Singel
Lieutenant Governor
Chairman, Pennsylvania Emergency
Management Council and
Emergency Response Commission

MESSAGE

Arthur A. Davis
Secretary of Environmental Resources
Commonwealth of Pennsylvania

In the all-too-short history of environmental protection, the 1980's began with the enactment of the federal Superfund law. Even recognizing the federal government's disappointing record of actual cleanups of hazardous waste sites, I believe this law commendably foreshadowed a focus at the national and state levels on environmental problems beyond pollution of the atmosphere and surface waters.

At the end of this decade, both state governments and the federal government are now challenged by a renewed national commitment to environmental protection. That commitment is driven by the public's realization that no natural resource — our air, our waters above and below ground, our coasts and wetlands, our forests and recreational lands, and our prime farmlands—is beyond the reach of environmental degradation, in spite of our most conscientious efforts to date. Pollution from hazardous substances seems especially heinous in the eyes of the public.

In Pennsylvania, the urgency of dealing with hazardous wastes led Governor Robert P. Casey and our General Assembly to create a pioneering state law to clean up sites not eligible for the federal Superfund. That law also provides us with the significant funding needed to match the state's share of a federal cleanup under Superfund. But beyond meeting these simple objectives, Pennsylvania's Hazardous Sites Cleanup Act of 1988 also faces the fact that we live in a society that is going to continue producing and disposing of hazardous materials, and that it is not possible to segregate our past from our future. As a result, Pennsylvania is committed to encouraging the establishment of environmentally sound treatment and disposal sites that, unlike commonly accepted practices of the past, acknowledge and minimize the long-term risks to the environment from wastes now known to be hazardous.

In one law, tools and funds to deal with the legacy of the past are linked to an enhanced activism to assure the safety of our future—a combination unique to Pennsylvania. The reason for this linkage is also simple. To create stable funding for an unprecedented number of cleanups, we required support from both the state's citizens and its businesses. Under the Hazardous Sites Cleanup Act, our businesses are bearing a special tax, increased manufacturing costs, and redoubled liability for past disposal practices. Our citizens are also contributing their tax dollars, and will ultimately experience greater costs of goods whose manufacture requires hazardous substances or results in hazardous wastes. We needed to show our citizens that we are able to clean up existing hazardous waste sites. We also needed to demonstrate our ability to prevent the creation of more such sites, and to address emergency situations where hazardous substances are released accidentally into the environment. At the same time, we needed to show our businesses that we are committed to assuring they will have adequate places in the future to treat and dispose of hazardous wastes that cannot be minimized or recycled. Pennsylvania's law does all of this, and, we hope, improves on the federal experience in encouraging business interests to become partners in voluntary cleanups.

With the passage of the 1988 Hazardous Sites Cleanup Act, a new chapter has begun in Pennsylvania's own environmental history. The commitment of the Pennsylvania Academy of Science to developing information in the complex field of hazardous materials and waste management is long-standing, worthy of note in its own right, and a much-needed, timely complement to governmental action. Thus, I applaud and welcome the publication of another timely volume on *Management of Hazardous Materials and Wastes: Treatment, Minimization and Environmental Impacts* by The Pennsylvania Academy of Science.

Arthur A. Davis
Secretary
Pennsylvania Department of
Environmental Resources

INTRODUCTORY REMARKS

Heinz G. Pfeiffer, Manager
Technology and Energy Assessment
Pennsylvania Power and Light Company
Two North Ninth Street
Allentown, PA 18101

The problems with waste disposal have continually become more severe during the last decades while increasing attention has been focused on them. Population growth and increased individual consumption have generally been blamed for the increasing problem exacerbated by the chimerical search for a zero risk society. Often overlooked is the contribution of the change in industrial and institutional operations. A packinghouse once had the proud motto, "We use all of the pig but the squeal." The present low cost of energy as compared to human labor has encouraged operations that are material intensive and minimize labor input. The result has been throw-away devices and containers in hospitals, restaurants, and our present day activities rather than washing and collecting the containers for reuse. With our new technologies of robotics and information handling, we should be able to reduce the waste streams and minimize both material and labor input at the same time.

The national problem with existing hazardous waste sites, often called "Superfund Sites", has been much better defined since 1984 when the Pennsylvania Academy of Sciences issued their book, *Hazardous and Toxic Wastes.* During the same period, very little progress has been made in the disposal of the hazardous wastes that are being continuously generated as a direct consequence of our civilization. Much of the lack of progress has been due to confusion in the minds of the public and their elected officials as to sound solutions. A reassessment of the problem, an evaluation of experience with the mitigation efforts, and guidance for the development of future solutions is very timely. This volume answers this need very well.

Perhaps the most difficult decision involving hazardous waste is what constitutes a hazardous waste and at what level need it be controlled. At some level, many of the common things we eat and breathe are hazardous, whether of natural or man-made origin, and eliminating all of these risks is impossible. The questions that face society are how serious is the risk and what level can be tolerated. We tend to focus on large single-source problems while many of our day-to-day functions may place us at higher risk. Living in some of the urban environments probably has more effect on overall health and longevity than many of the substances that are being controlled very strictly.

This book strikes a good balance between solutions that are available to us now and innovative ideas that may serve us in the future. Never was the Elizabethan adage, "Be not the first on whom the new is tried nor yet the last to lay the old aside", more appropriate. The public is bombarded on the one hand by those that would minimize the problem and go on as we have in the past, and on the other hand, by those that oppose every suggested solution by promising illusionary alternatives. In order of their general desirability, the solutions are: 1) changes in process and product that minimize the waste production, 2) reuse in other processes, 3) destruction of the waste, and 4) disposal in a secure depository.

The Environmental Protection Agency has a major objective, the minimization of waste, and their booklet EPA/530-SW-87-026 expresses the benefits very well:

> "Waste minimization is one of the few areas where national environmental goals and industry's economic interests clearly coincide. For generators, the benefits include reduced costs, liabilities, and regulatory burdens associated with hazardous waste management. For the general public, waste minimization pays off in an improved environment."
>
> "Waste minimization clearly provides opportunities to deal more efficiently and effectively with wastes that are hazardous to human health and the environment. These opportunities are unique in that they provide immediate financial rewards to industry, increased waste management flexibility to generators, and reduced pressures on the nation's existing treatment and land disposal capacity. Now is the time to investigate and take practical steps toward waste minimization, before major commitments are made for treatment and disposal options. Over the longer term, the benefits of source reduction and recycling will be key incentives for generators to integrate waste minimization techniques into their overall hazardous waste management programs."

Process and product changes are sometimes resisted, but many businesses have been leaders in finding solutions once a problem has been identified and in overall improvements that reduce costs even before the savings in waste disposal. Two examples of whole industry revolutions that were successful are the total replacement of beryllium phosphors by calcium- based phosphors in fluorescent lights and the phasing out of freons in aerosol cans. In both cases, the manufacturers involved did the research and provided the solution when the problem was defined.

The destruction of hazardous materials has been an area of major controversy since the publication of *Hazardous and Toxic Wastes.* Incineration, done properly, has been demonstrated to be a successful way to destroy wastes. Public experience with actual incinerators in the past has often been bad, and the fears of the problems that those old incinerators had has been a major factor in the public reluctance to accept the new technology. The question of "how clean is clean" must be addressed if destruction methods are to be accepted; seldom is the public given the data that compares the ultimate emission level from an incinerator with that from the alternative landfill.

One option of waste destruction that is often overlooked is the conversion of the waste into less toxic forms by chemical reactions and stabilization. In this form, it can often be used for constructive purposes or used as construction fill.

Secure landfill is the final option and is a sequel to the other methods; reduction of the waste stream or incineration still leave a residual material that must be disposed of. One of the objections to incineration is that toxic materials are concentrated, but the same argument applies to landfill. The leaching and ultimate rupture of liners is inevitable, and most people are not aware of the many emissions from landfills. Also, biodegradable has become a popular phrase without considering the volatile compounds that the biodegradable materials often give off. All of the methods of waste disposal have their role but none is perfect. The minimization of the amount of waste material that has to be left for future generations should be our prime objective.

ACKNOWLEDGMENTS

The publication of this book was aided by contributions from The Pennsylvania Power and Light Company, Allentown, Pennsylvania; Pennsylvania Emergency Management Agency; and Pennsylvania Department of Environmental Resources.

Management of Hazardous Materials and Wastes: Treatment, Minimization and Environmental Impacts. Edited by S. K. Majumdar, E. W. Miller and R. F. Schmalz.

Chapter One

HAZARDOUS AND TOXIC SUBSTANCES: AN OVERVIEW

ROBERT F. SCHMALZ,[1] E. WILLARD MILLER,[2] and S.K. MAJUMDAR[3]

[1]Professor of Geology
The Pennsylvania State University
University Park, PA 16802

[2]Professor of Geography and
Associate Dean for Instruction, Emeritus
The Pennsylvania State University
University Park, PA 16802

[3]Professor of Biology
Lafayette College
Easton, PA 18042

INTRODUCTION

The United States generates approximately 4.5 billion tons of solid wastes annually, or roughly 100 pounds per person each day (Table 1). Hazardous industrial wastes and radioactive wastes together make up approximately one percent of the total. Nationally, nearly 1,500 times as much hazardous wastes are produced as compared to the low level radioactive wastes. In Pennsylvania, which is one of the largest generators of low level radioactive wastes in the country, the ratio is approximately 500: 1 (U.S. Department of Commerce, 1988; Eliassen, 1969).

HAZARDOUS AND TOXIC WASTES

Hazardous and toxic wastes are the "garbage" of a highly technological society. They are produced by many segments of our society. Of these, industry is by far the largest source, generating these wastes to manufacture fuel, paper, plastics, clothing, rubber, paints, insecticides, medicines and a host of other products that are required in our daily lives. Of the industrial producers of hazardous wastes, chemical and allied products generate 60 percent of the total, followed by machinery, primary metals, paper and allied products, fabricated metal products and stone clay and glass products (chapter 2)*. Unfortunately, it has become evident that hazardous wastes from these products may adversely affect human health and cause serious environmental damage when the wastes are not properly disposed of (chapters 20, 21 and 24).

In order to control hazardous wastes, Congress in 1976 passed the Resource Conservation and Recovery Act. This Act directed the Environmental Protection Agency (EPA) to develop regulations to protect human health and the environment from improper solid waste management practices (chapters 11, 12, 13 and 14). The EPA on May 19, 1980 published a list of approximately 363 chemical (organic and inorganic) substances that are considered hazardous. These substances cover a wide range of industrial sources. Of the 4.5 billion tons of solid wastes, the EPA estimated that 57 million metric tons can be identified as hazardous.

DEFINITION AND PROPERTIES

The Department of Environmental Resources has defined a hazardous waste as "a solid waste, or combination of solid wastes, which because of its quantity, concentration, or physical, chemical or infectious characteristics may (a) cause, or significantly contribute to an increase in mortality or an increase in serious irreversible, or incapacitating reversible illness; or (b) pose a substantial present or potential hazard to human health or the environment when improperly treated, stored, transported, or disposed of, or otherwise managed."

The EPA has also proposed that a hazardous waste be identified by testing

TABLE 1

Solid Waste Generated Annually in the United States

Source	Millions of tons	
Agriculture	2,340	52%
Mineral industries (mining & milling waste)	1,620	36%
Industrial (non-hazardous)	225	5%
Municipal (domestic)	180	4%
Utility	90	2%
Hazardous industrial	45	1%
Low-level Radioactive	3.15	0.0007%
TOTAL (Approximate):	4.5 billion tons 100 lb/capita/day	

(Sources: Eliassen, 1969; U.S. Department of Commerce, 1988)

*Appropriate chapters from this book are identified for detailed information.

it to determine if it possesses any one of four characteristics. If it does it will be subject to regulation under the Resource Conservation and Recovery Act. The four characteristics are (Regan, 1985):

> Ignitability, which identifies wastes that pose a fire hazard during routine management. Fires not only present immediate dangers of heat and smoke but also can spread harmful particles over wide areas.
>
> Corrosivity, which identifies wastes requiring special containers because of their ability to corrode standard materials, or requiring segregation from other wastes because of their ability to dissolve toxic contaminants. A waste is corrosive if it is liquid and has a pH less than or equal to 2 or if it is a liquid which corrodes steel at a specified rate.
>
> Reactivity (or explosiveness) which identifies wastes that, during routine management, tend to react spontaneously, to react vigorously with air or water, to be unstable to shock or heat, to generate toxic gases such as cyanide or sulfides, or to explode.
>
> Toxicity, which identifies wastes that, when improperly managed, may release toxicants in sufficient quantities to pose a substantial hazard to human health or the environment.

The Environmental Protection Agency uses some tests identical with those used by the Department of Transportation (chapters 18 and 19) in its program to control transport of hazardous material. The EPA may add to or delete hazardous wastes or characteristics identifying hazardous wastes. Any changes are published in the Federal Register, and the public provided an opportunity to comment on the proposed changes.

TYPES OF ENVIRONMENTAL POLLUTION

Each year wastes discarded into the environment range from common household trash to complex materials in industrial wastes, sewage sludge, agricultural residues, mining refuse and pathological wastes from institutions such as hospitals and laboratories (chapters 2 and 15). Each of these has the potential for environmental degradation. The Environmental Protection Agency has on file hundreds of documented cases of damage to health and the environment from the improper management of hazardous wastes.

The vast majority of cases involve the pollution of groundwater due to improperly sited or operated landfills and surface impoundments. There are many types of groundwater pollution. To illustrate, within the Denver, Colorado environment, a 30 square mile area was contaminated from the disposal of pesticide wastes from unlined disposal ponds. The pesticide wastes that were sprayed on the land between 1943 and 1957 were not detected until the 1980's. Decontamination will take several years and cost $80 million. In another example, the groundwater supplies of Toone and Teague, Tennessee were contaminated when chemicals leaked from a nearby landfill. When the landfill closed in 1972 there were over 300 drums leaking pesticide wastes. It was considered too costly to remove the chemical and the town had to secure water from more distant sources.

Hazardous wastes can contaminate rivers, lakes, and other surface water. Many examples of this can be given. A recent example is the Ashland oil spill (chapter 34). Another example is found in an area about 25 miles south of Louisville, Kentucky. In a seven-acre site, which has come to be known as The Valley of the Drums, about 17,000 drums of toxic chemicals were stored on the surface as well as in many subsurface sites. In 1979 when the EPA analyzed the soils of the drainage area they identified some 200 organic chemicals and 30 metals. The entire water supply of the region was polluted and the area suffered long-range damage.

Hazardous wastes are also a possible pollutant of the atmosphere. As an example, in 1972 waste containing hexachlorobenzene (HCB) was disposed of in a landfill. The HCB is an organic compound that contains chlorine. As the HCB decomposed, chlorine was released into the atmosphere. As it accumulated on the surface, the cattle in an area of 100 square miles had to be destroyed with direct economic loss of $380,000.

A major danger from a hazardous waste site is its potential to ignite and/or explode. This occurred in 1978 in Chester, Pennsylvania where a single site had between 30,000 and 50,000 drums of industrial waste. The fire forced the closing of the Commodore Barry Bridge over the Delaware River and 45 firemen required medical treatment for lung and skin irritation from chemical fumes. The disregard for the danger of hazardous waste was demonstrated at this site. Drums of waste were stored within 20 feet of a natural gas storage tank and liquified natural gas tanks were about 100 yards away.

The effects of hazardous waste on health has often been shown. For example, a serious illness occurred to a family in Albuquerque, New Mexico after eating a pig that had been fed corn treated with a mercury compound. In another case, sickness occurred to a family in Indiana who drank milk from the family's cow that had been grazing in a pasture fertilized with the city of Bloomington's sewage sludge. The sludge had a concentration of polychlorinated biphenyls (PCBs). The Bhopal incident is an example of the effects of an industrial toxic chemical plant accident on human health and environment (chapter 33).

Hazardous wastes can also cause disease by direct contact. The classic example of this situation is the Love Canal of Niagara Falls. As drums of waste corroded, the chemicals percolated through the soil into yards and basements forcing the evacuation of over 200 families in 1978 and 1979. About 80 chemicals, some suspected to be carcinogens, were identified.

More than 8,000 chemicals are currently in use, and many of these are toxic and genetically damaging and may affect humans as a teratogens (cause birth defects), carcinogens (cause cancer), mutagens (cause DNA damage) and clastogens (induce chromosome damage). The Environmental Protection Agency recommends various evaluation methods for testing mutagenic and carcinogenic chemicals (chapter 21 and 24).

The enforcement of the safe and healthful working conditions and standards is the responsibility of the Occupational Safety and Health Administration (OSHA), and these are discussed in chapters 22 and 23.

MAGNITUDE OF THE HAZARDOUS WASTE PROBLEM

About 255 to 275 million metric tons of hazardous wastes generated annually are now under federal and state regulations. Some states have stricter definitions for hazardous waste than the federal program, which regulate about 40 million tons annually. Hundreds of millions of tons of federally exempted hazardous waste are disposed of in sanitary landfills that pose substantial risks in the future.

Land disposal is used for as much as 80 percent of regulated hazardous waste, some of which may remain hazardous for centuries. Inappropriate disposal of hazardous waste on land creates the risk of contaminating the environment, particularly groundwater, which may create health problems (chapters 20, 21, and 24).

Uncontrolled and careless disposal of industrial wastes became a concern in the mid-1970's. Before Congress passed the Resource Conservation and Recovery Act, (RCRA) few states had regulatory programs dealing with hazardous wastes. In implementing the RCRA, the Environmental Protection Agency discovered that the disposal of hazardous waste decades earlier had left undetermined but possibly very large amounts of dangerous substances in and on the land. In addition, wastes were leaking from many hazardous waste sites.

In an EPA survey of 350 uncontrolled hazardous waste sites in 1980 there was substantial evidence that a health problem existed. At that time the 350 sites were the only ones with detailed information. Since then, 15,000 additional uncontrolled sites have been investigated by the EPA. The EPA's Emergency and Remedial Response Information System found that from these sites at least 168 communities had contamination of groundwater and reservoirs; at least 468 wells had contaminated drinking water and had to be abandoned, and there were 108 other adverse incidents including damage to human health, natural habitat, fish and livestock, crops, sewer systems and soils.

In an unreleased EPA study it was reported that there were 80,263 sites in the nation with contaminated surface impoundments such as pits, ponds and lagoons. Ninety percent of these pose a potential threat of groundwater contamination. In another study of 954 sites contamination of drinking water was found in 29 percent. Leaking of toxic substances from waste landfills is thought to be the major contributing factor in these cases.

The magnitude of the hazardous and toxic waste problem is only becoming apparent in the 1980's. It is now recognized that the national data base is incomplete, and in some respects even inaccurate. Accurate data on the use of deficient waste management methods are not available. Use of land disposal methods varies greatly from state to state. For example, in Texas, 95 percent of hazardous waste is land disposed but in Massachusetts only 7 percent is land disposed, and all of that is sent to other states for disposal. Within the United States, 418 sites have been selected for the National Priority List from more than 80,000 sites. Moreover, the inventory of hazardous sites in the nation is still incomplete, and the severity of the problem is still unknown. It will require many years to complete the data base alone, and no estimates can reasonably be made for the removal of the potential number of hazardous sites.

While the past presents a formidable problem, policies must be developed to control present-day problems. An important way to reduce threats to public health and the environment from hazardous wastes and to lessen the cost of waste management is to reduce the amount of waste generated. Generators of waste can accomplish this by segregating wastes more carefully, recycling them, or sometimes changing the manufacturing processes. Whether this is done depends on the economic costs and savings involved. The federal program at present does not provide direct incentives for any of these approaches. Emphasis has not been on reducing waste generation but rather on its management. The program in the future must consider not only management but also the means to reduce the amount of hazardous materials produced (chapters 8, 9 and 10).

TECHNOLOGICAL MANAGEMENT STRATEGIES

Once hazardous wastes are generated, they can be managed by one of two broad categories of technologies (Table 2):

1. Disposal through containment or disposal on land or in the ocean
2. Treatment by one or more steps to reduce the hazard level of the waste

Containment technologies hold waste in a manner intended to inhibit release of hazardous components into the environment or keep releases at acceptable levels. These technologies include landfills, surface impoundments and underground injection wells. With most containment options, it is likely that releases will occur in the future. In fact, some landfills are designed to transfer material to the ground on a regulated basis. Disposal techniques, such as land treatment (spreading waste on the surface) or ocean dumping, rely on naturally occurring processes to reduce the hazardous level (chapter 5).

TABLE 2

Comparison of Some Hazard Reduction Technologies

	Disposal		Treatment		
	Landfills and Impoundment	Injection Wells	Incineration and other Thermal Destruction	High Temp Decomposition	Chemical Stabilization
Effectiveness: How well it contains or destroys hazardous characteristics	Low for volatiles, questionable for liquids	High, based on theory, but limited field data available.	High based on field tests.	Very high, commercial tests.	High for many metals. Based on Lab tests.
Reliability issues	Siting, construction and operation.	Site history, geology, well depth, construction & operation.	Monitoring uncertainties with respect to destruction and removal efficiency. Incinerability.	Limited experience, onsite treatment, operational simple.	Some inorganics still soluble, surrogate for weathering.
Environmental media most affected	Surface & Groundwater	Surface & Groundwater	Air	Air	Groundwater
Least compatible wastes	Liner reactive, highly toxic, mobile persistent and bio-accumulative.	Reactive, corrosive, highly toxic, mobile and persistent.	Highly toxic & refractory organics high, heavy metal concentration.	Some inorganics	Organics
Costs: low, medium, high	Low-medium	Low	Medium-high	Medium-high	Medium
Resource recovery potential	None	None	Energy and some acids	Energy and some metals	Possible building materials

Source: *Technologies and Management Strategies for Hazardous Waste Control,* 1983, Washington, D.C. Office of Technology Assessment, Congress of the United States, p. 21.

Treatment technologies reduce the hazard level directly or facilitate reduction in other ways by changing the physical or chemical nature of the waste, by separating waste constituents, by reducing the waste volumes, by incineration, by reducing the concentration of hazardous substances in the waste or by innovative biotechnologies (chapters 3, 4, 6 and the Appendix). The treatment technologies include chemical, thermal and biological processes. Management strategies for hazardous materials and waste are important and discussed in chapters 7, 15, 16 and 17.

As the waste materials have grown, there has arisen the needs for regulations and laws to protect, not only the environment, but people as well, because the disposal of waste products is usually unsightly and smelly, and public resistance is developing to having these degrading activities in many communities. While many of the problems are pragmatic, others are emotional in community resistance to establishing new sanitary landfill sites. These social issues are covered in several chapters in part 5. The community right to know law—SARA Title III, and trends in environmental education in Pennsylvania are discussed in chapters 12 and 25, respectively.

RADIOACTIVE WASTE

Although exact figures are not available (in part for reasons of national security), the total quantity of high level radioactive waste generated annually in the United States must be several million tons of which military heavy metal waste (uranium and thorium processing waste and trans-uranic elements) makes up approximately 1.1 million tons, and spent fuel generated by commercial (civilian) power reactors represents slightly less than 1,550 tons. The remainder is primarily made up of uranium and thorium mine and mill tailings and processing waste. The federal government has assumed responsibility for disposing of all high level radioactive waste, both military and commercial, and is currently carrying out final design studies for a mined geological waste repository at Yucca Mountain in the Nevada Test Site, near Beatty, Nevada. Until that facility is ready to receive waste, probably no earlier than the mid- to late-1990's, all civilian high level radioactive waste will be held in temporary storage. Aspects of radioactive waste management are detailed in Chapters 7 and 16.

Responsibility for disposing of commercially generated low level radioactive waste, however, was delegated to the states in 1980 by the Low Level Waste Policy Act (P.L. 96-573). Despite the small quantity of waste involved and the relatively small hazard it poses compared to many of the highly toxic or carcinogenic chemicals in hazardous industrial waste, the management of low level radioactive waste has become the focus of an extended and sometimes bitter public debate. Widespread public opposition to the establishment of low level waste disposal facilities stems, in part, from a fear of the unknown, aggravated by the acknowledged role of nuclear radiation as a cause of cancer and genetic abnormalities and compounded by the fearful power of nuclear weapons and the accidents at Three Mile Island and Chernobyl (chapter 32). These factors, together with a real concern for the potential environmental impact of radioactive waste disposal give rise to the "Not In My Backyard" (NIMBY) phenomenon and suggest that effective management of low level waste will prove to be more of a social-political problem than a serious technical challenge.

The rapid growth of the nuclear-based industry in the United States after 1945 occurred despite our national reluctance to confront and resolve the problem of disposing of the resulting radioactive waste. The magnitude of the problem was not fully appreciated at first, in part because the early epidemiological studies at Hiroshima and Nagasaki tended to underestimate the health hazard of radiation exposure (RERF, 1987). Initially, the federal government assumed responsibility for management of all radioactive waste, but with the growing application of nuclear technology in medicine, research and industry, and especially with the development of a significant private nuclear power industry, this proved no longer practical. The federal response was to encourage the development of commercial low level waste disposal facilities at several locations during the early 1960's and to restrict the use of government disposal facilities to the military and a limited number of government contractors. By the late 1970's, containment failures had been reported at commercial disposal sites near Sheffield, Illinois, West Valley, New York, and Maxey Flats, Kentucky (Cartwright, 1982; Grant, 1982; Kelleher, *et al.* 1979). These problems, together with political pressure brought by the three states in which active radioactive waste disposal facilities were located (Nevada, South Carolina, and Washington), forced Congress to review radioactive waste disposal policies and to address the increasingly urgent problem of disposing of spent fuel from commercial power reactors (Friedman, 1985; Dornsife, 1985). After extensive deliberation, Congress enacted the Low-Level Waste Policy Act (1980), the Nuclear Waste Policy Act of 1982 (P.L. 97-425) and the Low Level Radioactive Waste Policy Amendment Act of 1985 (P.L. 99-240). Under the terms of these Acts, commercial radioactive waste was divided into two categories: (1) high level waste comprising uranium and thorium mill tailings, spent reactor fuel and alpha-emitting trans-uranic elements (TRU's), and (2) low level radioactive waste comprising of all radioactive waste not specifically included above. The federal government assumed responsibility for all high level nuclear waste, but delegated to each state the responsibility for disposing of low level radioactive waste generated within its borders. Under the terms of the Low Level Waste Policy Act (P.L. 96-573) the states may develop their own disposal facilities or neighboring states may join together in "compacts" to develop disposal facilities for the exclusive use of the member states. By encouraging the development of compacts, it was apparently the intention of Congress to avoid proliferation of disposal sites, but if this was the case, the plan was not wholly successful.

Many states were eager to form compacts with the three states which already had operational disposal sites, while states which generate large volumes of waste (New York, Texas, California, Pennsylvania, Massachusetts) were unwelcome everywhere. One state (Texas) elected to develop its own site, although plans to reserve use of the facility for Texas alone may be challenged as a restraint of interstate commerce in violation of Article I, Section 10 of the Constitution. Other large generators chose to form compacts with nearby states whose waste production was very small. This in effect gives the host state (usually the larger generator) control of a facility for its (almost) exclusive use and precludes legal challenge since the Low Level Waste Policy Act specifically allows Congress to exempt such compacts from the Constitutional prohibition. Consequently, today there are more than a dozen low level radioactive waste disposal facilities planned in the United States, whereas just three disposal facilities were adequate to meet our needs ten years ago.

CHARACTERISTICS OF LOW LEVEL RADIOACTIVE WASTE

There are four major commercial sources of low level radioactive waste: nuclear power plants, industry, medical diagnosis and treatment, and research laboratories.

By far the greatest volume of waste is produced by nuclear power plants. This material consists of contaminated filtering media and spent ion-exchange resins (used to purify the cooling water in the reactor), evaporator "bottoms", housekeeping trash (protective clothing, clean-up materials, etc.) and contaminated tools or irradiated reactor components. The average specific activity of reactor waste is commonly as low as 0.04 Curie/cubic foot and except for irradiated components, rarely exceed 0.2 Ci/cubic foot. By far the most abundant isotopes (in Curies) are cobalt-60, strontium-90, zinc-65 and cesium-134 and -137, with minor amounts of tritium, arsenic-76, chromium-51, cobalt-58, manganese-54 and silver-110m. The longest-lived of the major isotopes is cesium-137, with a half-life of 30 years (Choppin & Rydberg, 1980).

Industrial generators represent a very diverse group, and the wastes they ship for disposal are no less varied. Radio-isotopes are used in the production of radio-pharmaceuticals, for non-destructive testing and research, and in the manufacture of consumer products ranging from plastic food wrap to furniture and floor coverings. The bulk of the industrial waste is housekeeping trash and stabilized analytical liquids (scintillation fluids and laboratory waste). Contaminated tools and hardware, resins, and incinerator ash make up most of the remainder. The specific activity of industrial waste is as high or higher than reactor waste (Pollog, 1983). The principal radionuclides found in the industrial waste stream are tritium, cobalt-60, cesium-137, iridium-192 and nickel-63, together with up to forty short-lived nuclides in quantities of 500 millicuries or less. Sealed radiation sources used in radiographic examination of welded joints and heavy castings are normally returned to the supplier to be recharged, and do not enter the industrial waste stream (Singh, 1987).

Medical waste, including waste generated by medical research, diagnosis, and treatment, represents a substantial volume, but is generally of very low specific activity (0.001 Ci/cubic foot). It is made up of compacted trash, stabilized liquids, laboratory waste and minor quantities of exchange resins. The principal isotopes are tritium, carbon-14, iodine-125 and technetium-99m (Singh, 1987). More than forty additional isotopes are used in very small quantities in specialized research, diagnosis and treatment. Because the most widely used isotope (technetium-99m) has a half-life of only 6 hours, much of the hospital waste can be stored on site for 12-24 months and then discarded with ordinary trash (Vince, 1985).

The most diverse waste stream is that generated by government and academic research laboratories. The waste volume is extremely small, less than 4% of the total volume shipped for disposal, and its specific activity is very low. At least thirty radio-isotopes are widely utilized in research programs, only two of which, tritium and carbon-14, are employed in significant quantity.

LOW-LEVEL RADIOACTIVE WASTE MANAGEMENT

The volume of low-level radioactive waste can be substantially reduced by compaction, incineration or evaporation. Such volume reduction procedures must play a critical role in LLRW management (as they must in the management of all forms of solid and liquid waste). Because the cost of transporting and disposing of LLRW is both high and volume related ($50 or more per cubic foot), there is a strong economic incentive for generators to develop and implement volume reduction measures, and formal legal requirements are probably unnecessary.

Safe means of handling, storing and transporting low-level radioactive waste have been developed in the United States and elsewhere, and the transport of such wastes is strictly regulated and monitored.

Finally, four basic types of waste disposal facility have been developed for the management of low-level radioactive waste. After nearly thirty years practical experience we can be quite confident that any one of these facility designs, enhanced as necessary to raise public confidence and meet local environmental and legal requirements, will provide secure confinement of radioactive materials and afford the essential isolation of the waste from the biosphere (chapter 16).

SELECTED REFERENCES

Cartwright, K. 1982. A geological case history: Lessons learned at Sheffield,

Illinois. in: Yalcintas, 1982, volume 2, pp. 67-82.

Collins, John P. and Walter Saukin, eds. 1981. *The Hazardous Waste Dilemma, Issues and Solutions,* New York: American Society of Civil Engineers, 329 pp.

Cope, C.B., W.H. Fuller and S.L. Willetts. 1983. *The Scientific Management of Hazardous Wastes,* New York: Cambridge University Press, 488 pp.

Choppin, G.R. and J. Rydberg. *Nuclear Chemistry: Theory and Applications.* Pergamon Press, New York, 1980. 574 pp. + 14 appendices, index.

Dornsife, W.P. 1985. Classification of Radioactive Materials and Wastes. Chapter 1 in: Majumdar, S.K. and Miller, E. Willard, *Management of Radioactive Materials and Wastes: Issues and Progress,* The Pennsylvania Academy of Science, Easton, PA, 1985, pp. 2-9.

Eliassen, R. 1969. *Solid Waste Management.* Office of Science and Technology, Executive Office of the President, Washington, D.C.

Friedman, R.S. 1985. Political Considerations of Nuclear Waste Policy. Chapter 15 in: Majumdar, S.K. and Miller, E. Willard, *Management of Radioactive Materials and Wastes: Issues and Progress.* The Pennsylvania Academy of Science, Easton, PA 1985, pp. 203-215.

Grant, J.L. 1982. Geotechnical measurements at the Maxey Flats, Kentucky, low-level radioactive waste disposal site—lessons learned. in: Yalcintas, 1982, volume 2, pp. 151-166.

Hazardous Wastes: DER Answers Your Questions. 1981. Harrisburg, PA: Department of Environmental Resources.

Isman, Warren E. and Gene P. Carlson. 1980. *Hazardous Material,* Encino, CA: Glencoe, 445 pp.

Kelleher, M, J. Lyons, K. Tong, W. Greenman and T. Waldman. 1983. Report: Investigation of a Low Level Waste Burial Site at the Western New York Nuclear Services Center. Power Authority of the State of New York, Albany, New York.

Low Level Waste Policy Act, P.L. 96-573, 94 Stat. 3347, December 1980.

Low Level Radioactive Waste Policy Amendments Act of 1985, P.L. 99-240, 99 Stat. 1842, January, 1985.

Majumdar, S.K. and E.W. Miller, Eds. 1985. *Management of Radioactive Materials and Wastes: Issues and Progress.* Pennsylvania Academy of Science, Easton, PA 1885, 406 pp.

Nuclear Waste Policy Act of 1982, P.L. 97-425, 96 Stat. 2201, December, 1982.

Pierce, J. Jeffrey and P. Arne Vesilind, eds. 1981. *Hazardous Waste Management.* Ann Arbor, MI: Ann Arbor Science, 190 pp.

Pollog, T.E. 1984. *Pennsylvania Low-Level Radioactive Waste Management Survey.* Pennsylvania Department of Environmental Resources, Harrisburg, PA, 1984. 13 pp. + 2 appendices.

Regan, Raymond W. 1984. Classification and Properties of Hazardous and Toxic Wastes, in: S.K. Majumdar and E.W. Miller, *Hazardous and Toxic Wastes:*

Technology, Management and Health Effects. The Pennsylvania Academy of Science, Easton, PA, 1984, pp. 2-11.

RERF. 1987. *U.S.-Japan Joint Reassessment of Atomic Bomb Radiation Dossimetry in Hiroshima and Nagasaki, Final Report.* Radiation Effects Research Foundation, Hiroshima, Japan, 1987. Volume 1, 430 pp.

Singh, K.N. 1987. *Pennsylvania and Maryland Low-Level Radioactive Waste Management Survey—1986.* Commonwealth of Pennsylvania, Department of Environmental Resources, Harrisburg, PA 1987. 17 pp. + 3 appendices.

Technologies and Management Strategies for Hazardous Waste Control. 1983. Washington, D.C.: Office of Technology Assessment, Congress of the United States, 55 pp.

U.S. Department of Commerce. 1988. *Statistical Abstract of the United States, 108th Edition,* U.S. Department of Commerce, Bureau of the Census, Washington, D.C. 1988.

Vince, M.A. 1985. Handling of Radioactive Materials and Waste in Hospitals. Chapter 12, in: Majumdar, S.K. and Miller, E. Willard, 1985. *Management of Radioactive Materials and Wastes: Issues and Progress.* The Pennsylvania Academy of Science, Easton, PA 1985, pp. 157-173.

Wilkes, Angella S. 1980. *Everybody's Problem, Hazardous Waste,* Washington, D.C.: Environmental Protection Agency.

Yalcintas, M.G. (Ed.), Proceedings of the Symposium on Low-Level Waste Disposal Volume 2. Arlington, Virginia, June, 1982. Oak Ridge National Laboratory, Oak Ridge, TN. (NUREG/CP-0028 CONF-820674).

Management of Hazardous Materials and Wastes: Treatment, Minimization and Environmental Impacts. Edited by S. K. Majumdar, E. W. Miller and R. F. Schmalz.

Chapter Two

HAZARDOUS WASTE GENERATION IN PENNSYLVANIA

BRIDGET HOFMAN

Division of Emergency and Remedial Response
Bureau of Waste Management
Pennsylvania Department of Environmental Resources
P.O. Box 2063
Harrisburg, PA 17120

INTRODUCTION

Since the mid-1970's state governments and the federal government have been working to gather needed information on hazardous waste. It has been a slow and arduous process. Unfortunately, information on the past problems of hazardous waste disposal is surfacing as quickly as information can be gathered, and solutions can be developed, for the safe management of that waste today. This chapter will focus on hazardous waste generation. The data on hazardous waste generation is only one of the critical pieces of information our society needs to proceed with identifying workable solutions to the safe management of hazardous waste.

A WORD ON DEFINITIONS

Hazardous waste is a subset of all types of waste generated in Pennsylvania. While the federal government has chosen to focus its attention and substantial resources almost exclusively on hazardous waste, Pennsylvania is faced with developing regulations and program initiatives to deal with all types of waste. Figure 1 depicts the three types of waste defined by the Pennsylvania Solid Waste Management Act (Act 97), and shows the relative percentage of each. As can be seen in Figure 1, hazardous waste composes a relatively small portion of all waste types generated in Pennsylvania. It should be clearly noted here that hazardous waste does not include low-level radioactive waste. These materials are covered under separate statutes and separate regulatory provisions both at the state and federal levels. Pennsylvania's Solid Waste Management Act of 1980

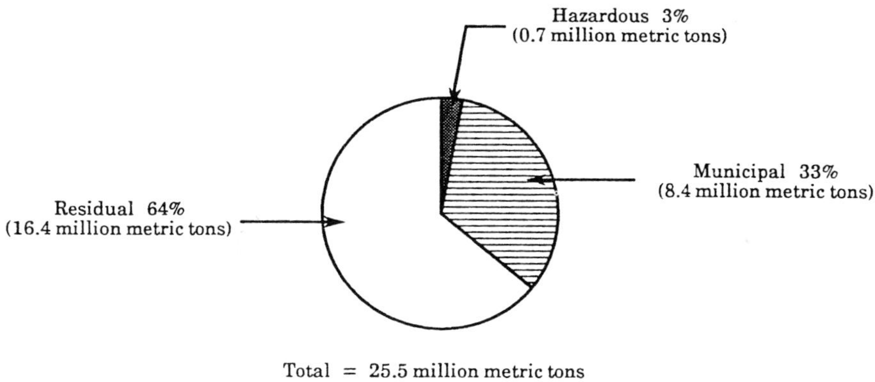

FIGURE 1. Solid Waste Generation in Pennsylvania.

defines hazardous waste as discarded materials which because of their quantity, concentration, or physical, chemical, or infectious characteristics, may:

1) Cause or significantly contribute to an increase in mortality or an increase in morbidity in either an individual or the total population; or
2) Pose a substantial present or potential hazard to human health or the environment when improperly treated, stored, transported, disposed of, or otherwise managed.

The U.S. Environmental Protection Agency (EPA) has identified four characteristics that define discarded materials as a hazardous waste: ignitability, corrosivity, reactivity, and toxicity. Many materials have been tested and are included on EPA's hazardous waste lists as either waste from specific sources, or waste from non-specific sources. For materials that are not specifically listed, it is the generators responsibility to test its waste for the four characteristics listed above. Mixtures of hazardous and non-hazardous waste are also defined as hazardous waste. All hazardous wastes have been assigned a four character hazardous waste number. The Pennsylvania Department of Environmental Resources (DER) has adopted the EPA lists of hazardous waste by reference.

The task of defining hazardous waste has become a science of its own. The few paragraphs above only scratch the surface of what has become a very complex definitional issue. Readers interested in the further details and nuances of this topic are referred to DER regulations at Chapter 75.261, EPA regulations at 40 CFR 261, and numerous background documents at EPA's Headquarters and Regional Offices.

SOURCES AND AMOUNTS OF HAZARDOUS WASTE

Hazardous waste is generated by many entities in the Commonwealth. These generators range from large industrial establishments to households.

Large Quantity Generators

Large quantity generators (LQGs) are the single largest source of hazardous waste generation in the Commonwealth. LQGs are required to obtain a U.S. EPA identification number and report on the management of that waste to the Pennsylvania Department of Environmental Resources (DER). By definition, any establishment which generates more than 1000 kg of hazardous waste in a calendar month is a LQG. LQGs are predominately industrial establishments. We estimate that there are 2,400 LQGs generating hazardous waste in Pennsylvania. The steel industry figures prominently in the Commonwealth's waste generation statistics. Summarized below in Table 1 are the major contributors to LQG waste in the Commonwealth. We currently estimate that LQGs in Pennsylvania generate 680,000 metric tons of hazardous waste annually.

Small Quantity Generators

Generators of between 100 kg and 1000 kg of hazardous waste in any calendar month are termed small quantity generators (SQGs). Small quantity generators are predominately small businesses and the smaller manufacturing and service firms. Currently, we estimate that there are 8,800 SQGs in Pennsylvania. Types of waste produced by SQGs include: spent solvents (32%), heavy metal waste (13.5%), and assorted other wastes, such as inks, paints and corrosives. Major contributors to SQG waste in the Commonwealth are listed below in Table 2.

The term very small quantity generator (VSQG) is used to represent those generators which produce less than 100 kg per month of hazardous waste. DER

TABLE 1

Predominate Hazardous Waste Generating Industries (LQGs) in Pennsylvania (by SIC Code)

SIC	Industry Group	Percent of Generators in SIC Group	Percent (Quantities) Waste Generated
33	Primary Metals	7%	50.0%
28	Chemicals & Allied Products	8%	12.0%
36	Electrical & Electronic Machinery	7%	10.0%
37	Transportation & Equipment	3%	1.4%
29	Petroleum & Refining	1%	1.2%
35	Machinery, except Electrical	6%	1.1%

TABLE 2

Predominate SQG Sources in Pennsylvania

Industry/Business Group	Percent Contribution
Vehicle Maintenance	78.7%
Metal Manufacturing	7.5%
Printing/Ceramics	2.4%
All others	11.4%

estimates that there are 20,400 VSQGs in Pennsylvania. The industry groups representing VSQG are similar to those represented by SQGs.

We currently estimate that SQGs generate 9,597 metric tons of hazardous waste annually, while VSQGs generate 3,316 metric tons, for a combined SQG/VSQG total of 12,913 metric tons/year.

Superfund Waste

There are currently (October 1988) 97 sites on EPA's National Priorities List (NPL) of sites. It is expected that up to 150 sites could be on that list by the year 2,000. In addition, Pennsylvania has enacted legislation that provides for a State program to cleanup hazardous sites. This program could add an additional 100 sites or more to the list of state/federal government priorities. Many factors effect the likelihood of waste being shipped from these sites; or in effect being "generated" from cleanup sites. These include the size and types of sites on the NPL (or state equivalent), state and federal policies regarding permanent on-site remedies, the size and cost of the cleanup alternative, and viable technological options. At this time, it is believed that approximately 18,000 metric tons/year of hazardous waste may be generated from hazardous site cleanups.

Household Hazardous Waste

At this point in time, waste generated from households is by regulatory definition *not* a hazardous waste. However, based on a technical definition, many waste products from the home may test as hazardous. Only gross estimates are available for rates of household hazardous waste generation. However, based on several studies (such as household hazardous waste collection days) an average of 7 -8 lbs/year/household appears to be a good average. The population of Pennsylvania is estimated at 11.8 million, or approximately 4.2 million households. Based on these averages, we can estimate household hazardous waste generation in Pennsylvania to be 15,000 metric tons/year.

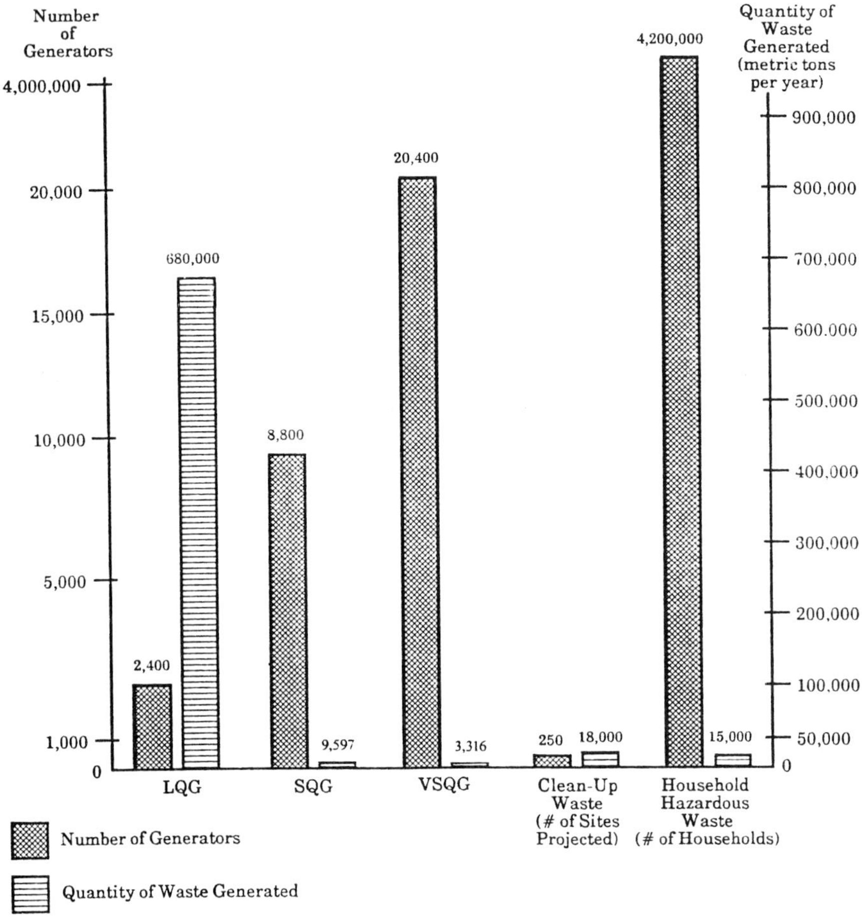

FIGURE 2. Generators of Hazardous Waste in Pennsylvania and Respective Contributions (1985).

Comparison of Waste Generators and Waste Quantities

Figure 2 provides an interesting comparison of numbers of entities producing waste and the amounts of waste they generate.

In order to arrive at the total amount of hazardous waste generated annually in the Commonwealth, a summation of contributions from the various sources is in order. As can be seen below in Table 3, total hazardous waste generation from these sources in Pennsylvania is estimated at 725,913 metric tons per year.

Variability of Data Sources

A comment on the reliability of the data as presented above is in order. Few

TABLE 3

Relative Contributions to Hazardous Waste Generation in Pennsylvania

Generator Type		Quantity of Waste (in metric tons)
Large Quantity Generator		680,000
Small Quantity Generator		9,597
Very Small Quantity Generators		3,316
Hazardous Waste Site Cleanups		18,000
Household Hazardous Waste		15,000
	TOTAL	725,913 metric tons per year

people realize that the data sources used by the government to present statistics range from actual reported data to literature searches and scale-up calculations.

Large quantity generators and owners/operators of hazardous waste management facilities (treatment, storage and disposal) have been required to report information on hazardous waste since 1980. However, none of these entities have been required to report the amount of waste that is actually generated. LQGs are required to report waste shipped off-site, i.e., manifested waste, as well as waste managed on-site in permitted units. The owners/operators of commercial treatment, storage, and disposal facilities are required to report waste received (from in-state and out-of-state), waste managed, e.g., treated, as well as waste shipped (or re-shipped) off-site. In order to arrive at statistics on hazardous waste generation, the raw data must be verified and massaged until a degree of consistency can be reached. A particular drum of hazardous waste can be stored on-site; shipped to a transfer facility; re-shipped to a treatment facility, with the treatment residue finally shipped to a destruction or disposal facility. Each of these operations will be counted as one activity and reported to DER. DER must then remove all "double counted" shipments, including shipments from out-of-state, to arrive at in-state waste generation totals.

Data on the other main sources of waste is less reliable. For example, the data on SQGs and VSQGs presented above is the result of a survey of business establishments likely to represent a significant number of small quantity generators. Data on household hazardous waste and hazardous site cleanup waste resulted from a literature search and scale-up of existing data, respectively.

Permit-by-Rule Facilities

One factor that has added considerable confusion to the waste generation numbers game is the categorization of hazardous wastewaters. Federal regulations, and many state regulations, allow discharges of hazardous wastewaters, which are regulated through Clean Water Act permits, to be excluded from the definition of hazardous waste. Some state regulations, such as Pennsylvania's, however, do require that these dischargers report annually to the regulatory agency. The decision as to whether to count this waste, which is usually many times greater in quantity than the "core" hazardous wastes, depends on the use of the statistic.

In Pennsylvania, most of the summary statistics gathered on waste generation and management have been used as input to the Hazardous Waste Facilities Plan. Since a primary purpose of the Plan is to identify types and capacities of facility that are needed, there is no reason to report wastewater quantities which are currently, and will continue to be, managed in units permitted under existing laws.

This particular waste type can be the single largest source of discrepancy between otherwise like states. It can often account for 90% or more of the total waste quantity. Some data prepared on a national level has not taken this discrepancy into consideration. In 1985, Pennsylvania records show that approximately 25 million metric tons of hazardous wastewaters were managed via NPDES, pre-treatment, or elementary neutralization permits. This compares to only 680,000 metric tons generated by the states large quantity generators. Combining this information would show that 97% of *all* hazardous waste (including hazardous wastewaters) generated in the Commonwealth are managed by way of Clean Water Act/Clean Streams Law permits. Since these wastes, by definition, must be managed on-site prior to discharge to retain their "permit-by-rule" designation, they are usually set aside during any discussion on hazardous waste generation.

WASTE TREATABILITY

The final topic to be discussed in this chapter is waste treatability categorization. Treatability categories are very important for any initiative regarding facility planning or facility siting. As discussed earlier, the U.S. EPA designated waste as hazardous based on its ignitable, corrosive, reactive, or toxic characteristics. This was done without any particular attention given to its form, e.g., aqueous inorganic liquid vs. organic sludge. These distinctions become critically important in identifying the types and capacities of state-of-the-art facilities that will need to be designed and built to manage these wastes.

While obviously some of these facilities currently exist, there is a general belief that new and advanced technologies will provide for better destruction of the hazardous components in the waste, thereby providing better long-term protection to public health and the environment. Destruction of hazardous molecules in the waste through the advanced, albeit expensive, treatment technologies, rather than "belt and suspenders" containment systems, should be our long-term goal.

Table 4 provides a summary of large quantity generator waste in Pennsylvania in 1985 divided into Waste Treatability Categories. The categories used in

TABLE 4

Hazardous Wastes Generated in Pennsylvania (1985)
(All quantities are in metric tons)

Waste Treatability Category	Off-Site Management	On-Site Management	Total
A. Aqueous Liquid Wastes			
1. Corrosive, with metals			
a. Pickle liquor	245,200	—	245,200
b. N.O.S.	63,400	—	63,400
2. Contains cyanide	7,800	—	7,800
3. Contains hex-chrome	22,300	—	22,300
4. Corrosive, N.O.S.	63,400	8,100	71,500
5. Reactive, N.O.S.	500	—	500
6. Organics, N.O.S.	19,800	—	19,800
7. Aquaous Wastes, N.O.S.	—	11,200	11,200
Subtotal	**422,500**	**19,300**	**441,800**
B. Organic Liquid Wastes			
1. Non-halogenated solvents	13,800	2,900	16,700
2. Halogenated solvents	9,200	—	9,200
3. Solvents, N.O.S.	9,600	500	10,100
4. Oils	—	—	—
5. Combustible liquids, N.O.S.	—	—	—
6. Organic liquids, N.O.S.	—	—	—
Subtotal	**33,400**	**3,400**	**36,800**
C. Solids and Sludges			
1. Non-halogenated organic constituents	700	—	700
2. Halogenated organic constituents	300	—	300
3. Oily solids and sludges	7,000	11,200	18,200
4. Paint residues	200	—	200
5. Combustible solids and sludges, N.O.S.	—	—	—
6. Solids and sludges with organics, N.O.S.	8,300	100	8,400
7. Heavy metal sludges			
a. Metal-finishing wastes	27,300	2,000	29,300
b. Pickle liquor lime sludge	—	—	—
c. Other	5,100	—	5,100
8 Solids with metals			
a. Electric arc furnace dust	55,400	27,600	83,000
b. Other emission control dust	10,200	—	10,200
c. Other	25,500	1,500	27,000
9. Corrosive solids and sludges, N.O.S.	2,300	—	2,300
10. Reactive solids and sludges, N.O.S.	100	—	100
11. Solids and sludges with inorganics, N.O.S.	—	—	—
12. Solids and sludges, N.O.S.	—	—	—
Subtotal	**142,600**	**42,400**	**185,000**
D. Other Wastes			
1. Ignitable	3,500	—	3,500
2. Corrosive	200	—	200
3. Reactive	—	—	—
4. Toxic	100	—	100
5. Other wastes, N.O.S.	—	—	—
Subtotal	**3,800**	—	**3,800**
E. "Hybrid" Categories			
1. Non-halogenated solvents: form not known	200	—	200
2. Halogenated solvents: form not known	1,900	—	1,900
3. Oil/solvents: blended in fuel	—	900	900
4. Inorganics: form not known	—	—	—
5. Combustible: form not known	—	—	—
6. Organics: form not known	2,800	—	2,800
7. Metal-finishing wastes: form not known	1,200	—	1,200
8. Metal wastes: form not known	4,400	—	4,400
Subtotal	**11,400**	**900**	**12,300**
TOTAL (Rounded)	**614,000**	**66,000**	**680,000**

N.O.S.: Not Otherwise Specified Note: "—" signifies < 100

Table 4 are those used in Pennsylvania's hazardous waste facilities planning documents. The table depicts 30 core categories and 13 other or hybrid categories. Many states in the northeast, and elsewhere across the country, have gone to great lengths to transform their existing manifest and annual report data into these types of waste treatability classifications. Unfortunately, each state took a slightly different approach to fit their own needs, e.g., their own definition of hazardous waste. This has made it very difficult for state authorities to compare generation and management data across state lines.

As a result of the need to better compare like information across state boundaries, the U.S. EPA and numerous state agencies, in cooperation with the National Governor's Association, has proposed a list of 27 treatability categories, several of which are optional. These are presented in Table 5. As can be seen, many of these have a close match to the existing state system. The U.S. EPA has promised to provide a generic methodology that all states can apply to existing data to arrive at consistent waste treatability classifications.

CONCLUSIONS

In the area of large quantity generation, our ability to summarize state-wide data has been improving. New reporting forms developed by EPA, together with more sophistication of state level computer data processing, will help us develop

TABLE 5

Proposed Standardized Waste Treatability Classifications

Organic Liquids	*Inorganic Sludges*
Waste Oil (optional)	Sludges with Metals
Halogenated Solvents	Sludges without Metals
Nonhalogenated Solvents	Unknown
Solvents, not otherwise classified	
Organic Liquids with Water	*Inorganic Solids*
Pesticides	Solids with Metals
	Solids without Metals
Organic Solids/Sludges	Unknown
Oily Sludge (optional)	
Halogenated Solids/Sludges	*Contaminated Sand, Soil, Clay*
Nonhalogenated Solids/Sludges	*Mixed Organic/Inorganic Waste*
Dye and Paint Sludges, Resins	Liquids
Unknown	Solids/Sludges
Inorganic Liquids	*Other*
Solutions with Metals	Explosives
Solutions without Metals	Mixed Radioactive/Hazardous
Unknown	Gases
	Lab Packs
	PCBs (optional)

greater confidence in our data over the next 3-5 years. While data on other important, but less significant, sources of waste generation is just becoming available, it is expected that this too will improve over the next 3-5 years. Efforts underway at the national level, mostly as a result of the state hazardous waste capacity assurance program required by the Superfund Amendments and Reauthorization Act of 1986, will be the greatest impetus for consistency from state to state. Better data will allow citizens, industry, and state governments to be better informed and, therefore, make confident decisions on future waste management practices and facility needs.

REFERENCE

All the data used in this chapter are currently in the form of numerous draft or final task reports prepared for DER by consultants for the Phase II Hazardous Waste Facilities Plan, or from various internal sources.

Management of Hazardous Materials and Wastes: Treatment, Minimization and Environmental Impacts. Edited by S. K. Majumdar, E. W. Miller and R. F. Schmalz.

Chapter Three

HAZARDOUS WASTE TREATMENT

C.A. McMenamy
Vice President and General Manager
Scott Environmental Technology
Plumsteadville, PA 18949

Recognition of the environmental impact of hazardous wastes has been a recent phenomenon in the United States and in many other parts of the world. Over the past 100 years we have developed and applied technologies to produce chemicals not to destroy them. All too often the decision has been made to bury chemical wastes rather than to convert them into forms that will not affect the environment or the health of those exposed to them.

The subject of leaking landfills containing chemical wastes has filled the media for the past ten years. The Love Canal hazardous waste dump became worldwide news in 1968 and focused attention on the problems resulting from chemical wastes in landfills.[1]

Historically hazardous wastes along with other materials were buried without treatment in more or less secure landfills. The high cost of treatment as compared to the low cost of landfilling caused those involved to select for landfill disposal. The generating organizations were often unaware or insensitive to the impact of the buried materials on human health and the environment. Adverse health and environmental effects were seldom well defined and understood by those making the decision as to the method of disposal.

Cost conscious manufacturers have always looked for the lowest cost raw materials, the most economical process, the highest production efficiency and the least expensive method of disposal.

January 1983 regulations in California were the first prevent land disposal of five categories of hazardous wastes. The intent of the California program was to encourage recycling, treatment and thermal destruction of the wastes that presented the greatest long-term risks when disposed of on or into the land.

Gary A. Davis[2] reported that the California program is based on two conclusions:

1. Land disposal of many hazardous wastes, particularly those that are highly toxic, persistent, mobile, and bioaccumulative, pose unacceptable long-term risks to public health and the environment.
2. Technologies do exist and are commercially available for recycling or detoxifying the wastes that present the greatest hazards when disposed of into the land.

   At present these technologies are not being used to any meaningful extent because of the availability of inexpensive land disposal.

The initial classes of materials selected in the California program, with the date of the phase out of land disposal were:

Cyanide Wastes	June 1, 1983
Toxic Metal Wastes	January 1, 1984
Strong Acids	January 1, 1984
PCB Liquids	January 1, 1984
Halogenated Organics:	
Liquids	January 1, 1985
Solids and Sludges	July 1, 1985

The U.S. Environmental Protection Agency in January 14, 1986[3] published the framework of the Congressionally mandated land disposal restrictions (LDR). Proposed rules for the first third of the wastes to be covered by these restrictions were published May 17, 1988.[4] The "first third" wastes included distillation bottoms from aniline production (K083), decanter tanks tar sludge from coking operations (K087) and wastewater treatment sludges from the mercury cell process in chlorine production (K106). The alphanumeric designation refers to RCRA identification number for the waste.[5]

In addition to restricting the land disposal of certain hazardous wastes, Congress also restricted the treatment and storage of hazardous wastes. The residues (nonwastewater) and the wastewaters have total composition limits for specified contaminates and maximum concentrations for the TCLP leachate. The TCLP procedure is designed to define the leachability of the waste. The TCLP test provides some assurance that the buried waste even when contacted by ground water will not leach chemicals from the waste into the water that can consequently migrate from the site.

The discussion of hazardous waste treatment must include a discussion of two important considerations: source and composition of the waste. The following questions must be answered:

1. What is the source of the waste?
2. Is it clearly defined?
3. What is the variability of the composition?
4. Can new previously unidentified components occur in the waste?

5. Can the interaction of the waste result in composition changes and potentially unsafe conditions, e.g., heat, pressure build up?

At one end of the spectrum of definition of a waste is the undefined waste, a drum of waste with no identification at an abandoned waste site. At the other end of the spectrum is a process stream such as spent pickling acid (K062) with well defined composition and little variability.

The waste of unknown composition presents a completely different treatment strategy than the defined waste. If it is judged transportable, (unlikely to detonate) it could be thermally treated in a rotary kiln incinerator facility designed to safely open and incinerate drums of hazardous wastes.

The spent pickling acid on the other hand can be treated in a physical/chemical treatment process designed for that specific material. The pH of the stream is adjusted to precipitate the iron as the hydroxide. The resultant precipitate is filtered and the waste water is recycled or treated with lime to precipitate sulfate in calcium sulfate. The precipitated hydroxide can be fixed and the precipitated calcium sulfate can be separated and sent to a landfill if it passes the TLCP test.

RCRA regulations list the spent pickling acid as K062, as of the listed RCRA hazardous wastes from specific sources: "Spent pickle liquor generated by steel finishing operations of facilities within the iron and steel industries (SIC codes 331 and 332). K062 is listed because of the possible hexavalent chromium and lead content.

The overall decision making process in hazardous waste management starts with four considerations shown in figure 1:

If the answer to the four questions in figure 2 is no, or the material from a site clean up is involved, second tier decisions must be made, these are shown in figure 2.

Selection of the method of treatment will depend on the composition and the variability of the waste. (Biological and thermal methods are considered elsewhere in this text.) The general methods used in physical treatment, chemical treatment and solidification are shown in figure 3.

The method used to select the method of treatment usually results in a number of steps. Process research and development is required followed by lab trials and pilot plant verification of the efficiency of the developed process. This

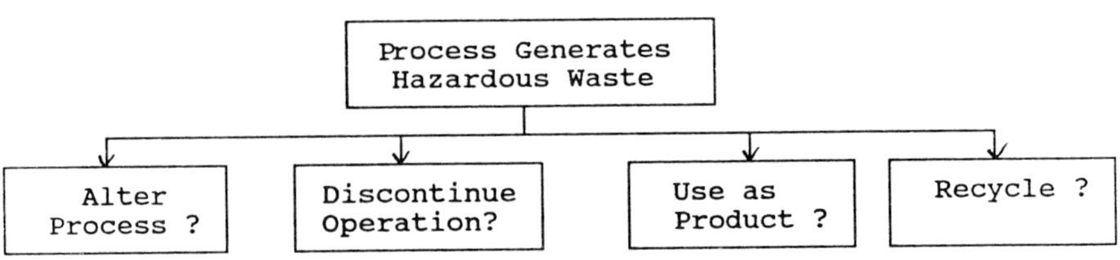

FIGURE 1. Four Considerations in Hazardous Waste Management.

is followed by selection and justification of the capital equipment necessary to treat the waste.

Most waste treatment processes are limited in their application. If the composition of the waste changes, the method may require modification.

The new proposed EPA LDR regulations not only restrict landfilling of certain wastes but also contains proposed treatment technology and limit the concentration of specified components.

One of the listed wastes is wastewater produced from the mercury cell process. In this process chlorine is produced by electrolytic decomposition of a

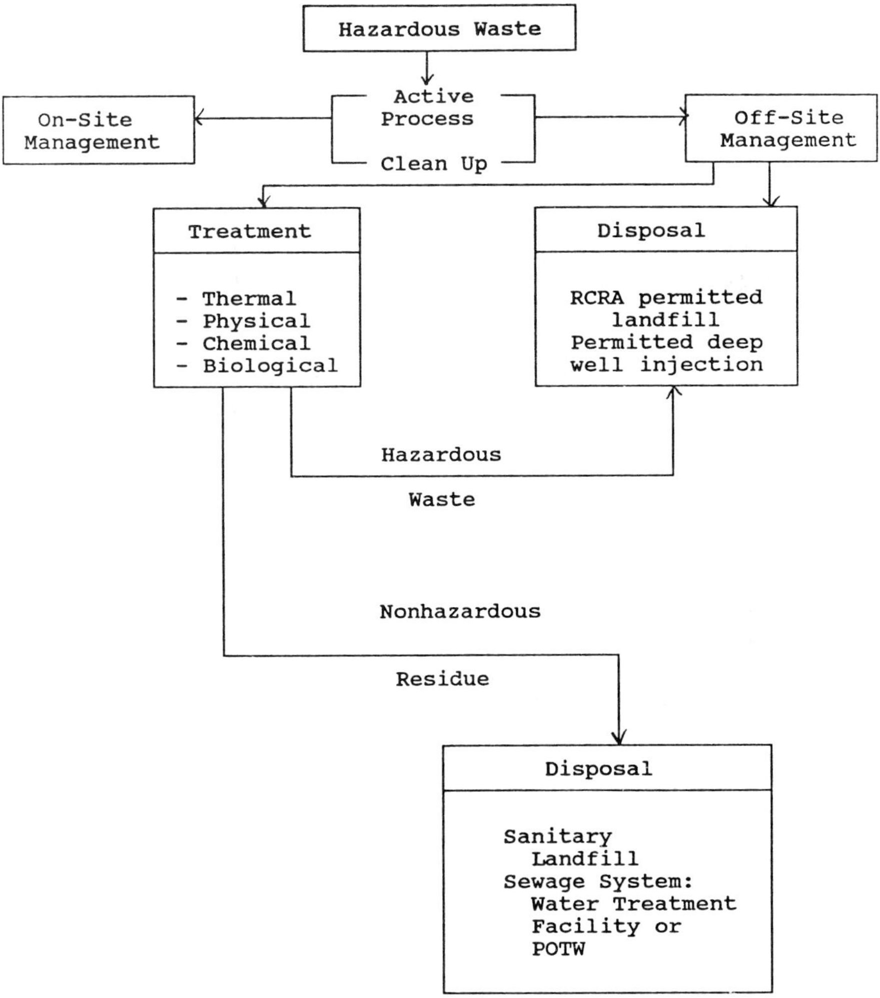

FIGURE 2. Second Tier Decisions.

Physical Treatment	Chemical Treatment	Solidification
Sedimentation Filtration Evaporation Distillation Flotation Absorption	Neutralization Precipitation Oxidation	Processes: Silicate-cement Lime Thermoplastic Organic Polymer Encapsulation

FIGURE 3. General Methods Used in Physical and Chemical Treatment, and Solidification.

	Max Hg level for any grab sample: Total	TCLP
K106 nonwastewater	630 mg/kg	0.028 mg/L*
K106 wastewater	0.030 mg/L	-

* Max in test leachate following the TCLP test procedure.

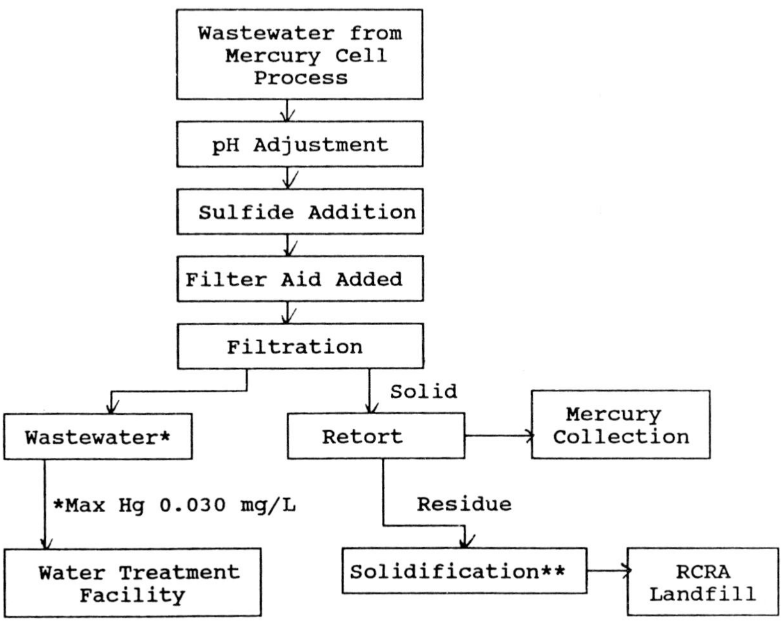

**Max Hg 630 mg/kg

Max TCLP 0.028 mg/L

FIGURE 4. Maximum Mercury Levels in the Wastewater and Nonwastewater (residue).

saturated solution of sodium chloride. Wastewater from the process contains both soluble and elemental mercury.

The BDAT (best demonstrated available technology)[6] for treatment of the wastewater involves chemical precipitation of the mercury preliminarily as a sulfide followed by sedimentation or filtration. This collected residue is listed as waste K106, "wastewater in treatment sludge from the mercuric cell process in chlorine production," the waste treatment process is shown in figure 4.

The untreated K106 waste typically consists of 60% water, 3% mercury sulfide, 2% other metallic sulfides and diatomaceous earth or other filler aid.

The EPA proposed treatment for K106 is retorting, the sludge is heated to volatize the elemented mercury and to decompose the mercuric sulfide. The vaporized mercury is condensed and collected for resale. The retort residue is treated in a stabilization process.

The hazardous waste processing for the mercury cell wastewater involves a number of steps. This is typical of most treatment processes particularly those involving inorganic materials. Organic materials on the other hand can be treated in a liquid injection or rotary kiln incinerator of the proper design.

Physical Treatment

A wide variety of physical treatment processes are used for hazardous waste treatment. Sedimentation and filtration are commonly used to separate solids from liquids. Sedimentation basins are often used in the treatment of waste water. The separation of chemically coagulated floc provides an economic method of liquid solid separation. Filtration can be accomplished by the use of microscreens (20 μm openings), diatomaceous earth filters, sand filters (0.4 mm sand, 150-400 mm thick) or mixed media filters. Filtration of sludges is accomplished by sand beds or vacuum filters.

Distillation and evaporation can be used to separate a waste stream into two or more factions. In waste oil recovery the lighter hydrocarbons can be evaporated from the waste and recovered. Stripping operations involve purging air through the waste and collection of the vaporized materials. Air stripping has been used to remove trace concentrations of volatile solvents from contaminated groundwater. The solvents can be collection by on carbon.

Absorption on activated carbon also permits concentration of organics dissolved in water. Leachate from the Love Canal site, for example, was adjusted to a neutral pH and fed to a clarifier. Suspended solids and immersable organics were allowed to settle as a sludge and disposed of separately. The aqueous phase was passed through activated carbon at a rate of 760L/min. The carbon was contained in two polypropylene bags. The effluent was monitored for breath-through. Concentrations of a wide variety of organic chemicals were reduced.

Chlorobenzene was reduced from 50,000 $\mu g/L$ to less than 12 $\mu g/L$, benzene from 28,000 to less than 10 and phenol from 2,400 to less than 10. Effluent water from the adsorption was sent to the municipal water treatment facility.[7]

Chemical Treatment

Chemical treatment is used to convert the components in the waste to a form that can be separated to a less toxic or corrosive compound. Chemical reactions are limited to specific reactants usually a concentration within a certain range. Mixed composition wastes may produce side reactions that will consume the reactant or produce undesired reaction products. The addition of soluble sulfide to precipitate heavy metal sulfides, if added to an acidic waste could produce highly toxic hydrogen sulfide.

Cyanide can be converted to cyanate using chlorine or sodium hypochlorite. Hexavalent chromium can be reduced to less toxic trivalent chromium by the controlled addition of sulfur dioxide.

The chemistry of hazardous waste treatment is often not clearly defined because of the reactions often take place in a highly diluted phase. Commonly contaminants are present at low ppm and ppb concentrations. The efficiency of the treatment must be demonstrated and sufficient analytical work done to define the limiting conditions.

Acidic material such as the pickling liquor can be effectively neutralized. When sulfuric acid is used the sulfate can be precipitated as calcium sulfate and collected. The concentration of toxic components must be considered before the calcium sulfate is landfilled. Acids and alkaline material can be neutralized and a variety of methods can be separated by precipitation as sulfate, sulfide or compounds as hydroxide. The separated metallic compounds can be solidified. The cost involved in design, purchase of capital equipment and start up can be substantial. Consider the fact that a single glass lined 1000 gallon reaction vessel for neutralization will cost about \$60,000 and the associated equipment another \$20,000.

An Integrated Facility

The Kommunekemi hazardous waste treatment facility in Denmark[8] provides an example of chemical and physical waste treatment. Wastes are collect at over 300 collection stations, where small business and household hazardous wastes can be taken and 21 transfer stations for collection of individual wastes. The wastes are taken to a central facility that uses the four principal plants for treatment and a controlled landfill:

1. An oil treatment plant.
2. An inorganic treatment plant.
3. An incinerator plant.
4. A water treatment plant.
5. A controlled landfill.

Oily wastes in a primary treatment process are separated into oil, water and sludge. The oil is filtered and in a stripper, the volatites are separated and collected. Volatable hydrocarbons are also separated in the primary treatment. Sludge from the primary treatment, the waste oil filtered material along with contaminated water go to the incinerator. The collected oil is reused and can be used as a support fuel for the incinerator plant.

The inorganic plant processes spent pickling baths, electroplating wastes and sludge containing metal for water treatment processes.

Sodium hypochlonite is added to oxidize the cyanide containing waste. Hexavalent chromium is reduced to trivalent chromium and bivalent iron is oxidized to trivalent iron. A pH adjustment precipitates a number of metals as hydroxides. The precipitated metals include iron, chromium, nickel, copper, aluminum and zinc. The precipitate is collected in a filter. The filter cakes are absorbed in a separate section of Kommunekemi's landfill near Klinthalm.

An estimated 53% of the received waste is incinerated (41% liquid, 12% solid), 32% is heated in the waste oil facility and 14% in the organic treatment facility. The remaining 1% is landfilled.

Solidification

The term solidification identifies those technologies that fix or encapsulate waste in a solid material resistant to leaching. Fixation techniques chemically and physically bind the waste with a solidification agent. Encapsulation techniques physically surround the waste in a protective envelope.

Cement based methods have been used successfully on sludges generated by precipitation of heavy metals. The high pH of the mixture tends to hug the metal in the unsoluble hydroxide or carbonate. Other waste components such as asbestos, latex, metal filings and ground plastic can serve to increase the strength and stability of the solid.

A number of other solidification methods, some proprietary, are available including lime based processes, thermoplastics (including polyethylene), organic polymer and encapsulation. One recent encapsulation process developed by TRW uses polybutadiene mixed with the waste as a forming agent. The polymer waste mix is formed into cubes and coated with polyethylene which is fused on the surface. Sealosafe, an English process, based on the use of a silicate based

		Waste Concentration	Leachate Concentration
Metal finishing chrome sludge	Cr	87,000 ppm	0.15 ppm
	Cu	5,200 ppm	0.05 ppm
	Pb	22,000 ppm	0.01 ppm
Tin production waste	As	16,200 ppm	0.25 ppm
	Pb	4,000 ppm	0.1 ppm
Acid neutralization waste	Hg	500 ppm	0.005 ppm

FIGURE 5. Sealosafe process effectively reduces the leachability. (Test Data).

agent has been reported to effectively reduce the leachability. Leachate concentration in the Sealosafe[9] bulk material compared to the leachate concentration is shown in figure 5.

Currently the TCLP test is required. Materials must pass the test before they can be landfilled. The solid portion of the waste is milled and extracted in a rotating (30 rpm) extraction apparatus for 18 hours. The extracting fluid is a buffered acidic solution. One of two extraction fluids are used, 4.93 pH or 2.88 pH. A qualification test is used to select the appropriate extraction fluid.[10]

A modification of the TCLP test was proposed in the May 24, 1988 Federal Register, p. 18792.[11]

Acceptance for Treatment

Once the treatment process is established and the acceptable variability of the waste is established it is critical to set up an acceptance process. This is achieved by the acceptance analysis establishing and analyzing waste for Key Parameters.

The hazardous waste shipment is first compared externally to its manifest, discrepancies must be resolved or the shipment rejected. The shipment is then inspected for damage, leaking and other visible signs of problems. A sample(s) is taken following a written procedure and analyzed for Key Parameters. The data is evaluated and a proceed or reject decision made, again a discussion may be required. Finally having passed the manifest check, the visual check and the Key Parameter analysis check, the waste is accepted for treatment. The block diagram is shown in figure 6.

CONCLUSION

Slowly but surely we are moving away from land disposal of untreated hazardous wastes. The Congressionally mandated land disposal restrictions will require treatment and will encourage waste minimization.

The treatment of wastes is a new area of endeavor. Large industry has been actively addressing the hazardous waste problem. The problem of those without the ability to effectively manage wastes remains a service problem. Contamination of ground water in the area of sanitary land by household hazardous wastes also is a serious problem. We should look to the Danish system as an example of the way to collect wastes and transport them to a treatment center.

Paul MacClennan in an 1988 article entitled, "The Lessons of Love Canal Fall on Deaf Ears" point out the painfully slow response of the hazardous waste problem. The article was written on the tenth anniversary of the first emergency Love Canal Declaration. He points out Love Canal has become a legend around the world—a symbol of mankind's disregard of a natural environment and after ten years many questions and concerns remain.

The U.S. General Accounting Office in a July 1988 Report to Congress titled;

Hazardous Waste, New Approach Needed to Manage the Resource Conservation and Recovery Act[13] concluded EPA's approach to studying hazardous wastes has suffered from low priority and changing approaches. GAO feels EPA's progress in developing regulations to control hazardous wastes has been limited. In the 1984 RCRA amendments, Congress specified 76 deadlines. As of April 1988, 66 of the 76 deadlines had come due and EPA had completed action on less than half of them.

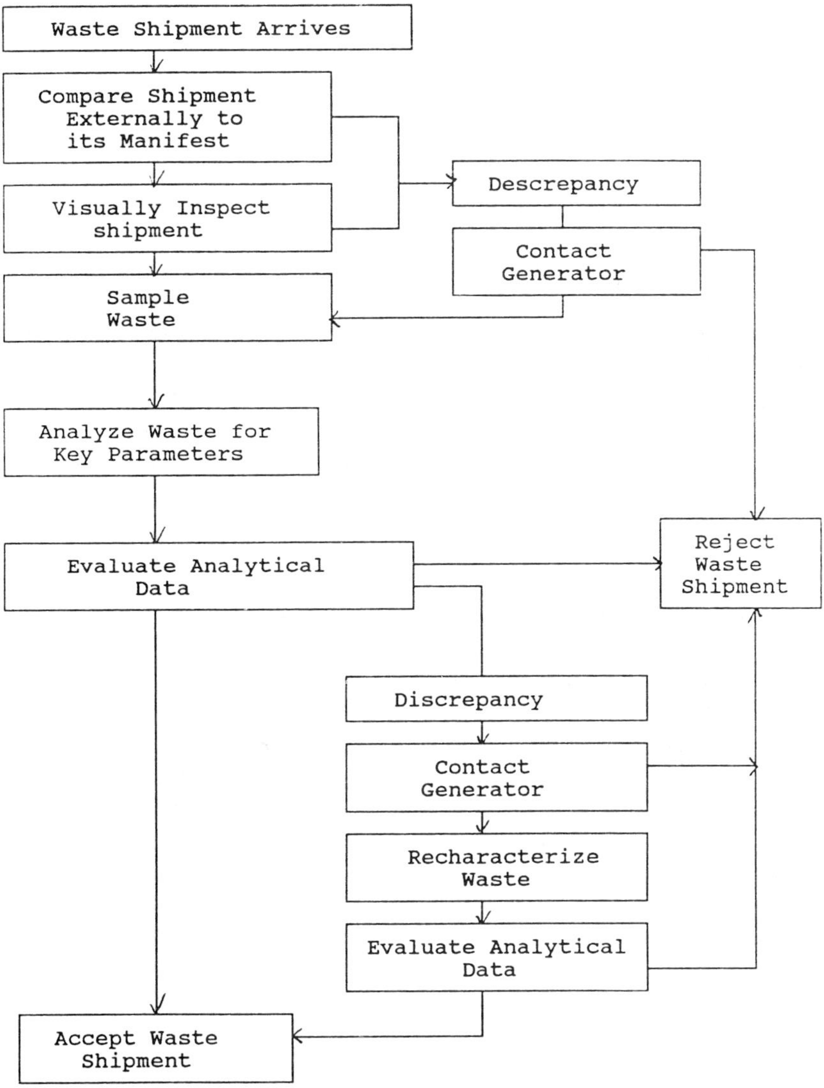

FIGURE 6. Waste Shipment Screening Procedure.

A worldwide investment of more than $115 billion is needed to carry out effective removal of deposits of hazardous wastes according to the Secretary General of ISWA 88![14] ISWA is the International Solid Wastes and Public Cleansing Association. Well designed and managed treatment facilities are needed.

We can hope that there is a far greater recognition of the fact that the cost of treatment of hazardous wastes is a small fraction of the cost of remedial action once the waste escapes from a landfill into the groundwater. In those cases real clean up is often impossible and the resultant contamination will be passed on to succeeding generations.

REFERENCES

1. Buffalo Courier Express, Buffalo, New York, Extreme Caution Urged in Probe of Love Canal, August 10, 1978.
2. Davis, G.A., Discouraging Dumping: The California Example, in Beyond Dumping, Quorum Books, Westport, Connecticut, 1984.
3. Environmental Protection Agency, Federal Register, November 7, 1986 [51 FR 40572].
4. Environmental Protection Agency, Federal Register, May 17, 1988 [53 FR 17578].
5. Environmental Protection Agency, Code of Federal Regulations, Subpart D—Lists of Hazardous Wastes [50 CFR 261.31, 261.32].
6. Environmental Protection Agency, Federal Register, May 17, 1988 [53 FR 17601 - 17603].
7. Environmental Protection Agency, Management of Hazardous Waste Leachate, SW-871, September 1982, p. 6-24.
8. Kommunekemi, a/s (descriptive 19 page brochure), Lindholmnez 3, DK-5800 Nyborg, Denmark.
9. Christensen, D.C., Wahanya, W. A Solid Future for Solidification/Fixation Processes, in Toxic and Hazardous Waste Disposal, Vol. 4, Ann Arbor Science, Ann Arbor, Michigan (1980).
10. Environmental Protection Agency, Toxicity Characteristics Leaching Procedure (TCLP), Code of Federal Regulations, [50 CFR 268 Appendix I].
11. Environmental Protection Agency, Federal Register, May 24, 1988, [53 FR 18792].
12. Mac Clannan, P., The Buffalo News, Buffalo, New York, July 31, 1988.
13. U.S. General Accounting Office, Hazardous Waste, New Approach Needed to Manage the Resource Conservation and Recovery Act, [GAO/RCED-88-115, July 19, 1988].
14. ISWA 88 News, November 1987, Dabofa, V. Farimagsgade 29, DK 1606 Copenhagen V. Denmark.

Management of Hazardous Materials and Wastes: Treatment, Minimization and Environmental Impacts. Edited by S. K. Majumdar, E. W. Miller and R. F. Schmalz. © 1989, The Pennsylvania Academy of Science.

Chapter Four
HAZARDOUS WASTE INCINERATION

ARTHUR B. NUNN, III
Vice President and General Manager
Scott Environmental Services
Plumsteadville, Pennsylvania 18949

INTRODUCTION

Perhaps no other facet of the general field of hazardous waste processing and disposal is as controversial and as poorly understood as that of incineration. People view the practice of incineration of hazardous waste as anything from the great solution to all waste disposal problems, to the most devastating and dangerous operation ever practiced by mankind. The true value of incineration lies, of course, somewhere between these two extremes. A great deal of knowledge has been obtained during the past ten years or more about the applications, advantages and limitations of incineration of hazardous waste as well as that of municipal and hospital wastes. Along with this expanded knowledge base have come many technological improvements in the various incineration processes which are available. In spite of these advances, and in spite of an ever increasing level of regulatory control over incinerator operations, the general public remains opposed to, or at the very least, leery of the practice of waste incineration. This has resulted in a political environment which has made it increasingly difficult to permit, construct, and operate incineration facilities. This is, in effect, closing the door on one of the most promising means of waste disposal currently available.

INCINERATION APPLICATIONS

It is estimated that between 200 and 300 million metric tons of hazardous waste are generated in the United States each year. Any estimate of how much of this waste could be disposed of by incineration would be dependent upon the definition of what is considered to be incinerable. Historically, a waste has not been considered to be incinerable unless its heat content was at least 4000 to 5000 BTUs per pound. Such a heat value would allow such a waste to sustain combustion without the use of auxiliary fuels except during start-up. Because of this, a positive economic benefit can be obtained by way of heat recovery and hence, reduced fossil fuel costs for an industrial facility.

By the early to mid 1980s, the definition of what is incinerable had been broadened considerably to include wastes which are not capable of sustaining combustion. This broadening occurred because of the desire on the part of industry throughout the United States to destroy as much waste as possible rather than to face the long term liability associated with any other means of hazardous waste disposal. Wastes which are now considered appropriate for incineration include almost any waste which is hazardous because it contains toxic organic compounds regardless of their concentrations and regardless of the matrix in which it occurs. Even contaminated water and soil, as well as many other solids and liquids which contain trace levels of organic compounds, are now being routinely incinerated.

There are only two classes of wastes which are not generally believed to be appropriate for incineration. One such class of waste is any waste, whether liquid or solid, which is hazardous only because it contains toxic heavy metals. Incineration cannot destroy heavy metals. It can change the form in which they exist in the waste to a potentially less toxic form, but it cannot remove the metals themselves. The metals will be found in the incinerator bottom ash or fly ash. Therefore, except for some advantage due to volume reduction, there is no true environmental benefit from the incineration of these types of wastes.

The other class of waste which cannot be routinely incinerated is any waste with a very high halogen content. Wastes with chlorine and/or fluorine contents in excess of approximately 30% are very difficult to incinerate due to the highly corrosive gases which are formed during the incineration process. These gases can be very damaging to the incinerator refractory and other materials of construction. As a result, only incinerators which are specially designed for this type of service can be used to incinerate high halogen wastes.

PRINCIPLES OF INCINERATION

Incineration can be defined in many ways. In practice however, incineration is a carefully engineered and controlled process of destroying organic materials by way of high temperature thermal oxidation. By such a process all organic waste materials are converted to carbon dioxide and water. When properly designed and engineered, incineration systems are easily capable of destroying solid and liquid hazardous organic substances with an efficiency of at least 99.99%, and often well in excess of 99.9999%. The key to proper incinerator design and operation is known as the three "T"s: Time, Temperature, and

Turbulence. Time refers to the amount of time that a substance is actually exposed to the high temperatures of the incinerator. This is referred to as the residence time. The longer any organic molecule is exposed to the high temperature of an incinerator, the more complete the destruction of that molecule will be. In the past, a residence time of 0.25 seconds was considered standard. By today's standards, however, most new incinerators are designed to operate with residence times in excess of 2.0 seconds.

Temperature is a measure of the amount of heat to which the hazardous organic substances are exposed during the incineration process. In the past, it was standard practice to operate incinerators at a temperature of approximately 1400 degrees Fahrenheit. By today's standards, most new incinerators are designed to operate at temperatures in excess of 2000 degrees Fahrenheit. As with residence time, the higher the temperature, the more complete is the destruction that the incinerator will produce.

Turbulence is a measure of the amount of mixing of the waste in the incinerator with the hot gases around it. The greater the turbulence, the greater the heat transfer from the incinerator flame to the waste materials. High temperatures and long residence times are of little benefit if the waste remains in an unmixed pile in the bottom of the unit with only the outer edges being burned. As a result, turbulence is very important in ensuring that proper mixing occurs and that all of the waste is fully exposed to the destruction of the incinerator.

The proper design of a given incinerator system will involve a careful balancing of time, temperature, and turbulence. Turbulence should always be maximized. Time and temperature however, can often be traded off against each other. When an incinerator has a very long residence time (greater than 3 or 4 seconds, for example) it can destroy waste very effectively at much lower temperatures (as low as 1200 degrees Fahrenheit). Similarly, if an incinerator operates at a very high temperature, it can be designed to have a lower residence time without a negative impact on combustion efficiency.

INCINERATION TECHNOLOGY

There are many different types of incinerators available today, including the following:

- Fixed hearth
- Fluidized bed
- Liquid injection
- Rotary kiln
- Cyclonic
- Auger
- Infrared
- Plasma.

Of these, the four most common types are the fixed hearth, fluidized bed, liquid injection, and rotary kiln. Each of these four incinerator types will be discussed in the following sections.

FIXED HEARTH

Fixed hearth incinerators are primarily designed for the combustion of homogeneous granular solids, irregular bulky solids, and heavy tars. These incinerators can also be used, however, for the combustion of pumpable liquid organic and aqueous waste materials.

A typical fixed hearth incinerator is depicted in Figure 1. As presented in the figure, the incinerator consists of primary and secondary chambers with the bulk solids, granular solids, and tars fed through a ram feed system on the front end of the primary chamber. The feeding of waste material to the incinerator is a batch process. Waste material is loaded into a feed chute, which is separated from the internal portion of the primary chamber by a hydraulically operated door. When waste is fed into the unit, this door opens and a ram is activated, which pushes the waste material into the incinerator. The ram is withdrawn and the door is closed. Inside the incinerator, the waste material is exposed to the high temperatures of the combustion zone as it sits on a grate through which combustion air is pumped into the unit. This combustion air, which is pushed up through the waste, is referred to as underfired air, and serves the dual purpose of providing oxygen to sustain combustion as well as disturbing the bed of waste material and thereby increasing turbulence. Each time the waste feed ram is activated and new waste is charged into the primary chamber, the partially combusted waste which is in the unit is pushed farther down toward

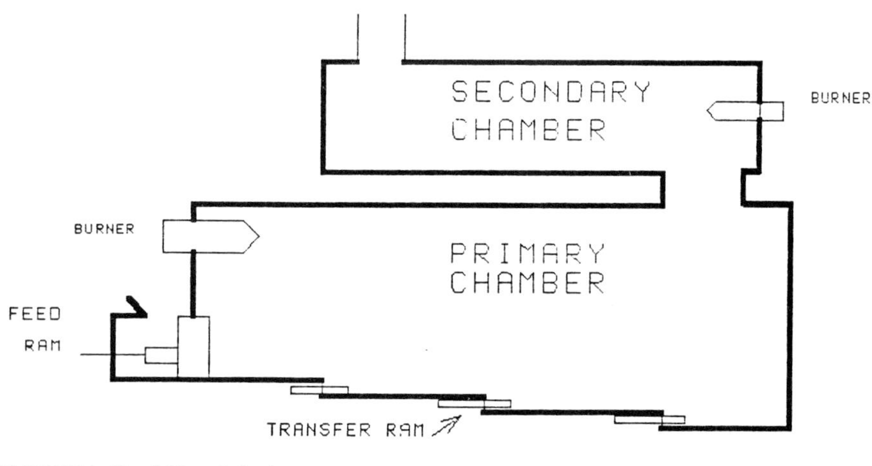

FIGURE 1. Fixed Hearth Incinerator.

the back end of the chamber, which also improves turbulence. This is further enhanced by the presence of steps and transfer rams in the bottom of the chamber, which combine to ensure that the waste material is turned over several times as it is passed through the incinerator. By the time that it reaches the back end of the unit, all organic material is completely combusted and only ash remains. This ash is removed from the chamber by way of an ash discharge ram. The total residence time of the solid waste material in the primary chamber of a fixed hearth incinerator is usually in excess of two hours.

The secondary chamber of a fixed hearth incinerator serves the purpose of completing the combustion process. As small unburned particles are elevated from the bed of waste materials by way of the underfired air, they are carried into the secondary chamber where they are combusted. Similarly, as organic vapors are released from the waste in the primary chamber, they are passed through the secondary chamber where they are combusted. The temperature in the secondary chamber is maintained by one or more burners which are firing fossil fuel and/or high BTU content liquid organic waste. The temperature in the secondary chamber is generally maintained at or above 2000 degrees Fahrenheit as opposed to a temperature in the primary chamber of 1400 to 1800 degrees Fahrenheit. Turbulence in the secondary chamber is maintained by designing the entry of the hot gases into the secondary chamber at an angle, so as to induce a swirling action in the chamber. This is often further enhanced by placing the secondary chamber burner(s) at an angle rather than simply shooting directly down the middle of the chamber. The residence time of organic gases in the secondary chamber is usually in the area of one to two seconds.

Fixed hearth incinerators tend to be rather small, and as a result, are usually found in industrial facilities where only a small amount of hazardous waste is generated and incinerated. These incinerators are often used for the disposal of general plant trash (non hazardous) as well as hazardous waste.

FLUIDIZED BED INCINERATORS

Fluidized bed incinerators offer somewhat more flexibility overall than fixed hearth incinerators in terms of the types of waste materials that they can burn. These incinerators can be used to destroy homogeneous granular solids, heavy tars, aqueous sludges, organic sludges, aqueous liquids, and organic liquids. The only type of waste that they are not designed to burn is irregular bulky solids. A typical fluidized bed incinerator is shown in Figure 2. As shown, the incinerator consists, basically, of one large vertical cyclindrical chamber. At the bottom of the chamber is a metal plate with holes, through which combustion air is passed. A bed of sand or some other type of granular inorganic material sits on top of this plate. As combustion air is forced through the holes in the plate, the bed of sand becomes suspended, or fluidized. The combustion of waste materials actually takes place in this fluidized bed of sand. The bed is fluidized and preheated by way of a preheat burner that is located above the bed and is pointing down onto the bed. After the bed has been preheated to a temperature of 1200 to 1400 degrees Fahrenheit, organic waste material is fed into the bed where it is mixed with the hot sand and the hot air of the combustion zone. Pumpable wastes, such as organic and aqueous liquids and sludges, as well as tars are injected through nozzles directly into the center of the fluidized bed. Granular solids are dropped into the bed from a solids feed chute directly above the bed. If they are high enough in heat content, these wastes will sustain combustion in the fluidized bed. If additional heat is needed, it can be supplied by auxiliary fossil fuel which is injected into the bed through a separate nozzle. Organic waste is primarily destroyed in the fluidized bed. Some organic vapors, as well as fine solid particles, are released from the bed, however. The combustion of these materials is completed in the open area above the bed. This area is referred to as the free board zone, and it serves exactly the same purpose as that of the secondary chamber of a fixed hearth incinerator. Because of the way in which the combustion vapors rise from the fluidized bed, the air is very turbulent in the freeboard zone. The freeboard zone is quite large; therefore the residence time of the gases in this zone is quite long with times up to five seconds.

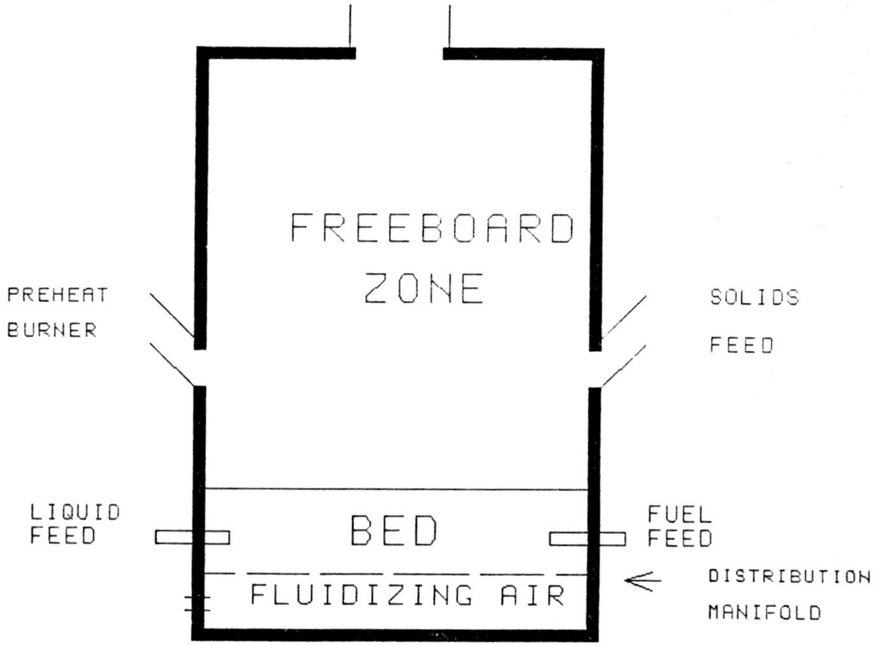

FIGURE 2. Fluidized Bed Incinerator.

Because of the long residence time offered by fluidized bed incinerators, these units can be operated at lower temperatures than other types of incinerators. Operating temperatures seldom exceed 1700 degrees Fahrenheit. The ability to operate effectively at this low temperature is further supported by the great amount of turbulence offered by fluidized bed incinerators. Heat transfer is optimized by the direct, intimate contact of the waste to be destroyed with the hot sand of the fluidized bed.

For all of their advantages, fluidized bed incinerators offer some distinct disadvantages as incinerators because of the extremely critical level of control needed to operate these types of units properly. Combustion of some sludges and solids must be carefully controlled in order to ensure that materials which are too heavy are not charged to the bed. Charging too heavy a load would result in the waste falling to the bottom of the bed which would result in clogging of the holes in the plate thereby ending fluidization of the sand. If this occurs, the incinerator must be shut down and manually cleaned. Similarly, the temperature of the incinerator must be carefully controlled to ensure that it does not get too high, which would cause the sand to melt and slag out on the plate. This too would end fluidization and cause a very difficult manual cleaning with possible replacement of the air distribution manifold.

LIQUID INJECTION INCINERATORS

Liquid waste incinerators are very common throughout the petroleum and chemical industry for the on site destruction of liquid waste materials generated at these plants. These types of incinerators are limited in their adaptability in that they cannot be used to incinerate any type of solid or sludge waste. In order to be burned properly in a liquid injection incinerator, a waste must not only be pumpable; it must be able to be very finely atomized. This may limit the ability to burn some very viscous materials even if they are pumpable. In spite of the limitation on the types of waste they can burn, liquid waste incinerators are very popular where only liquid wastes are present.

A typical liquid waste incinerator is shown in Figure 3. These incinerators can be mounted either horizontally as shown, or vertically. Vertical liquid waste incinerators can be either down fired (burner at the top pointing down) or bottom fired (burner at the bottom pointing up). In all cases there is a burner or a series of burners which atomize, mix, and burn the waste. In the case of low BTU content aqueous waste, the waste is injected through an atomization nozzle rather than a burner. The atomized waste is impinged upon the flame from another burner or burners. There may be as many nozzles and/or burners as there are different wastes to be burned in the incinerator. A series of burners and/or nozzles may be required to properly handle the different heat content, density, and viscosity of the different wastes to be burned. Burner and nozzle design is based primarily on the need to finely atomize the waste. It is generally desirable to atomize waste to a droplet size of approximately one micron to ensure adequate heat transfer and thus, complete combustion. Large droplets may simply pass through the incinerator and out the exhaust without being fully destroyed. Fine atomization prevents this from happening.

Most waste which is burned in a liquid waste incinerator is completely incinerated in the burner flame. Temperatures in the flame often exceed 3000 degrees Fahrenheit. Any unburned organic materials surviving the flame are subjected to temperatures of approximately 2000 degrees Fahrenheit as they pass through the remainder of the incinerator chamber. This is much like the freeboard zone of a fluidized bed incinerator. The residence time in this area is usually in excess of two seconds.

Liquid waste incinerators are available in many sizes. They are normally built in sizes ranging from 20 to 40 million BTUs per hour but they can be built as small as one million BTUs per hour and as large as 70 million BTUs per hour. These incinerators are adaptable in terms of the quantity of halogenated waste that they can burn. Liquid waste incinerators can be designed and constructed to burn wastes containing very high concentrations of chlorine and fluorine. This is a particularly attractive attribute to operators of some synthetic organic chemical manufacturing plants, because these plants often generate high halogen content wastes.

ROTARY KILN INCINERATORS

The most common type of incinerator found in commercial hazardous waste incineration facilities is the rotary kiln incinerator. This is because the rotary kiln is by far the most versatile type of incinerator available. This type of unit can burn tars, organic and aqueous liquids, sludges, homogeneous granular solids, and irregular bulky solids including drums, capacitors, and transformers. Rotary kilns can burn waste with a high fusible ash content as well as waste with a high halogen content. The average size of all rotary kiln incinerators

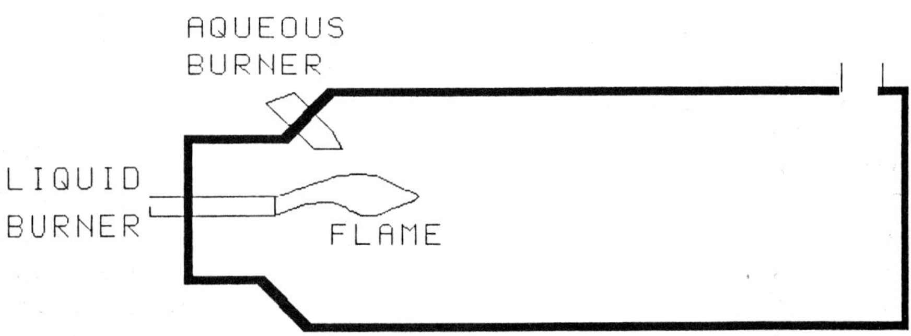

FIGURE 3. Liquid Waste Incinerator.

is approximately 60 million BTUs per hour, but they can be as large as 90 million BTUs per hour.

A typical rotary kiln incinerator is shown in Figure 4. The main body of the incinerator is a long cylindrical tube which is mounted on a slight downward incline from the front end where the waste is fed to the back end where the ash is removed. This cylinder rotates at a speed of one to five revolutions per minute. By this rotation, the solid waste, which is on the bottom of the cylinder, is constantly turned over which greatly increases heat transfer and turbulence. Liquid wastes and auxiliary fuels are fired through burners on the front end of the kiln with high atomization burners as discussed for liquid waste incinerators. Solid wastes are fed into the front end of the kiln by a solids feed chute similar to the ram feed system described for fixed hearth incinerators. The back end of the rotary kiln is often equipped with an auxiliary fuel fired burner, which helps to hold a high temperature in the kiln. This burner can be fired with a high BTU content liquid waste instead of fossil fuel if desired.

The rotary kiln is, by itself, a very efficient combustor of hazardous waste. The incinerator is made even more efficient, however, by the addition of a secondary chamber. This chamber is similar to the secondary chamber described for the fixed hearth incinerator. It is fired with auxiliary fossil fuel and/or liquid waste and the temperature in the chamber is maintained at a minimum of 2000 degrees Fahrenheit. Turbulence is maximized by pushing the exhaust from the kiln into the secondary chamber at an angle to induce swirl.

In addition to a high degree of turbulence and very high temperature, rotary kiln incinerators also often have very long residence times. Solids often exist in the kiln for more than 90 minutes. The total residence time for gases in the incinerator is often more than five seconds. The rotary kiln incinerator, in conclusion, is the most versatile of all available types of incinerators, and it is the only one of the four major types of incinerators that maximizes all three of the critical incineration factors of time, temperature, and turbulence.

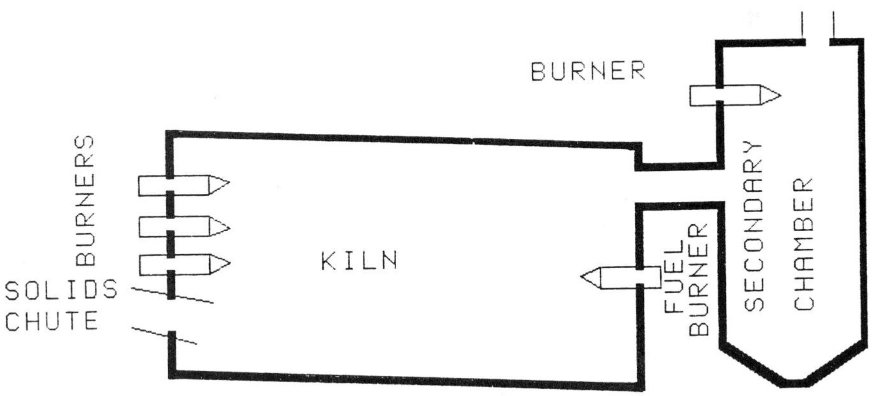

FIGURE 4. Rotary Kiln Incinerator.

EMISSIONS CONTROL EQUIPMENT

Air pollutant emissions control equipment is used on hazardous waste incinerators to control emissions of particulates and hydrogen halides. Emission control equipment is not used in any way for the control of unburned organic materials because the incinerators are so efficient that there is not enough organic material left in the exhaust to be able to be scrubbed. Most incinerators are controlled by a quench followed by a venturi and/or packed tower scrubber. The quench is used purely for the purpose of temperature reduction prior to further treatment of the gas. The venturi scrubber, if used, is present for the purpose of removal of particulate matter from the exhaust. The packed tower scrubber is used primarily for acid gas (hydrogen chloride, hydrogen fluoride) removal. Packed tower scrubbers use either pure water, or water with a low concentration of sodium hydroxide to remove the acid gases. Scrubber discharge water from both venturi and packed tower scrubbers is usually treated in on site wastewater treatment plants then disposed to the municipal wastewater treatment system in accordance with the plant's EPA or State wastewater discharge permit.

Other types of emission control systems may be used on hazardous waste incinerators. These include electrostatic precipitators, baghouses, dry scrubbers, and ionizing scrubbers. Some incinerators, which burn low ash waste and waste containing no halogenated materials, may not use any type of emission control systems.

REGULATORY REQUIREMENTS

Hazardous waste incinerators are regulated by the regulations of the Resource Conservation and Recovery Act except when burning polychlorinated biphenyls (PCBs). The incineration of PCBs is governed by the Toxic Substance Control Act. These regulations require that all waste which is burned in incinerators must be destroyed with an efficiency of at least 99.99% (99.9999% for PCBs). This destruction efficiency must be demonstrated in a very extensive trial burn test program which must be conducted under EPA supervision before a final operating permit can be granted. For commercial hazardous waste incineration facilities, this trial burn may have to be repeated every year.

Other requirements which are placed on hazardous waste incinerators pertain to emissions of particulates and hydrogen chloride. The particulate matter concentration in the exhaust from an incinerator must be no greater than 0.08 grains of particulate matter per dry standard cubic foot of exhaust with the exhaust corrected to a level of 7.0% oxygen. Some states have adopted more stringent levels of control for particulates with maximum allowable particulate concentrations as low as 0.015 grains per dry standard cubic foot at 7.0% oxygen. Federal regulations limit the amount of hydrogen chloride that can be emitted from a hazardous waste incinerator to 4.0 pounds per hour or, if over 4.0 pounds per hour, no more than 1.0% of the amount of hydrogen chloride present in the exhaust prior to the emission control equipment. Some states have also made this more stringent by adapting the EPA requirements to all hydrogen halides, not just hydrogen chloride.

Additional requirements placed on hazardous waste incinerators by the EPA and state environmental regulatory agencies involve continuous monitoring of incinerator operation. Continuous monitoring is required of the incinerator combustion air flow rate, the incinerator temperature, and the waste feed rate to the incinerator as well as continuous monitoring and recording of incinerator exhaust concentrations of carbon monoxide and oxygen. Some states are also requiring monitoring of exhaust concentrations of carbon dioxide and total hydrocarbons. All of these monitoring requirements are designed to ensure that the incinerator is operated at the same conditions as those when the trial burn test was conducted and passed. All monitors are calibrated frequently and checked randomly by regulatory personnel to ensure proper operation. Monitors are connected to an emergency waste feed cutoff system which will stop the flow of hazardous wastes to the incinerator whenever any one of the monitors measures parameters outside of the acceptable range established by the regulatory agency at the completion of the trial burn.

CONCLUSION

Hazardous waste incineration is a very complex process involving many variables. It is a process that is very environmentally sound for many types of hazardous waste. Many tests have been performed on hazardous waste incinerators since the late 1970s, and incinerators which are properly designed have proven to be very effective with destruction efficiencies well in excess of the 99.99% required by the U.S. Environmental Protection Agency. Hazardous waste incinerators are heavily regulated and closely monitored to ensure compliance with all applicable regulations. Furthermore, incineration is somewhat self complying, because, the more efficiently the incinerator is operated, the more economical that operation is. For this reason, incinerator operators strive to run their units in such a manner as to ensure the optimum destruction efficiency of all hazardous wastes which they burn.

Hazardous waste incineration is not the overall answer to all problems associated with hazardous waste. It is, however, an important piece of the puzzle. Currently, it is the only process available which completely destroys the waste rather than partially treating and/or permanently storing it in some location believed to be safe.

Management of Hazardous Materials and Wastes: Treatment, Minimization and Environmental Impacts. Edited by S. K. Majumdar, E. W. Miller and R. F. Schmalz.

Chapter Five

SANITARY LANDFILL TECHNOLOGY: ENGINEERING PRINCIPLES FOR IMPROVED DESIGN

JOHN MELHAM[1] and RAYMOND W. REGAN[2]

[1]President
Melham Associates
Harrisburg, PA
and
[2]Director
Office of Hazardous and Toxic Waste Management
The Pennsylvania State University
University Park, PA 16802

INTRODUCTION

Unquestionably, one of the most difficult problems of waste management confronting the nation today involves obtaining new sites for the secure long-term storage and final placement of municipal solid wastes (MSW). Although a frequently expressed viewpoint is that the sanitary landfill concept has outlived its usefulness and viability, several critical aspects support the contention that such an opinion follows from assumptive arguments based on past problems, rather than from objective facts and an open mind to engineering design practices.

Persuading arguments in favor of the landfill concept include:

- the lack of currently available options for some nonrecyclable materials in essentially all municipal solid wastes;
- the fact that other options such as incineration involve lengthy developmental and construction times, as well as their share of public opposition;
- the comparatively recent developments in landfill design engineering;
- established and proven monitoring equipment and procedures; and
- straightforward engineering requirements utilizing experience proven earth-moving and construction equipment and materials.

These strong points for the landfill concept should be kept in mind in evaluating solid waste disposal options. Equally, the landfill option should be objectively analyzed without undue prejudice from past uncontrolled 'dumps' and earlier attempts at controlled conditions. To rule out landfills as 'hopelessly unworkable' is to adopt a philosophy by which the airplane would never have flown, man would not have walked on the moon, and digital computing would consist of the animated use of five or ten extended fingers.

The purposes of this chapter include: (1) a basic definition of the traditional landfilling concept with the associated engineering principles for its design, (2) a discussion of the major shortcomings common to most sites chosen for landfill development, and (3) the presentation of a sanitary landfilling concept that may yield a far-higher factor of confidence by the public sector in this method of disposal.

PROCESS DEFINITION AND ENGINEERING PRINCIPLES

A basic definition of the sanitary landfilling process published by the ASCE[1] in 1959 was stated as follows:

"Sanitary landfill is a method of disposing of refuse on land without creating nuisances or hazards to public health or safety, by utilizing the *principles of engineering* (emphasis added) to confine the refuse to the smallest practical area, to reduce it to the smallest practical volume and to cover it with a layer of earth at the conclusion of each day's operation or at such more frequent intervals as may be necessary."

Several authors over the years have offered statements containing many of the basic elements of this definition[2,3,4] or by the lack of one, appear to indicate that such a statement is common knowledge.[5] Based on the writer's experience, a crystalized listing of the engineering principles for sanitary landfilling, along with a basis for their use, have been lacking. Therefore, such a list has been prepared (Table 1). Undoubtedly, this listing needs to be updated to address environmental impacts (surface and ground water protection, for example) as well.

Examples of poorly operated landfills (Figure 1) and well operated facilities (Figure 2) illustrate the need for continuing engineering planning during the operational phase of all sites, as has been stated[1] as follows:

"The apparent simplicity of the landfill method should not be construed as obviating the need for competent and continuing engineering planning and control. On the contrary, lack of engineering planning almost always results in dissatisfaction with the process . . ."

Without doubt, the public's visual perception of the local facility has an influence on their openess to future sanitary landfill developments in their area.

Addressing the issue of environmental impacts in general and the need for modern landfill designs to specifically mitigate ground water deterioration represent challenges for today's engineers, as indicated in the following sections.

CONVENTIONAL LANDFILL SITES

Traditionally, and perhaps up to the present day, landfills have been generally located on sites that are as flat as possible. Such sites were chosen because they facilitated the respective jobs of everyone directly concerned: the operator could easily oversee and manage the landfill; the engineering consultant could quickly make all volumetric calculations and plan sequential filling operations, and the refuse haulers would have easy site access and unloading capabilities. In the days before serious environmental concerns, these were the paramount considerations. Often the sites made available for landfills were low-lying areas which were unsuitable for either agriculture or real estate development. It is also worth mentioning that the preference for flat landfill sites apparently extended to locations where any level land was at a premium. As a general rule, most landfills have been located on sites with gradients of 10 percent or less.

The advent of environmental regulations did not alter the landfill siting

TABLE 1

Traditional Principles of Sanitary Landfilling

	Comment
1. Prepared site; developed on the basis of an engineering plan involving MSW placement, cover soil availability and surface grading;	Site selected primarily to meet the soils characteristics and groundwater level criteria of regulatory agency.
2. spread and compact the MSW in thin layers (two feet) on the working face;	Achieve compaction to 800-1000 lbs. per cubic yard.
3. develop a working face of compacted MSW at a horizontal-vertical slope of 3:1, with a lift depth of approximately eight feet;	Minimize exposed surface area and amount of cover soil needed daily.
4. spread and compact six inches of cover soil over the exposed MSW at the end of the day, or at more frequent intervals as warranted;	Prevent the emergence of insect larvae from the solid waste.
5. install intermediate cover soil at one foot compacted depth when a multi-lift operation is anticipated; and	Provide surface bearing capacity for vehicles unloading at site.
6. install a final layer of cover soil at two feet compacted depth.	Prevent the harboraging of animals into the solid waste.

FIGURE 1. View of Landfill Presenting a Poor Public Image.

FIGURE 2. View of Landfill Presenting an Improved Public Image.

preferences. Although it was obvious that new sites for any landfill purpose would have to be carefully chosen to minimize the potential for groundwater contamination, it has brought no change to the *type* of site being sought; although it would now have to be above flood plain levels, the desirable site may be viewed as having a flat-to-gently rolling topography.

The problem is that such sites actually *increase* everyone's burden from an environmental impact perspective, because flat sites share a common handicap in that they maximize conditions for water infiltration. Due to the shallow subsurface gradients infiltrating water tends to collect from higher adjacent land, as well as from within its own perimeter. This type of land surface acts as a recharge zone for the ground water table.

This natural hydrological function is the primary reason why such sites are undesirable for landfills. The high rates of infiltration, the seasonal and often substantial fluctuations in ground water levels, and soil dynamics due to moisture variations and hydrostatic pressures, all act on the landfill mass. Uncontained materials will sooner or later either combine with or be transported by water to form a leachate; waste containers, "impermeable" liners or similar protective measures will inexorably deteriorate under the presence of weathering. However careful the landfill operation attempts to be to minimize these problems, the simple truth of the matter is that the flatter topography is working directly against these efforts. The landfill presents a flat and "open" face to precipitation; infiltrating water from adjacent ground wants to seep into the landfill through the sidewalls of the excavation; water collecting in the pit has to be continually pumped to some discharge point; lastly, the tendency for groundwater to rise towards its preexcavation level as the site becomes landfilled, and the ensuing hydraulic pressure, is a continuing threat to the integrity of the liner. It is precisely these types of sites that have created the common (but by no means uneducated) opinion that "all landfills will leak".

While it is equally true that "all things built by man will eventually fail", the important aspect is not the certainty of failure, but the certainty that such a failure can be safely detected and repaired before damage has occurred. Here again, the flat sites chosen for landfills, even the newer generation "secure" landfills (Figure 3) are far less than satisfactory. Because of topographical constraints, leak detection systems are more compromises than what they should be. A common approach is to utilize vertical wells strategically placed within the landfill area to provide monitoring access to each "cell" or section of the facility. In the case of the current, state-of-the-art preferences for double liners, the monitoring well should provide detection that the upper liner has leaked. To achieve this, the well should penetrate the first liner to monitor the interstitial space between the two liners. These penetrations, however well sealed, constitute potential weaknesses particularly as the landfill settles and consolidates over several years. The potential problem is compounded by the detection accuracy desired: as the number of monitoring wells is increased to permit closer and quicker

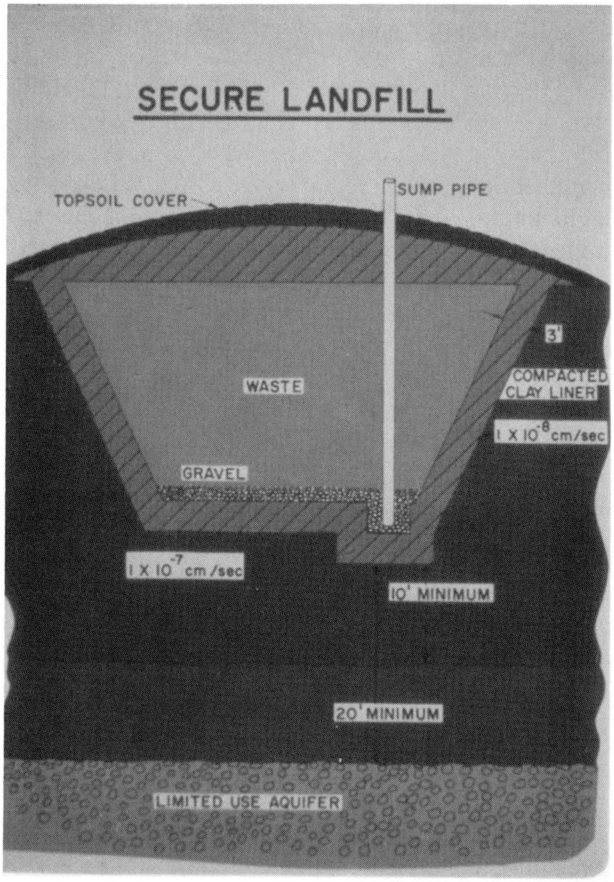

FIGURE 3. Design Elements for a Secure Landfill.

detection of leaks, the potential stress points are also increased. Over time, this has to mean increased probability of a leak around the monitoring well casing.

While the above discussion has centered on the inherent problems that the typical landfill site has regarding both surface and groundwater regimes, there are other environmental impacts which are exacerbated by their generally flat topography. Although these impacts are less critical to the natural environment than hydrological degradation, they are sufficiently important to the physical and human environments to create further negative (and usually hostile) attitudes of people within the impact area. These impacts include those that affect the perceptions of aesthetics, air pollution through airborne particulates, and increased ambient noise levels. All three of these perceived impacts are, in terms of severity, generally in an inverse relationship with the topographical complexity of the landfill site and surrounding area: the shallower and flatter the ground relief, the more potential impact. Visual prominence probably ranks second highest to the hydrological threat, in terms of site factors that contribute to the "not in my backyard" syndrome, and therefore should not be underestimated. And yet the typically flat sites chosen for landfills, with their often considerable sight distances, are in direct confrontation with this simple human preference. It is also why elaborate attempts to screen the landfill with some type of landscape buffer has limited effect in reducing opposition, because the buffer itself reminds the viewer what lies beyond. The extended sight distances of such flatter areas also increased the potential impacts of noise and airborne particulates generated by landfill equipment, although buffers can be effective in reducing these adverse effects.

The above summation of problems typically encountered with conventional landfill sites gives ample reason for the low confidence level on the part of the public towards current landfill practices. Although these problems can each be addressed through potential engineering solutions, the fact of the matter is that by not starting with more environmentally acceptable sites, the engineering becomes far more complex, involved, and prone to failure. Confronted then with an informed public, which is increasingly wary of overly complicated engineering solutions, the entire case for the landfill disposal method is usually not advanced. In this new age of environmental concern and accountability, the landfill concept will have to be more responsibly designed to address the problems discussed above, or it will cease to be a viable option.

AN ALTERNATIVE: THE STEEP-SLOPE CONCEPT

As the name implies, the alternative siting concept is axiomatically different from the conventional approach. The following section explains the concept in terms of the rationale behind site selection. To focus on the merits of this alternative, it is assumed that any landfill site will be located in a rural area.

The paramount concerns in site selection have nothing to do with operations convenience, traditional preferences, or even proximity to the sources of solid waste generation. Recognizing that without environmental acceptance there will be no landfill, the major criterion for site selection is to locate a site where topography, geology, and hydrogeology can be used to minimize or effectively eliminate the main nemesis of a landfill, its potential contamination of water. As indicated in the previous section, this can occur in three ways: surface infiltration from precipitation, lateral subsurface migration, and upward movement of the water table.

Steeper slopes, which shall be categorized here as slopes in excess of 15 percent, have an intrinsic advantage over shallower gradients in terms of controlling water intake of the soil. Because their runoff quotients are far higher than the flatter areas, the rates of infiltration for steep slopes are commensurately far lower. This fact, of course, is well documented by the many formulas (Mannings and others) for storm runoff calculations utilized by civil engineers.[7] Steeper slopes also permit easy interception and diversion of surface runoff draining from uphill areas, as the higher overall slopes permit great flexibility in diversion ditch design. Steeper slopes, also by virtue of their upland elevations, are exposed to higher wind regimes, and a positive effect on evaporation. And if such slopes are oriented in a southerly direction, there is a substantial increase in evaporation through solar gain, further minimizing infiltration. So, the first aspect of site selection is to locate potential steep-slope areas, with southern exposures, if possible. Often as not, such slopes have an additional advantage, as such slopes usually have been spared from agricultural or residential development.

Geology is also important. While much interest has been shown in finding suitable soil conditions, this is primarily of economic benefit rather than an essential requirement, as clay or other impermeable materials can always be imported to the site. Of equal or greater interest is to find "dry" lithology, heavy in aquicludes, and having no water bearing aquifers. Minor seeps between strata are acceptable, because these can be easily diverted by isolation trenches around the perimeter of the site. Steep-slope sites with geologic faults would be avoided, in keeping with most current state regulations.

Besides subsurface aquifers and seeps, the other hydrological concern is that of the watertable. Here again, the steeper slopes have an advantage, for such sites with these gradients are less effective as recharging zones and therefore have much lower ground watertables than flatter sites. Steep-sloped, upland sites are very common and numerous in many states, with watertables generally far too deep to ever present a problem for landfilling.

The steep-slope siting concept therefore utilizes hydrology, geology, and topography to provide maximum *natural* controls for decreasing water buildup in the landfill site. The other factors of concern usually can be easily addressed as well. Steeper-sloped areas are generally within a system of some form of hills and valleys, and therefore sites within the topographical system can be chosen that are effectively hidden by the surrounding hills. Hence, the potential problems of aesthetic impairment, dust and noise generation, and other similar concerns can be effectively mitigated. And because the natural elements of the environment are used to their full advantage, subsequent design and construction costs for environmental counter measures are reduced. Overall, the potential environmental acceptability is improved, because properly selected steep-slope sites address the environmental concerns far better than the traditional landfill sites.

ELEMENTS OF SITE DEVELOPMENT

The method of developing a landfill on a steep-slope site is envisioned to be heavily influenced by the topography, and therefore would differ significantly from that of more conventional flat sites. The excavation equipment and machinery are essentially the same, but it is the sequencing of their utilization that differs. Elements for the development of a steep-slope site follow.

Site road access would basically run along the general contours until it reached the elevation at which the landfill will be constructed. The excavation area is then planned, and this will extend perpendicular to, and uphill from, the access road. It is at this point that a fundamental difference with conventional sites becomes apparent: steep-slope sites will be more linear in plan, and will take the form of a uniform width excavation that follows along the access road. Maximum widths will be an inverse function of the slope of the existing ground: the greater the slope, the narrower the width of the excavation, because as the steepness of the slope increases, so does the requisite amount of excavation. Because steep slopes generally have shallow depths of soils, the excavation will be into the bedrock, and expenses increase with the depth of the cut. For simplicity here, we will assume that the "floor" of the excavation will be 10 feet or so above the level of the access road. The floor would be constructed horizontally flat; therefore, on a site with a slope of 5:1 (horizontal to vertical) or 20 percent, a nominal excavation width of 250 feet would entail an exposed uphill rock cut of some 50 feet. This is by no means unusual, as the contour strip mining of coal frequently involves rock cuts (referred to as 'highwalls') of 100-120 feet.

Machinery to be used is straightforward civil engineering equipment. A drill rig bores blast holes along a contour line uphill from the access road; these holes are drilled to the desired depth of floor of the excavation; charges (usually a mixture of ammonium nitrate and fuel oil) are placed, the holes are stemmed, and the charges detonated. Large front-end loaders and bulldozers then remove the fractured rock to a temporary storage area. Depending on the extent of the cut into the hillside, the blasting/loading cycle may have to be repeated a number of times until the maximum uphill extent of the landfill excavation is reached.

It is important to note here that while the width of the landfill is dependent upon topographical limitations, there is no such physical limitation on the length of the cut along the contour. Similarly, the "floor" itself of the landfill can be terraced up the hillside in two or three levels, thereby reducing the extent of the final cut along the "highwall". Therefore, there is no real limit to the size of the landfill, as it can expand both along the contour and uphill as the need arises.

The next important aspect of site development is that of excavation sequencing. Unlike some flat landfills, the entire area is not laid bare with one gigantic excavation. Rather, only a section at a time is cut, perhaps 100-200 feet along the length of the access road. Narrow cuts minimize impacts to the environment from temporary overburden storage and rainwater collection, and through associated cost savings, permits greater resources for environmental protection to be provided within each incremental cut.

Following this initial excavation of one section or "cell", the protective measures can be installed. The exposed highwall gives an excellent indication of any existing seeps, besides lithological confirmation. Generally, any seep that might be present can be easily directed, either along the highwall or under the floor of the excavation through the use of crushed rock 'chimney' drains. These drains can be easily permanently sealed from the landfill area itself, their downhill drainage exits would be accessible for periodic monitoring.

A comprehensive leak detection system would now be installed in the floor of the cell. Here again, the steep-slope concept provides a considerable advantage over the conventional flat sites, because the leak detection system relies on the infallible laws of gravity. After lining the floor with whatever synthetic or natural materials are chosen, the detection system, consisting of a network of pipes bedded in gravel, is installed. Four to six inch pipes would be laid horizontally to a slight pitch of one or two percent to drain towards the access road. Each pipe would exit the landfill into a small receiving sump; these sumps would be roughly at grade within the previously mentioned 10 foot high slope between the access road and floor of the landfill. Although these pipes would end in a uniform line of access sumps along the contours between the access road and the landfill for the first cell, they will have staggered beginnings in an order sequence uphill, as additional cells are constructed.

Not only does this type of detection system provide a high degree of locational accuracy of any problem occurrence, the simplicity of its design guarantees its continuing function for as long as the longevity of the pipe materials themselves. The horizontal configuration would guarantee continued functioning, whereas, vertical wells can be closed off by lateral ground movement. The system has the inherent longevity of its historic antecedent, the ancient Roman drainage tiles that continue to function today.

Two advantages over the conventional flat site's monitoring system discussed earlier are immediately evident: because the pipes are laid horizontally, there is no penetration of the overlying liner and therefore the liner's integrity is protected; secondly, the number of monitoring pipes, together with sequencing of the staggered uphill ends, permits the location of any leak to be accurately pinpointed.

Following the construction of the leak detection system and the underlying and overlying containment liners, the rest of the first "cell" can be developed. The highwall area of exposed rock is the next area of attention. Strictly speaking, if the dip of thetrata is down towards the inside of the excavation (and up towards the top of the hill), then it would be impossible for any chemical leakage to seep out of the landfill through the highwall. Practically speaking, however, it will probably be easier to line the highwall zone with impermeable material rather than debate about its redundancy in such cases. So, the design of the highwall area would include three general elements: an uphill diversion ditch to direct water away from the landfill, a drainage curtain or 'chimney drain' to divert seeps, and a membrane along the base to contain the landfill material and associate leachate contamination.

CONCLUSION

This discussion has put forth an option that promises to give new credibility to the landfilling of solid wastes. The many inherent advantages of the steep-slope concept span all technical aspects that are essential to the success of a landfill operation, including highest priority (environmental protection), middle priority (landfill economics) or lower-priority advantages (operations flexibility and local aesthetics). Disadvantages are minimized, such as the increased transportation costs associated with the potential greater distances of such sites from the communities generating the waste materials.

The potential number of environmentally acceptable areas, however, ensures that sites may be located very close to the network of interstate and other limited access routes.

Perhaps the greatest advantage of the concept is that by applying the natural laws to protect the environment, the steep-slope siting alternative may earn the public's acceptance, and restore their confidence that solid wastes may be more safely, and practically, disposed of in this fashion than any other currently available technique.

ACKNOWLEDGMENT

The genesis of this topic was presented as a term project report by John Melham, CE596D, *Advanced Hazardous and Industrial Waste Treatment,* The Pennsylvania State University at Harrisburg, Summer Session 1984.

REFERENCES

1. Am. Society of Civil Engr. 1959. *Sanitary Landfill.* Manuals of Engineering Practice No. 39, New York, 61 pp.
2. New York State Dept. of Envir. Conservation. 1971. *Sanitary Landfill, Planning, Design Operation Maintenance,* 32 pp.
3. Brunner, D.R. and D.J. Keller. 1972. *Sanitary Landfill Design and Operation.* U.S. Environmental Protection Agency (SW-65ts), Wash. DC, 59 pp.
4. Am. Society of Civil Engr. 1975. *Sanitary Landfill.* Manual of Practice No.

39, New York.

5. Tchobanoglous, G., H. Theisen and R. Eliassen. 1977. *Solid Wastes: Engineering Principles and Management Issues,* McGraw Hill Book Co., New York, 621 pp.
6. Pavoni, J.L., J.E. Heer, Jr. and D.J. Hagerty. 1975. *Handbook of Solid Waste Disposal, Materials and Energy Recovery.* Van Hostrand Reinhold Environmental Engineering Services, New York, 549 pp. (1975).
7. Fenn, D.G., K.J. Hanley and T.V. DeGeare. 1975. Use of the Water Balance Method for Predicting Leachate Generation from Solid Waste Disposal Sites. (SW-168) U.S. Environmental Protection Agency, Wash. DC, 40 pp.

Management of Hazardous Materials and Wastes: Treatment, Minimization and Environmental Impacts. Edited by S. K. Majumdar, E. W. Miller and R. F. Schmalz. © 1989, The Pennsylvania Academy of Science.

Chapter Six

INNOVATIVE BIOTECHNOLOGIES FOR DETECTION AND DISPOSAL OF HAZARDOUS MATERIALS AND WASTES[1]

BRUCE MOLHOLT

U.S. Environmental Protection Agency
Region III
841 Chestnut Street
Philadelphia, PA 19107

INTRODUCTION

The 1986 Superfund Amendments and Reauthorization Act (SARA) requires use of innovative technologies in the detection and disposal of hazardous materials and wastes. Whereas physical techniques such as gas chromatography may be modified to enhance detection of hazardous substances and such as incineration may be controlled to permanently destroy many hazardous waste constituents, in this Chapter I will only consider biological approaches to detection and disposal.

Several innovative biotechnologies are available for enhanced *detection* of hazardous substances which may be applied to hazardous materials and wastes. In this Chapter I will discuss biotechnologies which are appropriate for detection of carcinogens either alone or in complex mixtures. These biotechnologic approaches are of two sorts:

1. Short-term tests for mutagenicity, which have been available for a decade, but rarely applied to hazardous wastes.
2. Monoclonal antibodies, which react with high specificity and affinity to many hazardous waste constituents.

Biotechnologic strategies for *disposing* of hazardous wastes, by contrast, are in their infancy and have as yet relatively little data from actual field applica-

---

[1]The opinions expressed in this Chapter are those of the author and should not be construed as the official position of EPA.

tion. Hence, the discussion of disposal of hazardous waste constituents through biodegration by genetically engineered micro-organisms will be largely theoretical.

SHORT-TERM TESTS FOR DETECTION OF HAZARDOUS SUBSTANCES

The fact that carcinogenesis is initiated by mutagenesis of human oncogenes may be exploited in the use of short-term mutagenic assays for the enhanced detection of carcinogens in hazardous materials and wastes (Molholt, 1989). Whereas short-term tests have been frequently employed to screen chemicals for potential carcinogenicity, they have not been widely applied in the field for measuring the carcinogenic potential of hazardous wastes, despite the fact that these mutagenesis tests are quick, inexpensive and allow fairly good estimation of the total carcinogenic potency of a given waste stream. In that the total carcinogenic impact of a given hazardous waste site normally drives the risk assessment process, it would seem logical that such a facile approach as short-term testing would play a more major role in early assessment, perhaps even in ranking, of hazardous waste sites nationwide.

Short-term tests for carcinogenicity have improved vastly upon the Ames test which employed various histidine-requiring strains of *Salmonella typhimurium*. A recent, relatively complete breakdown of short-term tests by cost and accuracy in predicting carcinogenicity may be found in Table 2 of Lave and Omenn (1986), from which the following table is excerpted:

SUMMARY AND COMPARISON OF SHORT-TERM TESTS*

Test cell type	Carcinogenicity score**	Cost***
PROKARYOTE		
Ames (*Sal. typhim.*)	70	1600
E. coli recA	60	2000
B. subtilis recA	79	1100
EUKARYOTE		
Saccharomyces (yeast)	66	1800
Sister chromatid exchange	60	4000
Chinese hamster ovary (HPRT)	67	8700
Human foreskin (HPRT)	60	8700
Unscheduled DNA synthesis (HeLa)	68	7000

*Excerpted from Lave and Omenn (1986)
**Percent detection of carcinogenicity in 20 tests or more
***In 1988 dollars

In total some 250 different short-term tests for mutagenicity/carcinogenicity are available with 40-90 percent positive detection of carcinogens at costs of 550-16,000 dollars per test. Costs are definitely advantageous in comparison to *in vivo* assays for carcinogenicity where a given test employing hundreds of test animals and a team of pathologists over 3 years may cost $2.5 million or more.

The advantage for employment of short-term mutagenic tests in the bioassay of potential carcinogenicity of hazardous materials or wastes is twofold for complex mixtures:

1. Carcinogens may be detected among other organic chemicals which interfere in physical methods of analysis.
2. The total carcinogenic impact of many carcinogens may be quantified in a single test.

Some of these advantages (and techniques) are summarized in the recent monograph edited by Sandhu et al. (1987). In EPA Region III, we have proposed utilization of short-term tests to summarize carcinogenic risk at two hazardous waste sites, the Drake Chemical Site and Millcreek Site in middle and western Pennsylvania, respectively. An additional description of short-term tests and their utility in description and analysis of complex hazardous waste mixtures may be found in Waters et al. (1984).

A recent example of high detection limits for a probable human carcinogen in a waste site containing many species of organic acids is didactic. The waste site in question was known to contain trace amounts of bis(2-chloroethyl)ether (BCEE), however, groundwater samples were so contaminated with a complex series of organic acids that the detection limit by normal analytical instrumentation for BCEE was 20 ppb. As a fairly potent carcinogen, BCEE at only 30 pp*t* represents a 10^{-6} cancer risk by EPA risk assessment methodology (see Molholt, 1989). Hence, at the site in question, the detection limit for BCEE represented a $20{,}000/30 \times 10^{-6}$, or 6.7×10^{-4} individual cancer risk, which detection limit is untenable in terms of the public health protection mandate in SARA.

Since BCEE is mutagenic as well as carcinogenic, employment of short-term test systems for BCEE detection and quantification would enhance the sensitivity of risk assessment in the above-mentioned case. An additional method for enhanced detection of BCEE in the midst of competing organic acids is discussed below in the section on monoclonal antibodies.

There are several difficulties in the utilization of short-term test systems to quantify hazardous waste constituents which must be discussed before leaving the subject. First, the various short-term tests are not able to detect all carcinogens. Since mutagenicity is required, only carcinogenic initiators score positively and the test will not score carcinogenic promotors. (However, this point may be moot in light of the probable risk assessment model for promotors. See Molholt, 1988). Second, since many carcinogenic initiators must be chemically modified before they are mutagenic, short-term tests such as the Ames assay employ microsomal oxidases or other enzymes to convert procarcinogens to ultimate carcinogens capable of mutagenizing the test organism. These *in vitro* metabolic manipulations may not mimic those in vivo and alter apparent carcinogenic potency significantly. Lastly, if bacteriocidal or bacteriostatic agents contaminate a given hazardous waste mixture, short-term tests which employ bacteria will be impaired and give misleadingly low results. The same caveat pertains to other short-term tests which employ eukaryotic organisms when hazardous waste mixtures contain other toxins for the cell type in question.

Short-term tests for carcinogenicity have not yet been employed for screening carcinogens by any regulatory agency. This is unfortunate in that mutagens in short-term tests almost always are positive for both carcinogenicity and mutagenicity in higher animals, including, where there is adequate epidemiologic evidence, *H. sapiens.* It is ironic that the Chemical Manufacturers Association has persistently blocked attempts to utilize short-term tests in regulating carcinogens, accepting only experimental carcinogenic or mutagenic data from laboratory animals, and yet argue after much more expenditure than for short-term tests "of course, results in rats or mice have dubious significance in humans." (Cox, 1987).

IMMUNOLOGICAL DETECTION OF HAZARDOUS SUBSTANCES

Many chemical carcinogens are organochlorines, substituted benzene or benzidine rings, or polycyclic aromatic hydrocarbons each of which, when combined with a protein carrier, elicit an antibody response. The ability of these carcinogens to act as haptens may be utilized in enhanced immunological detection by linking their reaction with cognate antibodies to an indicator which is sensitive to the formation of antigen-antibody complexes. Many of these immunocomplex indicators are quite sensitive and colorimetric such that nanogram quantities of chemical carcinogen are easily detected.

An example of one such immunocomplex indicator which is useful in detection of carcinogenic chemicals is the ELISA test (enzyme-linked immunosorbent assay). Each antigen-antibody complex in the ELISA test activates an enzyme whose chromogenic substrate changes color in a cascade at each molecular event. Hence, enzyme catalysis multiplies the sensitivity of carcinogen detection many thousand-fold, allowing lowered detection limits from micrograms to nanograms or less, depending upon the antigenic strength and immunoglobulin avidity of the test system employed.

Utilization of immunological assays for carcinogens requires isolation of monoclonal antibodies for each carcinogenic chemical. These are conveniently selected from murine lymphocytes which elicit cognate immunoglobulins after antigenic challenge which are then hybridized with immortalized myeloma cells. The resulting hybridoma is an immortal cell line which perpetually synthesizes immunoglobulin with high affinity for a given carcinogen as hapten.

Enhanced detection of carcinogens by several orders of magnitude is required over present analytical techniques to quantify carcinogenic risk in the 10^{-6} range for many common carcinogens such as vinyl chloride, PCBs and chlorinated dioxins and furans (see discussion above). In addition, since immunological reactions are highly specific and capable of resolving even closely related isomers of the same compound, they are useful in detection of small quantities even in the presence of gross contamination, provided the contaminants do not denature proteins or contribute significantly to ionic strength.

Potentially, ELISA-based assays could be modified for consumer use, similar to the familiar "dip-sticks" used for detection of glucose in urine. It has been realized that much human exposure to carcinogenic substances occurs in the home and that contaminated indoor air frequently provides the vehicle (EPA, 1987). Just how much risk do you encounter in your own home? Wouldn't it be nice to have a convenient colorimetric monitor to check if certain contaminants were above a dangerous level?

For example, it appears that chlordane and heptachlor, the termiticides of choice for millions of American homes over the past 40 years, incur about the same carcinogenic risk nationwide as radon gas (Molholt, 1986). Since the risk for each homeowner is proportional to indoor air levels of chlordane and heptachlor, it would be desirable to monitor each home to which these termiticides have been applied. Inexpensive ELISA-based monitors for chlordane and heptachlor would allow each homeowner to decide whether or not they were at risk and, if so, the extent to which they should take corrective action. Similar monitors for other pesticides could be used to evaluate residues which remain on fruits and vegetables, especially those that are imported, by appropriate importers or regulatory agencies.

INNOVATIVE BIOTECHNOLOGIES FOR DISPOSAL OF HAZARDOUS WASTE

Microbial decomposition is a fundamental facet of natural recycling. Many saprophytic micro-organisms depend upon a wide variety of organic chemicals for their very existence and achieve evolutionary status through clever schemes for degrading the virtually indigestible. For example, only microbes (whether free living or in the termite gut) can digest cellulose, and do so with such ferocity that they contribute more atmospheric carbon dioxide from cellulose digestion than does mankind from all its burning and internal combustion.

Bacteria and fungi are especially adept at evolving wide varieties of catabolic enzymes for the digestion of unpalatable substrates. Perhaps this is because they were the first organisms to evolve in the hostile environment of primeval Earth. Marine bacteria exist today in superheated thermal vents of ocean trenches at 300 C. Fungi often contaminate solutions at pH extremes (under pH 1, over pH 13). As an additional facility for environmental hardiness, bacteria and fungi are strikingly mutable. For example, if *E. coli lac* deletions are alternately mutagenized and grown on lactose as a sole carbon source, a new lactose-cleaving activity evolves to replace the deleted β-galactosidase.

The Superfund Amendments and Reauthorization Act (SARA) requires use of innovative technologies for the destruction of hazardous waste constituents. In that many microbes are found which contaminate hazardous waste sites, it only seemed natural to include biodegradation among the schemes for hazardous chemical destruction according to the provisions of SARA. For example, many sites which contain trichloroethylene (TCE) are also found to contain lesser amounts of dichloroethylene (DCE) and vinyl chloride (VC, which is monochloroethylene), even though TCE was the only chlorinated ethylene disposed. The culprits are anaerobic bacteria which clip off chlorines from the ethylene backbone one at a time. This is a very dangerous situation in that each successive chlorine removal vastly increases the vapor pressure of the remaining ethylene until VC, with a vapor pressure of 3.5 atmospheres veritably leaps out of contaminated soil or groundwater. There have been homes impacted by VC gas from TCE biodegradation in which the lifetime cancer risk is *35* (each person breathing indoor air contaminated with VC at the level found will develop 35 cancers in a lifetime).

The impetus for biotechnologic engineering of organisms for biodegradation of hazardous wastes originated 20 years ago with the development of pseudomonads useful in cleanup of oil spills (Atlas, 1981). Of interest in the biotechnologic revolution, General Electric would go on to patent the plasmid-containing pseudomonad for petroleum hydrocarbon degradation isolated by Chakrabarty. Recently this precedent on patenting a microbial lifeform has been extended to a strain of mouse which is predisposed to cancer (the MYC mouse).

Several strategies have been developed for the derivation of microbes useful in hazardous waste biodegradation. Two major variations in microbe selection strategies are 1) isolation of naturally existing micro-organisms which have evolved within hazardous waste contamination, and 2) genetic manipulation of known voracious strains such that they catabolize the hazardous waste in question. Both major strategies of microbe selection have several substrategies which are summarized in the paragraphs below.

1) *Naturally existing microbes* may be isolated either as purified strains or groups of microbes which have proliferated at a hazardous waste site. Most often these are soil bacteria and fungi which are naturally able to digest synthetic organic chemicals. Oftentimes these naturally existing microbes are enriched by incubation, aeration and/or nutrient addition in contaminated soil samples brought into the laboratory setting.

2) *Genetically manipulated microbes* are generally well known genera such as *Pseudomonas* spp. which have a wide variation in catabolic capability and are amenable to mutagenesis, DNA transformation and laboratory manipula-

tion. Some strategies call for the mobilization of genes which encode catabolic enzymes into a more favorable host background. Others call for enzyme overproduction through gene amplification (viz. on plasmids), up-promotion or deletion of genes active in repression or feedback inhibition. Occasionally novel catabolic capabilities are created *de novo* through forced mutagenesis ala the evolution of lactase activity in *lac* deletions of *E. coli*. It is of advantage in terms of genetic manipulation (ease of gene isolation, mutation, transfer, etc.) if the selected catabolic traits are encoded by plasmid genes, and this often turns out to be naturally the case, as if Nature had been running a giant genetic manipulation laboratory (called Evolution?) all the while.

Reviews of selection strategies for biodegradation of hazardous wastes may be found in Alexander (1981), Kobayashi and Rittman (1982) and Amdurer et al (1986). Other heterotrophic bacteria which have proved useful in hazardous waste biodegradation (besides *Pseudomonas*) include *Achromobacter, Arthrobacter, Acinetobacter, Micrococcus, Vibrio, Brevibacterium, Corynebacterium* and *Flavobacterium.*

The pseudomonads continue to be the most useful microorganisms for biodegradation of synthetic organics and other environmental toxins (Ornston, 1971). Pseudomonads have been isolated by both natural selection and laboratory (genetic) manipulation which will biodegrade the following compounds:

1-octene	cyclohexane	camphor	DDT
1,7-octadiene	benzene	phenolics	
1,2-epoxyoctane	toluene	4-chlorophenol	
chloroacetate	alkyl benzene sulfonate	pentachlorophenol	
polyvinyl alcohol	p-hydroxybenzoate	catechol	
C18-carboxylic acid	dihydroxybenzoate	naphthalene	
acetamide	3-methylbenzoate	chloronaphthalenes	
diethylamine	3-methylmuconic acid	PCBs	
diethanolamine	3-chlorobenzoate	benzo(a)pyrene	
dimethylamine	p-dichlorobenzene	diazinon	
di-n-butylphthalate	1,3,5-trichlorobenzene	2,4,5-T	
dimethylnitrosamine	1,2,4-trichlorobenzene	dieldrin	

Many of the compounds digested by pseudomonads are carcinogenic, such as benzene, p-dichlorobenzene, PCBs, benzo(a)pyrene, DDT and dieldrin. In the case of PCBs, it must be noted that completion of biodegradation by pseudomonads is inversely proportional to the degree of PCB chlorination (Furukawa and Chakrabarty, 1982) (percent chlorination is indicated by the last two digits of the Aroclor number, e.g., Aroclor-1254 is 54% chlorinated).

Many hazardous waste constituents are organochlorines which were hailed as revolutionary chemicals due to their thermal stability and environmental longevity. The chlorine-carbon bond resists biodegradation and the ability of pseudomonads and other microbes to degrade this particular linkage has proved a breakthrough in bioremediation strategies for most hazardous waste mixtures (Rochkind et al., 1986). Unfortunately, in addition to being more resistant to biodegradation, increasing chlorination of organic chemicals also leads to higher octanol/water partitioning and increasing bioaccumulation. Hence, most of the world's animals, including *H. sapiens,* regardless of locality, contain part-per-million residues of PCBs, DDT and dieldrin and part-per-billion residues of polychlorinated dibenzodioxins in their fatty tissues.

The white rot fungus synthesizes the enzyme ligninase which, besides degrading lignin, a critical reaction in rotting wood, partially degrades many organochlorines, including the polychlorinated dibenzodioxins (Bumpus et al., 1985). A clever biotechnologic strategy would be to isolate the white rot fungus gene for ligninase, splice it by recombinant DNA technology into a bacterial plasmid and transform *Pseudomonas* to antibiotic-resistance encoded by the same plasmid. With the ability of many pseudomonads to degrade several organochlorines, the addition of ligninase should allow facile digestion of nearly every organochlorine constituent of hazardous waste mixtures, as well as many other toxic contaminants.

Many of the potential applications of innovative biotechnology have not yet been realized in biodegradation of hazardous wastes. There appear to be several reasons for this retardation of biodegradative technology development as compared to relatively rapid development of biotechnology in other related industries such as pharmaceuticals and agriculture. First, there is a public abhorrence of hazardous possibilities in both hazardous waste and biotechnologic arenas, such that they make for an uncomfortable marriage. Second, although there are several supportive initiatives by federal regulatory agencies, the clear road to approval of bioremediation strategies has yet to be paved. Lastly, whereas bioremediation promises to be enormously saving in terms of comparative cleanup costs vs. other technologies, the up-front research and development is both difficult and expensive. Each of these difficulties will be discussed more fully in the following paragraphs.

PUBLIC DISQUIETUDE OF BIOTECHNOLOGIC REMEDIATION

Ever since Love Canal the public has been finely attuned to imminent health endangerment from hazardous wastes. Unfortunately, in that innovative biotechnology is new, difficult to comprehend and has received some negative press, it, too, has been greeted with foreboding. However, whereas there is reason to fear exposure to hazardous waste constituents, which in virtually undetectable quantities can still manifest significant harm, there is little rational basis for that public fear which in some cases has accompanied biotechnologic research and development.

Genetically engineered organisms are created for a given purpose, e.g., complete digestion of tetrachlorinated dibenzo(p)dioxin (TCDD). There is no conceivable (or inconceivable) way for such an engineered organism to become an "Andromeda strain" which will suddenly adapt to digest humans, automobiles or skyscrapers instead of TCDD. The irrationality of this fear of biotechnology was recently displayed in vandalism of strawberry plants in the 1987 trial of "ice-minus" pseudomonads. The "ice-minus" strain has been deleted for a gene which controls cell wall formation in a paracrystalline array and normally, as a saprophyte of leaf surfaces, provides nuclei for frost condensation. Fear of "ice-minus" pseudomonads is thus akin to being afraid of a toothless strain of wolf. The "ice-minus" mutants are less able to thrive in the environment than normal pseudomonads.

Waste management through microbial degradation has, of course, dominated one arena of solid waste management, that of sewage treatment. Several extant hazardous waste management strategies which employ microbial degradation essentially are redesigned sewage treatment plants in terms of rationale and operation. Sewage treatment plants are able to handle low levels of hazardous wastes and satisfy discharge permits. Many microbial degradation schemes at hazardous waste sites call for completion of biodegradation after treatment on-site by nearby POTWs. Hence, in a rational world, there should be no more apprehension concerning bioremediation of hazardous waste sites than there is public fear of sewage treatment in the universally accepted manner.

FEDERAL SUPPORT OF BIOREMEDIATION

As mentioned earlier, SARA calls for innovative technologies in hazardous waste destruction at Superfund sites. However, in that biotechnology is federally regulated under programs other than Superfund, e.g., in EPA under the Toxic Substances Act (TSCA), it is necessary to receive clearance prior to application of genetically engineered micro-organisms for bioremediation of hazardous waste sites.

As authorized by SARA, EPA has initiated its SITE program (Superfund Innovative Testing and Evaluation) which accepted 22 innovative technology applications for trial in 1987, four of which employed micro-organisms. One of these was for biodegradation of PCBs, which may be important since there are an estimated 134 tons of PCBs in sediments of the upper Hudson River valley alone (Brown et al., 1987). The PCB project has not yet been funded (by mid-1988) and will cost an estimated $200,000.

In addition, EPA has proposed creation of a fund for the development of bioremediation demonstration projects, Biosystems for Pollution Control Initiative, which is proposed to begin in 1989 and will provide $12 million for such projects over a 3 year period. Obviously such a program is iffy during an election year and requires more than good faith by companies considering the development of novel bioremediation strategies. In the long run, since there are handsome profits to be made, it may be that a sounder mechanism of support for research and development of innovative biotechnologies for hazardous waste cleanup will be found within the private sector.

DIFFICULTIES IN BIOREMEDIATION

Most of the above discussion has applauded the technologic merits of biodegradation of hazardous wastes. However, the impression should not be left imbalanced: Innovative biotechnologies for hazardous waste remediation are fraught with enormous problems. Many of the problems considered below are those which should be addressed by research and development in this arena whether federally or privately funded.

Although finding micro-organisms which will degrade a given toxic chemical may be difficult, in general, a greater difficulty is encountered in the low level of these substrates under normal hazardous waste conditions. "Low level" in this sense is as perceived by potential microbial vultures eager for their carrion. Whereas part-per-million levels of TCDD are heinous for public health, this is not a very rich diet even for the famous white rot fungus!

Many strategies have been employed to get around low substrate concentrations, including, supplementation of waste sites with additional nontoxic nutrients, absorption of waste constituents to some matrix which provides microenvironments of increased concentration and physical removal of toxicants from contaminated media prior to bioremediation.

Even if toxic substrates are present at sufficient concentration to maintain microbial growth, at some interim time towards desired cleanup these same substrates will become growth limiting. Hence, additional strategies must be developed to keep micro-organisms growing until the desired cleanup level is attained.

SUCCESS STORIES

Although precarious, at least 65 companies are presently engaged in bioremediation of hazardous wastes. These companies are (as of mid-1988):

Advanced Biocultures Formulations
Advanced Mineral Technologies
American Cyanimid
American Genetics International
Amicon Corp
April Enterprises
Arco Performance Chemicals
Genex Corp
Groundwater Decontamination Sys.
Groundwater Technology
Haztech
Homestake Mining Co
Institute of Gas Technology
ITT Rayonier

Atlantic Research Corp
ATW Caldweld
Avasco
Battelle Memorial Institute
Biochem Technology
Bioclean
Bioscience Management
Biospherics
Biosystems
Biotechnica International
Biotechnology Unlimited
Biotrol
Cambridge Analytical Assoc
Cecos International
Chemical Waste Management
Celanese Chemical Co
Cytoculture International
Detox
Detox Industries
Dupont
Ecova
Enzon
Flow Laboratories
Fluor Corp
General Electric
General Environmental Science
Interbio
International Technologies
Ionics
Johnston Assoc
Keystone Environmental Resources
Lab Systems
Metropolitan Environmental
Microbe Masters
Microlife Techniques
Miles Laboratories
Motec
MycoSearch
Occidental Chemical
Ogden Environmental
Oxford Environmental
Polybac
Reidel Environmental Services
Rollins Environmental
Sybron Biochemical
Syntro Corp
Vertech Treatment
Westinghouse
Winston Laboratories
Worne Biotechnology
Zimpro

The above companies are engaged in one or more projects involving bioremediation of hazardous wastes. Additional companies employ bioremediation of their waste streams to achieve permitted discharge. Bethlehem Steel Co., for example, treats up to three tons per day phenol from its waste stream at the Sparrows Point, MD, plant through a bioreactor seeded with sewage sludge. The output two pounds of phenol which is discharged into the Chesapeake Bay represents a 99.97 percent reduction.

Occidental Chemical, through a subsidiary, has operated a leachate collection system at its Superfund Hyde Park Landfill since 1979. Heretofore, leachate was concentrated over activated carbon at a cost of $2 million per year. However, now through bioremediation the need for activated carbon has diminished by 96 percent which will save Occidental an estimated $20 million at this site alone.

At another Superfund site in California, Ecova is removing hazardous wastes from groundwater through a combination of bioremediation and air stripping. Microbes and nutrients are carefully tailored to remaining toxic chemicals in groundwater as it is being remediated. Ultimately the clarified waste stream is discharged through a POTW.

These are but three of hundreds of success stories using micro-organisms to remediate hazardous wastes and surely many more examples will be available in the near future. Despite problems encountered in virtually every type of hazardous waste situation, it is probable that a cost-effective bioremediation scheme may be found.

REFERENCES

REVIEWS

Alexander, M. 1981. Biodegradation of chemicals of environmental concern. Science *211*:132-138.

Amdurer, M., R.T. Fellman, J. Roetzer and C. Russ. 1986. Systems to accelerate *in situ* stabilization of waste deposits. Section 2. Biodegradation. pp. 74-124. Hazardous Waste Engineering Lab, EPA, Cincinnati, OH. EPA/540/2-86/002.

Chakrabarty, A. 1982. Biodegradation and detoxification of environmental pollutants. CRC Press, Inc., Boca Raton, FL.

Kobayashi, M. and B.E. Rittman. 1982. Microbial removal of hazardous organic compounds. Environmental Science and Technology *16*:170A-183A.

McCormick, D. 1985. One bug's meat. Bio/Technology *3*:429-435.

Omenn, G.S. 1986. Genetic control of environmental pollutants: A conference review. Microbial Ecology *12*:129-143.

SPECIFIC

Atlas, R.M. 1981. Microbial degradation of petroleum hydrocarbons: An environmental perspective. Antonie van Lieuwenhoek Journal of Microbiology and Serology 39:257-271.

Brown, J.F., D.L. Bedard, M.J. Brennan et al. 1987. Polychlorinated biphenyl dechlorination in aquatic sediments. Science *236*:709-712.

Bumpus, J.A., T. Ming, D. Wright et al. 1985. Oxidation of persistent environmental pollutants by a white rot fungus. Science *228*:1434-1436.

Cox, G.V. 1987. Quality assurance in risk models for regulatory decision making. Society for Risk Analysis Annual Conference, Houston, TX.

EPA. 1987. Unfinished business: A comparative assessment of environmental problems. Office of Policy Analysis, EPA, Washington, D.C.

Furukawa, K. and A.M. Chakrabarty. 1982. Involvement of plasmids in total degradation of chlorinated biphenyls. Applied Environmental Microbiology *44*:619-626.

Lave, L.B. and G.S. Omenn. 1986. Cost-effectiveness of short-term tests for carcinogenicity. Nature *324*:29-34.

Molholt, B. 1986. Risk assessment of indoor air pollution by termiticides. Society for Risk Analysis Annual Conference, Boston, MA.

Molholt, B. 1989. Risk assessment and management, *In* Majumdar, S.K., E.W. Miller and R.F. Schmalz (eds). *Management of Hazardous Materials and Wastes,* Pennsylvania Academy of Science.

Ornston, L.N. 1971. Regulation of catabolic pathways in *Pseudomonas.* Bacteriological Reviews *35*:87-116.

Rochkind, M.L., J.W. Blackburn and G.S. Sayler. 1986. Microbial decomposition of chlorinated aromatic compounds. U.S. Environmental Protection Agency, Cincinnati, OH. EPA/600/2-86/090.

Sandhu, S.S., D.M. DeMarini, M.J. Mass, M.M. Moore and J.L. Mumford (eds). 1987. Short-term bioassays in the analysis of complex environmental mixtures, V. Plenum Press, New York.

Waters, M.D., S.S. Sandhu, J. Lewtas, L. Claxton, G. Strauss and S. Nesnow (eds). 1984. Short-term bioassays in the analysis of complex environmental mixtures, IV. Plenum Press, New York.

Weber, A.S. and M.R. Matsumoto. 1987. Feasibility of intermittent biological treatment for hazardous wastes. Environmental Progress 6:166-171.

Management of Hazardous Materials and Wastes: Treatment, Minimization and Environmental Impacts. Edited by S. K. Majumdar, E. W. Miller and R. F. Schmalz.

Chapter Seven

WEST VALLEY, NEW YORK: NUCLEAR FUEL, NUCLEAR WASTE, NUCLEAR TECHNOLOGY

W. DEAN HOFFMAN

Deputy Project Manager
Westinghouse Electric Corp.
Low Level Waste Program
509 W. Capitol
Springfield, IL 62704

This chapter will discuss the history of the West Valley nuclear facility from its legacy as the world's first nuclear fuel reprocessing facility to its important role as a demonstration project for the nation's high level nuclear waste program. Included are discussions on how different types of nuclear waste are handled as well as a summary on the disposal areas located there.

NUCLEAR FUEL

In West Valley, New York, population 500, the nuclear age began in 1961 when the Atomic Energy Commission licensed the nation's first commercial nuclear fuel reprocessing plant. A key proponent of the facility was Governor Nelson Rockefeller who was motivated to bring jobs and technology to the depressed, rural western New York economy. New York State agreed to take responsibility for the final decontamination and decommissioning of the plant at the end of its operating tenure.

Concept of Reprocessing

The purpose of the facility was to reclaim reusable uranium and plutonium from spent nuclear fuel assemblies provided by utilities in the early era of commercial nuclear power. This fuel reclamation was achieved through a chemical or solvent extraction process. Fuel assemblies were first cooled in a storage pool, then remotely chopped and fed to a chemical process cell where the fuel pellets were dissolved in nitric acid. An organic solvent (kerosene) was added to separate out the uranium and plutonium which were shipped to other sites for fuel fabrication.

In essence, this was a recycling plant for nuclear fuel. The technology was not brand new. The government had been reprocessing nuclear fuel for defense purposes since the 1940s. However, commercial fuel reprocessing had not been attempted.

In 1963 New York State leased the facility to the W. R. Grace Corporation whose subsidiary, Nuclear Fuel Services (NFS), would operate the plant. Construction was finished in 1966 and fuel reprocessing operations began that year and continued for six years. Almost 650 metric tons of fuel were reprocessed. The Getty Oil Corporation acquired NFS in 1969 and foresaw a profitable business. Getty, seeking to expand the facility, shut the plant down in 1972. The closure was to be temporary but the plant never reopened for reprocessing.

Changing Regulations

The 1970s brought a rapidly changing regulatory climate to all industries including the nuclear business. Particularly stringent for NFS were new seismic regulations which impacted expansion costs greatly. Projected expenses to expand and meet new regulations ranged from $300 million to $600 million. The plant itself had only cost $32 million. Getty decided in 1976 not to make the investment. They announced to New York State, the property owner, that at the end of their lease in 1981, they would no longer operate the site.

What had begun with great promise ended with proven reprocessing technology, but unfulfilled commercial expectations, adverse media coverage, questions about plant safety and uncertainty over the future of the site. The focus of the issues concerning the site shifted from nuclear fuel to nuclear waste. The site was synonymous with controversy.

NUCLEAR WASTE

High Level Waste

High level waste, low level waste, transuranic waste, and spent fuel were all accumulated at the West Valley site. High level wastes were generated as part of nuclear fuel reprocessing. These liquids containing fission products were left over after the separation of uranium and plutonium from the fuel. The high level wastes were pumped to two underground storage tanks. One tank contains 600,000 gallons of high level waste while another contains 12,000 gallons of the liquid. The tanks are made from carbon steel and encased in concrete vaults. Each tank has a spare. The entire storage area is located ten feet beneath the ground and is monitored daily. These high level wastes will eventually be converted to glass for transport to a federal repository.

Spent Fuel

750 spent nuclear fuel assemblies were shipped to West Valley between 1972 and 1976 by six utilities with the intention they be reprocessed. Two of the utilities sold their fuel to NFS. New York State brought suit in 1982 to have all the fuel returned since it was never reprocessed. New York State argued the fuel owners were trespassing on the State, owner of the facility, by having fuel stored at West Valley. Court-ordered fuel shipping began in 1983 and 625 assemblies were returned in heavy casks to their utility owners. The remaining 125 assemblies were owned by NFS but now belong to the United States Department of Energy which will transport them to the Idaho National Engineering Laboratory for research purposes.

Transuranic Waste

Transuranic wastes are those wastes which contain prescribed concentrations of elements heavier than uranium 238. These wastes are found at various stages of the reprocessing cycle. So far during plant cleanup 1,000 cu. ft. have been segregated for eventual shipment to a federal repository.

Waste Disposal Area

NFS operated a 7-acre disposal area licensed by the Nuclear Regulatory Commission for radioactive wastes from plant operations. These included contaminated fuel hardware, obsolete equipment, and degraded process solvents absorbed in vermiculite. Approximately 140,000 cubic feet of radioactive materials were disposed of in 3 by 7 foot holes 50 feet deep. One half ton of irradiated fuel elements were also buried in this area which continues to be extensively monitored.

In 1983 a water sampling program led to the discovery of kerosene seeping from buried tanks in this disposal area. A two year cleanup program concluded with the removal of eight tanks of the solvent along with several tons of contaminated soil. The kerosene was stabilized in cement and the soil packaged in boxes. Disposal of liquids is prohibited under current regulations.

NFS also operated a commercial low level waste disposal facility for New York State. This 22 acre area was located adjacent to the NRC licensed disposal facility. This commercial landfill operated between 1963 and 1975 disposing of over 2 million cu. ft. of low level waste. The method used was shallow land burial utilizing 14 trenches and 12 acres which were covered and continue to be monitored.

The area is characterized by glaciated till which resembles modeling clay. It is highly impermeable; hence, the site was chosen for disposal. This asset became a liability in periods of heavy rain and melting snow. Because of its impermeability the soil held water in the trenches. This "bathtub effect" became a big issue in 1975 when water accumulated to such a degree that it surfaced in the disposal area. Millions of gallons were pumped from the trenches and treated before release.

The disposal area has remained closed since that time. New York State has completed additional remediation to the area by capping the trenches to reduce water infiltration. In 1987 a clay plug was inserted into a sand lens leading to one of the area trenches. This work has resulted in a successful curtailing of water accumulation in the area.

NUCLEAR TECHNOLOGY

The West Valley case provides considerable insight into nuclear waste technology. In the next section different technologies selected for managing various types of nuclear waste will be discussed. Different methods utilized for disposal area maintenance and remediation will also be assessed.

High Level Waste Solidification

For years liquid high level nuclear waste in the United States has been accumulated at defense nuclear fuel reprocessing plants. It is stored in shielded underground tanks. Obviously, the tanks are only a temporary solution. The federal government has sought permanent stabilization of these wastes and in 1980 chose West Valley as its demonstration project. Under Congressional Act the Department of Energy is responsible for converting 600,000 gallons of high level waste at the site to a durable solid suitable for shipment to a repository. The New York State Energy Authority is a partner in the project. West Valley Nuclear Services Company, a division of Westinghouse Electric Corporation, is the site operator.

Though the West Valley high level wastes are from commercial nuclear fuel reprocessing, the same technology will be applied at defense sites to solidify the wastes. Scientists and engineers at West Valley have elected to convert the waste to borosilicate glass. France has performed this process at its Marcoule site since 1978. The French dry the waste to a calcine powder then feed it to a melter for solidification into glass. The West Valley process will not necessitate drying the waste. The waste will be filtered, concentrated and converted to glass in a slurry fed ceramic melter. Borosilicate glass is chosen for its waste bonding qualities and durability. The molten mixtures will be fed into 300 stainless steel canisters for solidification. This will begin in the early 1990s. This basic technology will be applied at other high level waste sites in Washington, Idaho, and South Carolina. Eventually, all the solidified waste will be sent to a deep geologic repository.

Processing the West Valley high level waste is now ready to begin. The first step will be remote transfer of the liquid from the storage tank to a supernatant treatment system. This system has been installed in a spare tank in the waste storage area. Here the waste will be processed through a series of ion exchange columns filled with a zeolite clay which will extract the radioactive cesium from the liquid.

The resulting low level liquid is then transferred through underground piping to the liquid waste treatment system. The major components of the system are ion exchange filters and evaporation and concentration equipment. The concentrated waste is then stabilized in a cement solidification system. The cement system was designed to mix waste and cement with high shear mixers in a large container which fills 71 gallon square drums. Approximately 15,000 drums of low level waste will be produced. The drums will be stored in a specially designed vault called a drum cell. This engineered building features a steel shell which contains a 20 inch thick concrete shield wall and instrumentation for environmental monitoring. A remotely operated crane will control waste drum emplacement. An environmental impact statement will evaluate plans for capping this structure permanently.

These radioactive waste treatment systems will operate as a fully integrated network. Waste processing will reduce the volume of the liquid by 90 percent. The high level waste can then be converted to glass.

Transuranic Waste Inventory

These nuclear wastes containing prescribed concentrations of elements heavier than uranium 238 must be segregated as they are collected during the cleanup of the reprocessing plant at West Valley. The technology of segregation involves the use of a segmented gamma scanner which detects energy emitted from radioactive isotopes contained in waste drums. The data generated are used to inventory and classify low density wastes. For higher density materials a passive neutron assay system is utilized which detects the contents of a waste container on six sides. Transuranic wastes are boxed or put in drums for storage. They, like the high level wastes, will be sent to a federal repository.

Decontamination and Decommissioning (D&D)

Seventy percent of the former nuclear fuel reprocessing facility has been decontaminated and modified to support new waste processing systems. Over 350,000 cubic feet of space has undergone decommissioning. This is significant in that the West Valley nuclear facility provides a cost and experience data base for future D&D projects.

Miles of piping and hundreds of tons of hardware have been removed from plant areas. Most notable in this effort has been the development of remote controlled tooling for equipment removed. One tool in particular that will be applicable at Department of Energy and commercial nuclear facilities is a remote operated, crane suspended, high pressure sprayer. Using a water jet this equipment will clean contaminated dust and dirt from walls, remove paint, and even cut through several feet of concrete.

A bionic arm was developed to reach into areas hard to access in order to cut piping. A wheel mounted robot was refined and fitted with a vacuum to remove tons of debris in the plant's highly radioactive chemical process cell.

Low Level Waste Handling

New handling, packaging, and disposal techniques have been developed to manage low level waste associated with the cleanup. Engineers built a size reduction facility to torch cut vessels and hardware removed from the facility. A supercompactor was installed which compresses waste filled drums to a fraction of their original size maximizing storage space.

Sprung structure tents are being used to store low level wastes until an environmental impact statement is completed on site closure. These weatherized shelters provide temporary storage. Permanent disposal options will be examined in the impact study. Shallow land disposal has been suspended under a legal agreement until the study is complete.

Public Attitudes

An open public information program has been successful in communicating project work to the local public. Over 1200 Western New York citizens were polled in both 1986 and 1987 on their attitudes toward radioactive waste cleanup and nuclear energy. Ninety-five percent give waste cleanup a high priority and over forty percent think it is the most important environmental action of our time. Better than half of those surveyed think nuclear energy is necessary but another third see waste management as critical to its future. Regarding waste dispoal, people tend to favor centralized sites for both high and low level waste.

Age and education are important variables in predicting nuclear attitudes. As age increases so does support for nuclear energy and radioactive waste cleanup. As education level increases, support for nuclear projects decreases.

SUMMARY

A site which was home to a commercial failure and a major environmental issue has now become a tremendous opportunity for success. The demonstration of nuclear waste cleanup, solidification, and transport is essential to the government and the industry. The Department of Energy, New York State and Westinghouse face many challenges at West Valley. However, project officials and community leaders are confident those challenges will be successfully met.

The West Valley Demonstration Project is a model for nuclear waste cleanup. All types of nuclear waste are handled there. Many lessons have and will be learned. The technology developed will have a positive impact on the local environment and could have national impact. If successful, the project could change the way the United States perceives nuclear waste.

REFERENCES

1. Colglazier, Jr., E. W. (Ed.) 1982. *The Politics of Nuclear Waste.* Pergamon Press, New York.
2. Duckworth, J. P. 1983. Commercial Waste Disposal History of West Valley, N.Y. Presented at American Nuclear Winter Meeting. San Francisco.
3. Murray, R. L. 1985. *Understanding Radioactive Waste* (2nd Edition). Battelle Press, Columbus, OH.
4. Resnikoff, M. 1987. *Living Without Landfills.* Radioactive Waste Campaign. Faculty Press, Brooklyn, NY.

Management of Hazardous Materials and Wastes: Treatment, Minimization and Environmental Impacts. Edited by S. K. Majumdar, E. W. Miller and R. F. Schmalz.

Chapter Eight

HAZARDOUS WASTE MINIMIZATION —THE STATE PERSPECTIVE

GREGORY L. HARDER
Waste Minimization Specialist
Pennsylvania Department of Environmental Resources
P.O. Box 2063
Harrisburg, PA 17120

Everyone agrees that waste minimization is a good idea. Discrepancies arise when people try to decide how to encourage waste minimization, how to define waste minimization, and how to do it. Any time that waste is generated it can be viewed as a waste minimization opportunity.

WHAT IS WASTE MINIMIZATION?

It is vital to define precisely what is meant by the term "waste minimization". For discussion purposes here, the following definitions are used.

Waste minimization is the reduction of hazardous waste that is generated, or subsequently treated, stored, or disposed. Waste minimization consists of any source reduction, use, or reclamation activity undertaken by a generator that results in either the reduction of the quantity of hazardous waste, the reduction of toxicity of hazardous waste, or both, so long as such reduction is consistent with the goal of minimizing present and future threats to human health and the environment. Treatment which is intended solely to prepare the waste for disposal, such as dewatering, is not waste minimization.

Source reduction is the reduction or elimination of hazardous waste generation at the source, usually within a process. Source reduction measures include process modifications, feedstock substitutions, improvements in feedstock purity, various housekeeping and management practices, increases in the efficiency of machinery, use or reclamation within a process, or any action that reduces the amount of waste leaving a process.

Reclamation is the treatment of a material to recover a usable product, to recover distinct components as separate end products, or the regeneration of a material.

Use include the direct and effective substitution of waste for an ingredient or raw material used in an industrial process and the effective substitution for a commercial chemical product.

The Pennsylvania Hazardous Waste Facilities Plan (July 1986) establishes as a State policy a waste management hierarchy. The most preferred method is source reduction. The second is "reuse/recycling/recovery" which corresponds to use and reclamation. "Recycling" is a very imprecise term which means different things to different people. There seems to be no useful distinction between reuse and use. Recovery is similar to reclamation in meaning. Consequently, a comprehensive waste minimization strategy should place primary emphasis on source reduction, which is really avoiding the generation of the waste. Use and reclamation of waste after it has been generated should receive secondary emphasis.

WHAT ARE THE BENEFITS OF WASTE MINIMIZATION?

Waste minimization clearly provides a method for saving money by more efficiently using natural resources and energy. ". . . technology requires resources—water, air, land, raw materials, energy—which are to be found in the physical environment. This interaction between technology and the physical environment results in depletion of resources on the one hand and a build-up of wastes—pollution—on the other. The strategy which the modern manager must learn if he is to cope with this double problem is that of a non-waste technology, one which conserves resources, reduces pollution and saves money at the same time." (Royston, p. 44)

Some additional benefits include reduced costs for transporting, treating, storing, or disposing hazardous waste. Reductions in long term liability from reduced handling of waste are possible but difficult to quantify. There may be reduced regulatory burdens if less waste is generated or if regulated activities such as treatment or storage can be eliminated. Leadership is establishing a comprehensive waste minimization program which can result in an improved image and credibility with the local community and the general public.

HOW CAN I IMPLEMENT A COMPREHENSIVE WASTE MINIMIZATION PROGRAM?

Waste minimization can be conceptualized as occurring in three waves. The first is implementation of simple common sense techniques such as segregating wastes and repairing leaky equipment. The second is the "low hanging fruit" or technology which is available and needs only to be plucked and transferred to the minimization opportunity at hand. An example is the purchase of distillation apparatus to reclaim solvents. The third wave is developing new technology. (Harry Freeman, USEPA). This tiered approach is useful since it describes the order in which activities are implemented, the easier tasks first. Additionally, the initial waves are generally more cost effective with shorter pay-back periods.

The first step in developing a comprehensive waste minimization program is to obtain the commitment and support of top management. It is essential to include all the people involved in an industrial process in thinking about waste minimization. Everyone from product development engineers to production line workers can have input if they are initiated into an awareness of the principles, benefits, and basic techniques of waste minimization. For example, engineers developing a new product should consider the nature and quantity of waste that will be generated and how to reduce it. Production line workers may be aware of leaks, wasteful procedures, or other waste minimization opportunities.

The second step is to conduct a waste minimization assessment throughout the plant or production unit. After obtaining the commitment of top management the following steps are suggested.

1. organize the audit team
2. conduct research into known waste minimization technologies and techniques specific to the manufacturing process
3. review current waste generation sources, types, quantities, and disposal methods
4. collect existing process flow diagrams and data
5. conduct a facility inspection
6. identify and list potential waste minimization opportunities
7. perform technical and economic review of the options to prioritize them

(for additional information on conducting a waste minimization assessment see *The EPA Manual for Waste Minimization Opportunity Assessments* and the *Hazardous Waste Minimization Manual for Small Quantity Generators.*)

The third step is to implement the opportunities in order of priority.

PENNSYLVANIA'S WASTE MINIMIZATION STRATEGY

The Pennsylvania Department of Environmental Resources has developed a draft strategy on how to encourage hazardous waste minimization. A combination of these program elements is essential to the development and implementation of a successful statewide waste minimization program.

- — Collect Waste Minimization Information from Generators
- — Develop a Technical Assistance Program
- — Foster the Transfer of Technical Information
- — Conduct an Educational Outreach Program
- — Support Research, Pilot Tests, and Demonstration Projects
- — Support the Northeast Industrial Waste Exchange
- — Promote the Governor's Waste Minimization Award
- — Develop Waste Minimization Coordinator Position in Regional Offices
- — Streamline the Permitting/Approval Process for Legitimate Users and Reclaimers
- — Require Waste Minimization Plans from All Generators
- — Clearly Define What Waste Is/Is Not Through Regulations and Policy
- — Create Financial Incentives for Implementation Programs
- — Establish Fee/Tax Disincentives to Generate Waste

REFERENCES

1. Royston, Michael G. 1979. *Pollution Prevention Pays.* Pergamon Press Ltd., Oxford, England, pp. 44-45.

FOR FURTHER INFORMATION:

The EPA Manual for Waste Minimization Opportunity Assessments. United States Environmental Protection Agency, Office of Research and Development. April 1988.

Hazardous Waste Minimization Manual for Pennsylvania's Vehicle Maintenance Industry. Center for Hazardous Materials Research (412-826-5320 or 1-800-334-CHMR). October 1987.

Hazardous Waste Minimization Manual for Small Quantity Generators. Center for Hazardous Materials Research (412-826-5320 or 1-800-334-CHMR). April 1987.

Profit from Pollution Prevention: A Guide to Industrial Waste Reduction and Recycling. Monica E. Campbell and William M. Glenn. Pollution Probe Foundation. 1982. *Report to Congress: Minimization of Hazardous Waste.* United States Environmental Protection Agency. 2 volumes, October 1986.

Waste Minimization: Issues and Options. United States Environmental Protection Agency. 3 volumes, October 1986. (this and the previous reference available from the National Technical Information Service as a five volume set, order number PB87-114328, phone 703-487-4650).

Serious Reduction of Hazardous Waste: For Pollution Prevention and Industrial Efficiency. U.S. Congress, Office of Technology Assessment. September 1986. (U.S. Government Printing Office, Washington, D.C., OTA-ITE-317).

Waste Minimization: Environmental Quality with Economic Benefits. United States Environmental Protection Agency. 1987. (EPA/530-SW-87-026 available from RCRA hotline 1-800-424-9346).

Waste Minimization Assessments: A Step-by-Step Procedure. Lorton, Fromm and Freeman. December 2, 1987.

Management of Hazardous Materials and Wastes: Treatment, Minimization and Environmental Impacts. Edited by S. K. Majumdar, E. W. Miller and R. F. Schmalz. © 1989, The Pennsylvania Academy of Science.

Chapter Nine
HAZARDOUS WASTE REDUCTION

JOEL S. HIRSCHHORN
Office of Technology Assessment
United States Congress
Washington, D.C. 20510-8025

INTRODUCTION

Interest in reducing the generation of hazardous waste is increasing nationwide. Every study of the subject concludes that substantially more technically feasible and profitable waste reduction is possible with existing technology.[1-3] The OTA reports discuss a number of policy options which are consistent with these three approaches and these will not be repeated here. Every part of American society can play an important role in creating a new environmental protection strategy based on

— prevention instead of reaction,
— government technical assistance instead of regulation of industry, and
— a comprehensive or multimedia basis for action instead of the traditional piecemeal approach.

Much of the available data is misleading. Improving industrial efficiency by cutting waste production is variously called waste reduction, source reduction, pollution prevention, or waste minimization. There are no standard definitions. Companies often claim waste reduction credit for activities such as incineration that *follow* generation of a waste, rather than only those that *avoid* waste creation, handling, movement, and management. Survey results and published papers show that probably 75 percent of companies use a definition of waste reduction or some other term which gives them credit for improved waste management and pollution control. 3M's Pollution Prevention Pays and Dow Chemical's Waste Reduction Always Pays are more than slogans; they are simple statements of economic fact. These companies and others overcame both

---

(The views expressed here are solely those of the author and not necessarily those of OTA.)

internal obstacles to waste reduction and obstacles in the public sector. Not every company will be able to do so.

It is easy to miss the obstacles to waste reduction and, therefore, not understand why Federal policy is important and to overlook the negative consequences of relying on industry and the States to implement waste reduction. There is little public debate on Federal waste reduction policy. Industry fears burdensome regulation of waste reduction. Companies in the waste management and pollution control business fear a loss of markets. Environmental organizations worry about loss of support for established regulatory programs. The new interest in waste reduction may hide a serious national problem: nearly every part of American society is mentally locked into the established, institutionalized pollution control culture—or paradigm—which defines environmental protection in terms of what is done to wastes and pollutants *once they are produced.*

THE KEY INDUSTRIAL ACTION: THE WASTE REDUCTION AUDIT

People who have studied how reducing the generation of hazardous waste can best be implemented conclude that a critical initial action is to conduct a waste reduction audit or assessment (WRA). The WRA is a tool in overcoming the non-technical obstacles to waste reduction and in reaching sound technical and management decisions about waste reduction. When a WRA is used effectively the lack of technical obstacles to waste reduction or a means to overcome them are revealed.

Because waste reduction is an upstream activity, waste generators do not usually know as much about where, how, and why wastes are generated as they do about end-of-pipe pollution control and waste management. But a problem with WRAs is emerging. WRAs are getting unnecessarily complex, lengthy, and costly. Waste generators unfamiliar with waste reduction may not realize the proper role of WRAs. As a result, some waste generators may be deterred by the costs and difficulty of WRAs and may miss technically and economically feasible opportunities for waste reduction.

It is easy to lose sight of the principle purpose of a WRA. *A WRA should provide information for defensible, informed, and correct waste reduction management decisions. The risk, however, is that we may create a rigid WRA methodology which is inappropriate for the early phases of waste reduction and for some types of waste generation activities.*

THE AUDIT AND THE FOUR PHASES OF WASTE REDUCTION

A WRA is not like a law of physics. There are many different formulations of WRAs. However, there are a few basic elements of WRAs which are relevant to most waste generation situations. As described in OTA's 1986 report, these key WRA components are:

Step 1: Identify the amounts and kinds of hazardous substances in wastes and emissions, whether regulated or not.

Step 2: Identify the specific production sources of the wastes and emissions.

Step 3: Set priorities for waste reduction actions on the basis of costs, environmental concerns, worker health and safety, liabilities, and production constraints.

Step 4: Analyze and select technically and economically feasible reduction techniques.

Step 5: Compare the economics of waste reduction alternatives with current and future waste management or pollution control options.

Step 6: Evaluate the progress and success of chosen waste reduction measures.

It is also useful to think of four stages for waste reduction in order to assess the applicability and scope of WRAs:

Stage 1: Common sense waste reduction.
Stage 2: Information driven waste reduction.
Stage 3: Audit dependent waste reduction.
Stage 4: R&D based waste reduction.

As the names suggest, the third stage of waste reduction is where formal WRAs become critical. As discussed later, the first two stages can be effective without a WRA or can be enhanced by simple self-audit check lists.

COMMON SENSE WASTE REDUCTION

In the first phase, people in production exploit readily visible, common sense, low cost, and low risk waste reduction opportunities. At this stage there are no technological barriers to waste reduction. It is literally true that people can walk through industrial operations and, with waste reduction in mind, spot immediate reduction opportunities which can be implemented in days and weeks. The initial waste reduction efforts are small low-tech efforts, carried out quickly and probably with the largest percent return on investment. They often involve changing procedures rather than changing production technology or major equipment. Also, these early efforts require little study or engineering analysis, information, testing, or verification of feasibility. For example, reducing the flow of water in cleaning operations, halting the comingling of hazardous with non-hazardous waste, covering vessels which contain volatile chemicals, improving inventory controls, reusing slightly off-specification products, replacing water cleaning with mechanical methods, and installing closed-loop recycling of process waters or waste streams.

In the first phase, formal structural analysis can be replaced by intuition or by back-of-the-envelope calculations of costs and benefits. It is useful and instructive, however, for people to understand the evaluative thought processes they use at this early stage. By doing this, they can develop a formal methodology to assess more difficult and complex waste reduction possibilities. This is how the WRA concept was developed historically. The six steps of an WRA given above codify the logical process of reducing waste generation. But a lengthy formal WRA in the first phase would be a waste of time—like using a computer to add two and two. A simple self-audit check list would, however, be useful. This approach could stimulate waste reduction activity, educate production people, and provide a record of actions and results. The check list could use WRA steps 1 and 2, and combine the others into a brief description of the action taken and the results achieved. One page would be enough.

INFORMATION DRIVEN WASTE REDUCTION

The key difference between the first and second waste reduction stages is that production people need more detailed information to assess reduction opportunities and decide what to do. In stage one, the reduction opportunities spring from direct experience with a naked-eye observation of production. But in stage two the reduction opportunities are more subtle, but here too there are no significant technological obstacles to waste reduction. Various types of information can assure more systematic and comprehensive results.

Production people need to understand exactly why a waste is generated from a process, and its linkage to raw materials and product quality in order to figure out waste reduction opportunities. If production people become aware of many different ways to reduce certain types of process wastes, then they can see more reduction possibilities in their specific production operations. For example, it is helpful to know that a raw material change can be an effective way to reduce waste. But it is better to know about cases where an organic solvent was replaced by a water based product or by a mechanical technique.

This stage of waste reduction is relatively easy, quick, and low cost, but it does require more information about wastes, their production origins, and the types of waste reduction methods generally in use. The information can be obtained from many different sources: in-house information, professional articles and reports on waste reduction, conferences and workshops, discussions with other production people, or materials increasingly available from government waste reduction programs. The key point is that little effort or money is necessary to get a quantum improvement in waste reduction.

As with stage one, no lengthy formal WRA is necessary and the brief check list approach is probably best because the costs and benefits of the effort are self-evident or easily calculable. Outside consulting help, although strictly speaking not necessary in this stage, could be useful, if it were available at no cost, such as from a government program. Indeed, several States have found that on-site technical assistance programs provide quick payoffs because they provide the rapid increase in information which facilitates this stage of activity.

AUDIT DEPENDENT WASTE REDUCTION

Production people eventually will reach an end to waste reduction which does not require a formal analysis. The "low hanging fruit" gets picked. It is then that a formal detailed WRA becomes critical to achieving all the reduction of wastes and pollutants which is technically and economically feasible, and consistent with production requirements. At this stage some technological obstacles are identified.

Uncertainty about economic paybacks and technical feasibility becomes necessary and characteristic of this stage. More capital investment becomes necessary and risks increase. Even when the reduction action is profitable, payback periods may be long. Competition with other, traditional forms of capital investment increases. Testing and development in the laboratory and within production itself also increase. The time of production people becomes a significant factor to management. Effects on product quality require analysis. New wastes may be created which do not provide net environmental or worker health and safety improvements.

Without a formal WRA production people may incorrectly conclude that they have exhausted their waste reduction opportunities or they may even pursue projects which are either technically, economically, or environmentally ill advised. They may miss opportunities to reduce non-regulated wastes or relatively small waste streams which nevertheless pose substantial costs and liabilities. A big problem is that waste reduction responsibility may go to those people with regulatory compliance jobs who will focus only on regulated wastes. But history shows how the universe of regulated waste steadily increases, and Superfund liabilities cover all hazardous substances, even unregulated wastes.

Circumstances change and evolve as production people move up the waste reduction learning curve. Gradually, the costs and time of key production people to implement a WRA are justified. Waste reduction can become integrated with major process and product improvements. Moreover, there may come a time when outside consultants or people from other company operations can contribute uniquely to overcoming obstacles to waste reduction.

But any structured analysis can be embellished without end. And now WRAs run a risk of becoming unnecessarily complex and lengthy. *WRAs should not automatically become major engineering projects which involve outside consultants. A complicated WRA methodology and protocol needs caveats and shortcuts or it may blunt interest in waste reduction and reinforce the proclivity of production people to stay with traditional end-of-pipe waste management.*

Although the WRA is applicable to the third stage of waste reduction, not all such waste reduction is so capital intensive or technically complicated as to justify a traditional engineering approach. For example, third stage waste reduction for batch processes, service and other non-manufacturing operations, rapidly changing production processes, and relatively small waste streams do not routinely call for a major design and construction approach. For example, shifting from water cleaning of batch reactor vessels to ultrasonic cleaning requires buying commercial equipment, but no radical change in the process technology or equipment.

Waste generators, therefore, should consider a "short form" approach to a WRA which is less labor intensive and costly. The goal should be a series of questions for all six steps which stimulate thinking, break down initial biases, and give production people the responsibility to defend waste generation. It is necessary to prod production people and get them to define, in their own terms, the limits of existing information and resources. "I don't know" and "I'm not sure" answers establish the need for more detailed WRAs.

Moreover, the full EPA approach is expensive. It's cost for a major waste has been said by an EPA contractor to be about $20,000, more than many stage three waste reduction efforts justify. But that was for a WRA designed and implemented by an outside consultant at an early stage of WRA development. The emphasis thus far has been on designing WRAs probably for the most difficult waste reduction situations. Reducing the complexity and cost of WRAs should and probably will receive more attention by EPA.

It is not only a matter of cost. One of the many obstacles to waste reduction is the difficulty in getting the time and commitment of production people. A highly detailed WRA requires extensive time of production people, even before a waste reduction project can be justified, and it may seem that outside consultants are necessary.

But outside consultants do not always maintain a focus on true waste reduction. The underlying tendency throughout industrialized society to think in terms of end-of-pipe control or management also exists in the engineering consulting community. The result is often uncreative waste reduction analyses and a tendency to prefer waste management options as part of WRAs. In several WRAs already conducted by outside consultants, treatment solutions are selected because they fall under the aegis of waste minimization. *Whenever waste exists as a separate entity and is handled, transported, and managed, the solution is not true waste reduction.* Usually government regulation and uncertain liabilities remain, unless true waste reduction is used. When in doubt about whether true waste reduction is being practiced, ask if the activity or equipment physically requires a waste to operate.

If you want waste reduction stressed, use consultants who are experienced in production processes and not in pollution control. Unless consultants are process oriented, they may take their cues about reduction targets from company people who are reluctant to change existing systems. It is also necessary to be vigilant about consultants who offer a narrow, technical fix to a specific waste but do not address general obstacles to waste reduction within the company. These include: lack of supportive management policies, poor organizational support, and economic analyses which do not account for all short and long term direct and indirect costs of waste management.

If waste reduction is to achieve its full potential, then waste generators must pay attention to non-technical obstacles and to continue systematic and comprehensive waste reduction. A waste generator who is dependent on outside consultants has probably not designed an effective in-house waste reduction program. It is imperative to remember that no outside consultant is likely to know as much about a specific industrial process as in-house people. The objective should be to channel the in-house expertise toward waste reduction. The best consultant, therefore, is one who helps establish the organizational structure, attitudes, information base, and analytical methods which structure and sustain on-going waste reduction programs which respond to changing technologies, costs, and products.

It might be useful for companies to consider using an outside consultant when they first hit barriers to waste reduction after going through the first two phases. Thereafter, they should establish their own WRA procedures, possibly try a consultant again or use materials or services from a government agency to see if they can improve their WRAs.

R&D BASED WASTE REDUCTION

Lastly, there is a fourth stage of waste reduction which requires substantial R&D on process technology or equipment and possibly product composition or design. Many negative conclusions will result from activities in the first three stages of waste reduction. But R&D might provide a cost-effective new waste reduction solution. Some circumstances favor an R&D effort; these include: large waste streams from well established processes, wastes which seem inevitable unless a different product is designed, and wastes linked to major raw materials which also dictate basic processing technology.

One benefit of R&D is that a new product or process with a major waste reduction advantage could be marketed to others. For example, companies who have developed in-process recycling of used metalworking fluids and have found water based replacements for solvent based raw materials have begun to market them. Since waste reduction R&D requires careful and continuing economic analysis, economic spinoffs should be thought about as early as possible. The WRA audit can become a periodic effort for both the original situation and any new waste generating activity resulting from the R&D.

IS THERE A PROBLEM FOR CONGRESS TO FIX?

Moving to the Federal level, there is now evidence of Congressional interest in waste reduction but the legislative process has just begun. At this time, six bills have been introduced in the U.S. Senate and House of Representatives and Congress provided money for new waste reduction activities at EPA, including a States grant program, in the FY1988 appropriations. While differing in a number of details, none of these bills calls for traditional regulations to prescribe waste reduction actions for industry. Instead, these bills call for government to provide technical assistance to industry, of a kind already proven in several demonstration programs (e.g., Ventura Country, California and North Carolina). [Nothing said here is meant in any way to be an endorsement of or opposition to particular legislation. More bills may be introduced and the hearing process, which has not yet begun and which will bring to light new perspectives and information, typically results in significant changes in legislation.] Although there are many dedicated people working to put waste reduction on the national agenda, there has been no significant public support of these bills by either industry or public interest groups.

It must be remembered that Congress already took some action on waste reduction through its minimization actions in the 1984 reauthorization of the Resource Conservation and Recovery Act (Hazardous and Solid Waste Amendments Act of 1984). This resulted in the first waste reduction efforts at the Environmental Protection Agency, and these are still evolving. However, another important development was the following statement of national policy:

"The Congress hereby declares it to be the national policy of the United States that, wherever feasible, the generation of hazardous waste is to be reduced or eliminated as expeditiously as possible."

THE REAL NATIONAL PROBLEM

The new interest in waste reduction and pollution prevention may hide a serious national problem: nearly every part of American society is mentally locked into the established, institutionalized pollution control culture—or paradigm—which defines environmental protection in terms of what is done to wastes and pollutants *once they are produced.* Many people don't realize that pollution control often transfers pollution from one regulated environmental medium to another, and sometimes to a less or non regulated medium. Also, pollution control is based on the concept of safe or allowable levels of pollution which, because they are so difficult to set, means that many hazardous substances are left unregulated, while debate continues for years. Pollution control also pits economic and health benefits against one another.

COMPETING INDUSTRIAL PRIORITIES

There are plausible explanations for the lack of industrial support for a Federal waste reduction initiative. Ask two questions about a company: Does it have a successful waste reduction program? Is it in the waste management or pollutions control business?

The second question is more important than most people recognize. The steadily increasing national spending on the environment—now about $80 billion—helps companies understand the benefits of waste reduction. It is also a business opportunity for many of America's largest manufacturing companies.

When a company has a successful waste reduction program and is in the waste management business, there is no net advantage for it to foster a government program that would assist *other* companies to reduce their waste generation and, hence, shrink the waste management/pollution control market. Although some new business would be created for waste reduction consulting, there is little expensive hardware or engineering services to be sold. Waste reduction government intervention would create economic winners and losers.

When the company has a successful waste reduction program but is not in the waste management/pollution control business, a government program could reduce its competitive advantage relative to firms without a successful waste reduction program. Moreover, the company that has done it on its own may feel that it is unfair for the government to assist other companies with less initiative.

And companies without successful waste reduction programs? If a company is in the waste management/pollution control business it wants to maintain the market. If not, it sees no benefit in changing the status quo if its own environmental costs are relatively low. A company that hasn't recognized the economic benefits of waste reduction, will not see any purpose in a government waste reduction program.

For all companies, and particularly for those without a successful waste reduction program and no waste management/pollution control business, there is the understandable fear that *any* Federal waste reduction initiative, even if it is non-regulatory today, will lead to a traditional regulatory program tomorrow. Waste reduction, in other words, falls victim to industry's mistrust of any regulatory agency. (This is why some States have kept their waste reduction programs out of their environmental regulatory agencies.) Indeed, Congressional waste reduction action to date has been in the context of the Resource Conservation and Recovery Act regulatory program for hazardous waste. Few companies see a non-regulatory waste reduction program as making the regulatory programs easier to bear, or they give greater weight to their fears, as discussed above.

Industry has three concerns about waste reduction regulation:

—Requirements to cut waste generation might result in eliminating certain products and the waste reduction concept might be extended to unregulated wastes and to post-consumer wastes. The latter is particularly threatening because, from the industrial perspective, a company which is not generating a harmful waste or which is managing a waste in compliance with regulations, could be forced to change or drop an established, profitable product because of waste created late in the product's lifecycle.

—A company which voluntarily reduced its waste generation today-or-yesterday might tomorrow be *required* to meet some arbitrary additional level of waste reduction at great cost and difficulty. (This concern explains why some major companies are not revealing past waste reduction accomplishments; they are "banking" them in case they are needed to satisfy future regulatory requirements.)

—Some companies would like to keep the status quo because they have not suffered from pollution control regulations. Some large, low cost producers have increased profits because their price increases, unlike other, high cost producers in the same industry, have surpassed regulatory cost increases.

Lastly, industry has recently been funding university work in risk assessment and supporting expanded government activities in this area. Risk assessment can assist the troubled regulatory system by better allocating resources for traditional pollution control technology. And while no industrial money has been committed to public waste reduction programs, some leading companies, including several very active in waste reduction, have recently established a new trade association: the Coalition For Responsible Waste Incineration.

ESTABLISHED PRIORITIES AND SKEPTICISM BY PUBLIC INTEREST GROUPS

Early on, environmentalists accepted the pollution control strategy as expedient. Limiting, not eliminating, pollution *was* a practical first approach to solving the newly perceived and overwhelming problem. However, *organized* environmental interests and public policy have developed an attachment to the pollution control strategy. As Robert Reich has written, environmentalists like everyone else in the environmental regulatory arena have learned to play, maintain, and expand the game according to established rules. Indeed, many environmentalists move over to private industry or the government.

Expertise, priorities, and commitments are established. Partial wins seem preferable to gambling on a new strategy. Support of a Federal waste reduction initiative might detract from political interest in and, possibly, funding for existing regulatory programs. Waste reduction might even make the limits and faults of hard fought for regulations more visible. Moreover, there is a deep suspicion among environmentalists that a *non-regulatory* waste reduction program might give opportunities to industry to compromise regulatory programs without really eliminating pollution. Some environmentalists and people in government advocate the indirect approach of expanding the end-of-pipe regulatory program and improving its enforcement, thus increasing costs to industry, which will then turn to waste reduction to lower costs. But this strategy discounts industrial responses to rising regulatory costs other than waste reduction. These responses include plant relocation and closings, use of traditional pollution control technologies, litigation, lobbying to change laws and regulations, and, sometimes, even illegal waste disposal.

SHARED CONCERNS

There may be mutually reinforcing silence on Federal waste reduction policy. Simply put, private sector players (and even some regulatory bureaucrats) may have independently reached a similar conclusion: helping to create a Federal waste reduction program might crystallize the uncertain secondary impacts—outside their control—that we have discussed here.

Giving visible support to waste reduction legislation might lead to a snowball effect if others, particularly those who do not perceive negative consequences, join in. Silence seems safe, especially as long as concerned parties think that Congress will not pass waste reduction legislation. This is a rational strategy from their perspective, because waste reduction is environmental legislation and therefore must compete for Congressional attention with older, established environmental laws with active constituencies in government and the private sector. Also, the absence of good data can be used to defend the position that a Federal program is unnecessary or to delay action, for years, while information is obtained. But we see little chance of getting good data on true waste reduction for all of American industry for some years.

ADDRESSING CONCERNS ABOUT SECONDARY IMPACTS

Waste reduction supporters can adopt a strategy to broaden public support. First, three factors must be assumed: (1) the nation has not yet turned the corner on successful long term waste reduction and is not likely to under current public policy, (2) lack of private sector support results from perceived negative secondary impacts of a Federal program, and (3) overall, serious long term waste reduction is good for society as a whole.

Waste reduction proponents could tackle the concerns described earlier head on. For example, they might convincingly argue that applying traditional regulatory measures is technically infeasible and administratively impractical, because of the enormous number and diversity of waste generating situations. Waste reduction supporters can be unequivocally positive that technical assistance can work and will be made to work. Industry's concerns are valid because, in its report to Congress on the subject, EPA stated it is still studying the possibility of future regulation.

Regarding waste reduction's effect on the waste management/pollution control market, proponents of waste reduction can point to the large toxic waste site cleanup market and emphasize that we can never reach the zero waste level. Moreover, the bulk of the American industrial base could be convinced about the benefits they would receive from a major Federal waste reduction program.

Concerns of environmental groups could be allayed through a commitment to maintaining the regulatory program which, indeed, is necessary but not sufficient for a successful waste reduction program. A case might be made to increase regulatory penalties and liabilities over time as industry becomes able to reduce its waste generation and exit in part from the regulatory system.

More efforts to educate the general public about waste reduction and primacy over pollution control would also help environmental organizations to rethink their priorities. The public benefits of waste reduction are likely to offset its negative consequences and costs. For example, not assisting waste reduction could contribute to increased risks and costs associated with producing avoidable environmental wastes. If ten more cleanup sites are created a year because substantial feasible waste reduction is not done, then at a typical cleanup cost of $10 million a site another $100 million is added to the national cleanup bill, which OTA has estimated to be about $300 billion already. If industry cut 10 percent of its waste generation a year, now at over 500 million tons a year, then waste management costs of several hundred million dollars would be saved. In comparison, the costs of State and Federal technical assistance programs for waste reduction are not likely to exceed about $10 million to $20 million a year. Increased taxes on increased profits resulting from waste reduction savings could pay for such government activity.

CONCLUSION

As always, we need to learn from the past. The benefits of waste reduction were recognized in theory over ten years ago. But practice has not followed theory. Federal environmental policy has been part of the problem. And, even with new, strong Congressional interest in waste reduction, progress may be slow unless we understand why people and organizations cling to the old approach to protecting the environment.

It seems that the private sector does not support a Federal waste reduction program because they are worried about possible secondary impacts. Industry fears burdensome regulations of waste reduction. Companies in the waste management and pollution control business fear a loss of markets. Environmental organizations worry about loss of support for established regulatory programs. These fears, if not confronted, will handicap the public debate and impede development of Federal waste reduction policy and the adoption of a pollution prevention ethic.

REFERENCES

1. Office of Technology Assessment, *Serious Reduction of Hazardous Waste,* Wash., DC, 1986; *From Pollution to Prevention,* 1987. (Order from Superintendent of Documents, U.S. Government Printing Office, Wash., DC, 20402-9325; 202 783-3238. *Serious Reduction of Hazardous Waste:* GPO #052-003-01048, $12; *From Pollution To Prevention:* GPO #052-003-01071-2, $2.75. Summaries available free from OTA, Wash., DC 20510-8025.)
2. Sarokin, D.J., et al, *Cutting Chemical Wastes,* INFORM, New York, 1985.
3. U.S. Environmental Protection Agency, *Report to Congress: Minimization of Hazardous Waste,* Wash., DC, 1986.

Management of Hazardous Materials and Wastes: Treatment, Minimization and Environmental Impacts. Edited by S. K. Majumdar, E. W. Miller and R. F. Schmalz.

Chapter Ten

WASTE REDUCTION AT THE SOURCE: THE ECONOMIC AND ECOLOGICAL IMPERATIVE FOR NOW AND THE 21ST CENTURY

DONALD HUISINGH

University of Lund
TEM, Box 62
Asumsgatan 38
275 22 Sjöbo, Sweden

As environmental awareness and environmental concerns increased during the last two decades, regulatory responses have typically focused upon end-of-pipe, Pollution Control Solutions to achieve the environmental protection goals agreed upon by the society at that time. However it has been observed, repeatedly, in numerous nations that these Pollution Control Solutions, while they have helped in some instances, in others, they have often, at great expense, merely transferred the pollutant from the air to the water or to the soil and the food chain.

Fortunately, several leading industrial firms have been successful in substantially reducing or in eliminating many of their emissions by utilizing integrative system's approaches to Pollution Prevention and Waste Reduction, rather than by continuing to place sole reliance upon end-of-pipe Pollution Controls.

Governmental policy and efforts should increasingly be focused upon ways to stimulate and encourage industries to take creative paths to the Prevention of Pollution rather than continuing to insist upon the use of the costly and largely ineffective End-of-Pipe Pollution Control approaches they have mandated during the last two decades.

Governmental policy and financial resource allocations should increasingly put their money where they have said for ten years their priorities are: Pollution Prevention. To continue to focus 99.9% of environmental protection efforts upon Pollution Controls and only 0.1% upon Pollution Prevention is ecologically disastrous, economically foolish and ethically immoral.

This chapter emphasizes that some industrial firms are showing that Pollution Prevention Pays. They have found that in many situations, technically feasible, economically profitable and environmentally friendly pollution prevention approaches can be utilized to attain environmental quality goals. They have found that these approaches are better than continuing to place sole reliance upon end-of-pipe Pollution Control and Waste Management approaches. They have found that economic and ethical survival are based upon their utilization of processes that are designed with Nature in Mind.

INTRODUCTION

Centuries ago, the early Greeks showed their concern about the future of their city and of their environment by requiring their males to take an oath as they became official members of the Polis (city). This oath stated, in part,

> "I swear that I will strive to pass on to the next generation, a city that is better than the one I received."

In the intervening time similar sentiments have been expressed by many leaders. One example is a statement by President Theodore Roosevelt in his message to the U.S. Congress on December 3, 1907. He challenged Congress and the people of the United States by saying,

> "To waste and destroy our natural resources instead of increasing their usefulness will undermine the very prosperity which we are obligated to hand down to our children, amplified and developed."

He and other political leaders in many nations led the way in early environmental protection and conservation efforts during the first half of the twentieth century. Most of the focus at that time was upon conservation of soil and other natural resources.

However, in the years immediately after World War II, as people were struggling to recover from the ravages of war, they often disregarded environmental concerns as they eagerly embraced the promise implied by the slogan used by one chemical company,

"Better Life Through Chemistry"

While, in fact, most people of the world benefited from the new chemistry by having access to a vast array of new products that provided many lifestyle options not available to previous generations, public confidence in science and technology was repeatedly shaken as people became aware of the negative and unanticipated human health and environmental effects of some of those new products of chemistry that promised a better life.

Early warnings of unanticipated side-effects of some of these chemical products were sounded by several authors. Among them, Rachel Carson in her 1962 book, *Silent Spring*, alerted the public to the fact that chlorinated pesticides (such as DDT, even though it is an effective insecticide with comparatively low human toxicity) were found to bioaccumulate through the food chain and to threaten the extinction of species far removed in distance and time from the point of initial use. Swedish studies in the 1960s were among the first to document these types of food-chain effects.

In subsequent years, the public learned that many of the new chemicals used in society not only have environmental effects but also cause adverse health effects in humans, ranging from exzema and respiratory diseases to miscarriages, fetal anomalies and cancers of many types. Of all the maladies caused by toxic substances, the public is especially fearful of cancer, and of anything that may increase the probability of developing that dread disease in any of its humerous forms. Medical and environmental scientists estimate that as many as 60 to 80 percent of all human cancers are caused or are at least exacerbated by exposure to toxic and hazardous substances. Fatalities due to many types of cancers are increasing in most of the "developed" world, in spite of advances in the medical sciences that now permit early detection and treatment of some types of cancer.

During the 1970s and 1980s most governmental bodies responded to these "toxic substance" caused problems by enacting an array of environmental regulations. Most of the rules, laws and policies were developed in response to particular environmental crises and were usually medium specific. Consequently, air pollution control regulations were enacted to focus attention on controlling certain air pollutants. Water pollution problems triggered similar developments of pollution control legislation focused upon controlling the pollutants previously entering the water. More recently, hazardous waste problems have been addressed by yet another set of pollution control responses. As these medium-specific pollution-control laws were enforced, often the pollutants were kept from being released into the air, but were then transferred to the water where the pollutants once again caused problems. In other cases, pollutants were kept from entering the water directly because they were deposited on or in the land in pits, ponds, lagoons or landfills. To our chagrin, these substances then moved into surface water and groundwater, and from there, into the global food-chain. It is now estimated that the cost to "Clean-Up" the tens of thousands of hazardous waste dump sites throughout in the U.S. will be 100s of billions of dollars! Most other industrial countries are facing similar, very costly "clean-up" costs they also inherited from the past. As a result, leaders in many nations are increasingly realizing they must not only "clean-up" the results of unwise releases and deposits of materials in the past, but must also develop better, more integrative systems, capable of preventing the creation of thousands of additional hazardous waste contaminated sites for our children's children to "clean up." Certainly, I don't want to pass on such a heritage to my children! Do you?

Because of these and many other pollution problems, the public has become increasingly impatient with and distrustful of politicians and industrialists. In spite of having more and more pollution-control laws on the books than ever before and despite the expenditure of billions of dollars, pounds, francs, guilders and kroners on planning, building, and operating end-of-pipe pollution control devices, the public continues to hear of new Love Canals, Times Beaches, Bhopals, Chernobyls, or Sandoz's. Additionally, they continue to learn of new toxic substances in their air, water, food and in their homes. It is not surprising therefore, that people in many countries have been demanding that we relook at our basic environmental protection approaches and at the suppositions that underlie those approaches. They are saying, "there must be a better way!"

Emphasis upon prevention is a better way!

INTEGRATED PEST MANAGEMENT IN AGRICULTURE: A SYSTEM'S APPROACH TO PREVENTION

Fortunately there are promising "better ways," of planning and doing many of the things we have been doing. For example, within the agricultural industry, this type of reassessment has resulted in the development and implementation of integrated pest management programs that work more sensitively with the cycles of nature to prevent or minimize the build-up of pests beyond a certain economic threshold. The result is usually a decreased usage of pesticides and therefore, a reduction of the negative environmental impacts from pesticides and a substantial decrease in operating costs for farmers. Significantly, a number of studies have shown that the net profits for the farmer were substantially better using the integrated pest management approach than were those obtained using the calendar method of deciding when to apply pesticides. Furthermore, pesticide residues on the food were substantially reduced. Additionally, the risks of groundwater and ecosystem contamination from the pesticides were also reduced.

HOLISTIC HEALTH AND WELLNESS PROMOTION PROGRAMS IN HUMAN HEALTH CARE: A SYSTEM'S APPROACH TO PREVENTION

While some parts of the medical profession have traditionally placed high value upon prevention of diseases and illnesses, only recently has there been a more wide-spread emphasis, within the medical profession, upon disease prevention and wellness promotion to help people stay healthy, rather than placing such heavy reliance on the use of drugs and surgery for controlling or curing one's diseases once they have developed. This is resulting in dramatic decreases in the frequency and duration of hospitalizations and in impressive cost reductions in medical care in some countries.

Thus in both the agricultural and the health-care sectors, the truth of the old adage, An Ounce of Prevention is Worth a Pound of Cure, is being substantiated while economic, environmental and human health improvements are being documented.

POLLUTION PREVENTION AND WASTE REDUCTION IN INDUSTRY: A SYSTEM'S APPROACH TO PREVENTION

Fortunately, some industrial leaders are also developing and implementing innovative ways to reduce or prevent the production of pollutants generated within industry. Consequently, their firms have simultaneously realized economic and environmental benefits.

The 3M Corporation, for example, has been a leader in implementing what they call their Pollution Prevention Pays (3P) program. *Their 3P emphasis began in 1975 as an alternative to the costly, pollution-control, approach mandated by governmental agencies.* The aim of 3P is to prevent the production of pollutants at their source, rather than at controlling them at the end of the manufacturing process. The 3M Company has actively involved its employees, at all levels of industrial operations, in making suggestions of ways to prevent the production of all types of wastes. The program has elicited impressive employee suggestions for process changes which have saved money and prevented the production of pollutants. The creative ideas produced by the "Involved," employees help 3M to remain highly competitive through rapid advances in new and innovative processes, technologies and products.

What has been accomplished by this employee involvement in 3M's 3P Program? In the first eleven years of the 3P program, 1,200 employee suggestions for pollution prevention and waste reduction, were implemented. As a consequence, the company saved over $300 million in raw materials costs, waste disposal costs, waste management costs and energy costs. At the same time, the company eliminated the discharge, annually, of more than 90,000 tons of air pollutants, 10,000 tons of water pollutants, and 140,000 tons of sludge, and prevented the production of approximately one billion gallons of wastewater. Additionally, they saved 254,000 barrels of oil annually through increased operational efficiency. In other words, 3M has certainly found that Pollution Prevention Pays!

Fortunately, the positive experiences 3M has obtained and is obtaining by placing primary emphasis upon systems approaches to pollution prevention rather than relying solely upon pollution controls, have also been found to be true for many other companies world-wide. Industries are finding that pollution prevention and waste reduction are:

a. Technically Feasible
b. Economically Profitable
c. Ecologically Beneficial
d. Ethically Sound

The following statements by officials from Chevron and Monsanto illustrate the sentiments of many industrial leaders who are highly supportive of using waste reduction and pollution prevention approaches:

> "One of the keystones in our efforts has been and will be the reduction of volume of wastes produced."
>
> Chevron

> "In our existing chemical processes, we have a formal program to reduce the quantity of waste that we are currently producing, through equipment upgrading, more efficient use of raw materials, or different methods of process control."
>
> Monsanto

The leader of a Canadian industrial firm that switched over to the preventative approach, recently said: "I wish we had been 'forced' to look at systems approaches to waste reduction in this comprehensive way ten years ago. We are now *saving* $1,000,000/year in raw materials costs and in waste management costs. At the same time, our worker's environment and the stream's environment have improved dramatically. Productivity and product quality have improved and the firm's public image has also improved drastically. Just think what we could have done if we had started with this emphasis years earlier."

Governmental officials, scientists and technologists are also supportive, as is evident from the following four statements:

> "There needs to be a concentrated effort to make changes in manufacturing processes to reduce the generation of hazardous wastes requiring disposal."
>
> U.S. Environmental Protection Agency

"Elimination, rather than treatment or storage, is the optimal solution to the problems of industrial waste."

The Illinois Industrial Waste Elimination Research Center

"The ideal solution is to reduce waste at the source by changing the industrial processes so that hazardous by-products are not produced."

California Office of Appropriate Technology

"In-plant options are probably the most effective and economical means of managing hazardous wastes. . . . The committee strongly recommends a major commitment, both philosophically and in funding to approaches that prevent or eliminate hazardous materials from being discharged as wastes."

The United States National Academy of Sciences

In a speech at the "Commission of the European Communities" conference entitled, "Environment and Quality of Life," held in the Hague, November 4-7, 1980, Mr. C. van Veen, President of the Federation of the Netherlands Industry cautioned that the term Cleaner Technologies, is more appropriate that the term Clean Technologies. Mr. van Veen stated further, "It is very important for us to give the 'Cleaner Technology,' notion a precise connotation. We should not confine it, for example, to reducing the waste flow but make it comprise elements like:

—The most rational use of raw materials and energy;
—Reducing the amount of pollution effluents in the environment;
—Reducing the quantities of wastes produced during the manufacturing and during the use of the manufactured product."

Mr. van Veen's plea for a more comprehensive approach to the development of ways to reduce the production of wastes of all types throughout the life cycle of the product, is certainly commensurate with the activities of many industrialists that have implemented these types of approaches in their own firms.

Several governments, including the Netherlands government, are now firmly committed to the development and implementation of environmental policies and practices supportive of pollution prevention and waste reduction. This was clearly stated by Minister Pieter Winsemius in a statement made in support of the "Indicative Multi-Year Program 1985-1989 for Chemical Wastes." He said, "the two primary aims of this program are:

a. Prevention of the generation of chemical waste by stimulating research and formulating regulations. Research into non-waste technologies by process adaptation and by use of alternative resources will be continued.
b. Control the diffuse spreading of chemical wastes in the environment."

Winsemius then concluded his speech by saying, "In both aims, central issues are to control the waste streams qualitatively and quantitatively together with gaining insight into the chemical waste problem. In realizing the aims of this policy, *the prevention of the generation of chemical wastes will have the highest priority.*"

Certainly, the tone and intent of each of these people or organizations varies somewhat. However, one point is quite clear: all are agreed that increasingly there must be a comprehensive and cooperative focus by government and industry upon the development and implementation of cleaner technologies that are more energy and material efficient and which produce fewer and less toxic wastes throughout the product's entire life cycle.

POLLUTION PREVENTION AND WASTE REDUCTION CONCEPTS AND APPROACHES

While conventional pollution controls have emphasized "end-of-pipe" treatments or "out-the-back-door," practices to manage wastes produced, pollution prevention, waste reduction and cleaner technology approaches emphasize process-integrated ways to prevent or to minimize the production of wastes at the source. These integrated systems usually reduced pollutant production and discharges of pollutants to all media, to a greater extent and/or at a lower cost than is possible utilizing only "end-of-pipe approaches." Consequently, corporate profits are often enhanced, waste management costs are reduced and risks to the environment and to humans are decreased.

Successful corporate programs of this type have usually included the following steps:

A. *Perform a Waste reduction Audit*—This type of audit is designed to help industrialists to accurately quantify the flow of materials through each manufacturing and processing step. Sources of pollutants and wastes of all types are accurately identified and quantified. Once the nature, sources and present fate of all wastes have been identified, the feedstocks, the processes and the products are scrutinized in detail for opportunities for reducing waste. Alternative waste reduction strategies are developed and evaluated for their relative benefits for reducing the production of wastes and for reducing risks throughout the entire manufacturing system and throughout the product's life cycle.

B. *Develop Management Initiatives*—The most successful corporate pollution prevention, waste reduction programs in industries, have almost always utilized the following two management strategies in implementing their "Systems Approach," to waste reduction and pollution prevention:

1. They communicated the concepts and the details of the program to all employees in clear language. They also showed that they believed that all employees' creative inputs are essential for the identification of problems and for the development of solutions of them.
2. They provided Incentive Awards Programs, to stimulate and to reward creative pollution prevention and waste reduction initiatives from employees.

C. *Improve Housekeeping*—Many companies, once they performed a waste reduction audit, identified many simple, inexpensive, housekeeping changes they could make that would yield immediate reductions in wastes generated and in energy utilized, thereby improving the in-door and out-door environment and the economic profitability of the company at the same time.

D. *Material Substitution*—Many firms found ways, for example, to modify their systems so they could replace organic solvent based inks, paints or cleaners with less toxic, less expensive water-based materials. Other types of substitutions often involve process changes as well as materials changes.

E. *Redesign Equipment*—Redesign or replacement of equipment as well as optimization of process conditions can also be important factors in the reduction of wastes and in the prevention of pollution.

F. *Utilize Integrated Approaches to Effect Materials Reuse and Recycle*—Many new and improved membrane, filtration, adsorption, distillation and micro-processor techniques are making it increasingly feasible for industrialists to effect materials reuse; thereby permitting them to develop closed-loop or nearly closed-loop processes. (Many industrial engineers have commented that they have come much closer to "closed-loop," or nearly "closed-loop" processes than they believed would be possible, even a few years ago.)

The consequences of industrial utilization of these six steps are frequently dramatic reductions in wastes produced and in even more dramatic increases in profitability.

Certainly some processes are more readily improved than others. In some cases, current technologies or raw materials presently available present major obstacles to further waste reduction or pollution prevention. In those instances, joint research activities among governmental, academic and industrial groups should be organized and funded in order that new and improved processes or different types of products that avoid the problems inherent in those presently produced, may be developed. Research in these areas should be catalyzed and coordinated through some international organization such as UNEP, UNIDO or the new International Association of Clean Technology. Results of such research and technological development should be communicated internationally in usable forms so that all nations environments and economies can benefit. Certainly, the global interconnections of air and water pollution should be sufficiently evident to motivate such cooperation.

RESULTS OF 500 CASE STUDIES OF INDUSTRIAL FIRMS THAT HAVE IMPLEMENTED POLLUTION PREVENTION AND WASTE REDUCTION CONCEPTS AND TECHNOLOGIES

Based upon an evaluation of 500 case studies of industrial firms that have successfully implemented pollution preventing—waste reducing technologies, it is clear that the following general process modification areas are often used singly or in various combinations:

A. Replace chemical processes with mechanical processes.
B. Replace single-pass rinse processes with counter-current processes.
C. Replace single-pass processes with closed-loop processes.
D. Replace organic solvent-based inks, paints and coatings with water-based inks, paints or coatings.
E. Replace acidic or alkaline treatment processes with mechanical processes.
F. Install ion exchange, ultrafiltration, reverse osmosis, electrodialysis, adsorption or other processes to effect separation of components from the process stream and to return the useful ones to the processes from which they came.
G. Install new proprietary waste reduction-pollution prevention technologies.
H. Install decanters, conservation vents, floating roofs and scrap condensers to prevent or minimize the release of materials into the air or water media.
I. Recycle or reuse materials in various processes within the firm.

Implementation of these types of process modifications has resulted in 70 to 100 percent reductions of some air emissions, water emissions and/or the production of hazardous and non-hazardous solid wastes. Frequently, there is also a decrease in energy usage and a decrease in worker exposure to toxic substances.

According to the U.S. Office of Technology Assessment's 1986 report to the U.S. Congress, entitled, Serious Reduction of Hazardous Waste, 50% waste reduction can be effected today with presently existing technologies.

The waste reducing, pollution preventing, cleaner technologies, that can help to bring about such dramatic reductions in wastes produced, often pay for themselves in comparatively short times, usually less than two years and in many cases, less than one year. Table 1, characterizes ten specific, industrial waste reductions, and presents the percentage reduction of wastes produced and the payback period of the investments for the process changes. The very encouraging waste reduction and short payback results illustrated by these data were obtained due to improvement within one or more of the following areas:

A. Decreased raw material's costs;
B. Decreased waste management costs;
C. Deceased energy costs;
D. Improved product quality;
E. Enhanced productivity;
F. Decreased down-time;
G. Decreased worker health risks;
H. Decreased environmental hazards;

I. Decreased long-term liability for clean-up of waste materials that might have otherwise been buried in landfills;
J. Improved public image for the company.

The 500 case studies upon which these summary statements are based are from industrial firms from many different nations and many different industrial branches.

CONCLUDING RECOMMENDATIONS FOR GOVERNMENTS AND INDUSTRIES AND FOR SOCIETY AT LARGE

In the foregoing pages, a few of the benefits to industry and to the environment resulting from corporate focus upon pollution prevention and waste reduction are highlighted. In spite of the impressive results of these companies, the vast majority of industrialists have not switched to this approach for a number of reasons. Primary among them is that environmental authorities continue to enforce pollution control regulations by insisting that industry utilize end-

TABLE 1

Characterization of Waste Reduction Approaches

Industry	Method	Percent Reduction of Wastes	Payback Period
Pesticide production	Waste stream separation	100% of dust	10 months
Farm equipment manufacturing	Proprietary process	80% of sludge	2.5 years
Automotive manufacturing	Pneumatic cleaning process	100% of sludge, H_2O	2 years
Microelectronics	Vibratory cleaning process	100% of sludge	3 years
Paint, coatings	Pneumatic cleaning process	100% of solvent, paint	< 1 year
Leather-making	Ion-exchange, adsorption	99% of chromium	2.5 years
Pharmaceutical production	H_2O-base solvent replaced organic-based	100% of solvent	< 1 year
Organic chemical manufacturing	Adsorption, scrap condenser, conservation vent, floating roof	95% of cumene	1 month
Photographic processing	Electrolytic recovery, ion exchange, adsorption	85% of developer, 95% fixer, Ag, and solvent	< 1 year
Equipment manufacturing	Ultrafiltration	100% of solvent, oil; 98% of paint	2 years

of-pipe approaches to attain the environmental quality standards.

Some governments, in response to public pressure, and due to convincing industrial results based upon the utilization of Pollution Prevention approaches, are reluctantly and slowly beginning to devote some attention to integrated approaches to pollution prevention and waste reduction. However, the U.S. EPA's 1988 budget for pollution prevention and waste reduction is substantially less than one percent of the budget devoted to pollution controls.

This continues to be true, even though for the last fifteen years, the U.S. government and many European governments have stated that their preferred ranking for the management of hazardous and non-hazardous wastes is as shown below:

a. Waste prevention
b. Waste reuse within the production process
c. Waste reuse off-site
d. Utilization of the energy value of the wastes
e. Sanitary and/or secure landfilling of the wastes

Sadly, during those fifteen years, the first option, Prevention, did not receive much attention. While prevention was considered to be the best possible option, it was believed to only be possible in the long-run. In the meantime, 99.9% of the governmental efforts of almost every country continued to focus upon the end-of-pipe, waste management options which all agree are lower on the list of priorities. Consequently, the generation of wastes of all types steadily increased and the composition of wastes became more complex, thereby posing increasingly severe threats upon the quality of present and future environments and continuing to promise ever increasing waste management and clean-up costs for the future.

Governments must take seriously the results of the firms referred to earlier. They must increasingly do the following through their regulatory, research and educational efforts:

a. Develop and utilize coordinated, multi-media, regulatory strategies designed to foster the system's approach to waste reduction and pollution prevention. If governments do so, and industrialists take advantage of the benefits thus afforded, not only will the economic situation of most industries be improved; the cost of governmental regulatory action will decrease drastically. More importantly, the prime goals for ensuring environmental quality for us and for future generations will be more realizable. Truly sustainable industrial activities must increasingly be based upon an ecologically sound system's approaches to the prevention of problems rather than continuing to be based upon reacting to the problems after they occur.

b. Develop and implement policies and regulatory strategies that are multi-national or global in scope.

c. Develop environmental quality goals that are clearly defined and which have definite timetables for reaching those goals. (For example, some companies have set their own goals for 35% waste reduction within five years.)
d. Develop a mechanism to coordinate research efforts designed to achieve the environmental quality goals. (An International Association of Clean Technology (IACT), has been formed. It could, with support from governments, industry and academia, play a very significant role in coordinating the research and in disseminating the results of that research throughout the developed and the developing world.)
e. Develop and utilize "Product Design Criteria," to assist product designers plan new products that incorporate "Life Cycle" impact considerations into the product's design. Such design criteria would be selected to reduce wastes and minimize "Toxics Risks" in:

1. The production of the raw materials for the new products.
2. The production of the new product.
3. The consumer's use of the new products.
4. The end disposition of the new products.

If governmental environmental protection efforts emphasized these factors and if they were to invest 25 percent of their economic and intellectual resources during the next ten years on pollution prevention and waste reduction, instead of solely upon pollution controls, we would enter the 21st century with much greater economic, environmental and social health.

As we look to the 21st century, more and more people in all walks of life are realizing increasingly that they must focus upon longer-term responsibilities in such proactive and preventative ways.

Additionally, it is increasingly evident that legislative and regulatory efforts, as important as they are, are insufficient to ensure that we reach the goals essential to ensure a sustainable future for our children's children. Some believe that ethical factors that incorporate ecological principles must become integral to the thinking and decision-making of all industrial, governmental and academic persons. There is an encouraging indication that one group, the World Federation of Engineering Organizations, is taking these responsibilities seriously. This is illustrated by their Code of Environmental Ethics for Engineers, which was developed by a group of their environmental engineers. (A copy of this code is presented below.) This code makes it clearly evident that these engineers recognize that ethical values and ecological values are not abstractions that conflict with sound technological and economic decisions. Instead they see these principles as concrete essentialities for sustainable development. They realize that the Humans versus Nature relationship must be replaced with one of Humans with Nature.

I invite you to study their Code of Ethics and challenge you to develop your own Code of Ethics as a step in clarifying your values and responsibilities for your professional and personal decisions. I urge you to utilize those principles in your day-to-day decisions to increasingly focus upon proactive ways to prevent pollution and reduce waste for present and future generations.

CODE OF ENVIRONMENTAL ETHICS FOR ENGINEERS

The World Federation of Engineering Organizations Committee on Engineering and Environment, with a strong and clear belief that man's enjoyment and permanence on this planet will depend on the care and protection he provides to the environment, states the following principles.

TO ALL ENGINEERS

When you develop any professional activity:

1. Try with the best of your ability, courage, enthusiasm and dedication to obtain a superior technical achievement, which will contribute to and promote a healthy and agreeable surrounding for all men, in open spaces as well as indoors.
2. Strive to accomplish the beneficial objectives of your work with the lowest possible consumption of raw materials and energy and the lowest production of wastes and any kind of pollution.
3. Discuss in particular the consequences of your proposals and actions, direct or indirect, immediate or long term, upon the health of people, social equity and the local system of values.
4. Study thoroughly the environment that will be affected, assess all the impacts that might arise in the state, dynamics and aesthetics of the ecosystems involved, urbanized or natural, as well as in the pertinent socio-economic systems, and select the best alternative for an environmentally sound and sustainable development.
5. Promote a clear understanding of the actions required to restore and, if possible, to improve the environment that may be disturbed, and include them in your proposals.
6. Reject any kind of commitment that involves unfair damages for human surroundings and nature, and manage the best possible social and political solution.
7. Be aware that the principles of ecosystemic interdependence, diversity maintenance, resource recovery and interrelational harmony form the bases of our continued existence and that each of those bases poses a threshold of sustainability that should not be exceeded.

Always remember that war, greed, misery and ignorance, plus natural disasters and human induced pollution and destruction of resources, are the main causes of the progressive impairment of the environment and that you, as an active member of the engineering profession, deeply involved in the promotion of development, must use your talent, knowledge and imagination to assist society in removing those evils and improving the quality of life for all people.

Approved by the Committee on Engineering and Environment of the World Federation of Engineering Organizations, in the 6th Annual Plenary Session, New Delhi, 5th. November 1985.

Printed in Buenos Aires, Headquarters of the Committee, 1986.

SELECTED CITATIONS

1. Campbell, M.E., and W.M. Gleen. 1982. Profit from Pollution Prevention: A Guide to Industrial Waste Reduction & Recycling. The Pollution Probe Foundation, 12 Madison Avenue, Toronto, Ontario, Canada M5R 2S1. 404 pp.
2. Guyer, S. and D. Roe. 1986. Approaches to Source Reduction: Practical Guidance from Existing Policies and Programs. The Environmental Defense Fund, 2606 Dwight Way, Berkeley, CA 94704. 148 pp.
3. Hirschhorn, J.S., et al. 1986. Serious Reduction of Hazardous Waste. The U.S. Office of Technology Assessment, Washington, D.C. 20510-8025. 254 pp.
4. Huisingh, D. and V. Bailey, (Editors). 1982. Making Pollution Prevention Pay: Ecology with Economy as Policy. Pergamon Press, New York, NY. 156 pp.
5. Huisingh, D., et al. 1986. Proven Profits from Pollution Prevention: Case Studies in Resource Conservation and Waste Reduction. The Institute for Local Self-Reliance, 2425 18th St. NW, Washington, D.C. 20009. 316 pp.
6. Huisingh, D., L.R. Martin and J. Aberth. 1987. Toward Pollution-Free Manufacturing. The American Management Association, 135 West 50th Street, New York, NY 10020. 121 pp.
7. Huisingh, Donald and Lisa Gardner. 1987. "Waste Reduction Through Material and Process Substitutions: Progress and Problems Encountered in Industrial Implementation," in Hazardous Waste & Hazardous Materials. Vol. 4, Number 1, p. 23-33.
8. Kauders, R. 1986. Austrian Non-Waste Technologies. Schriftenreihe der Arbeitsgemeinschaft Umweltschutz der Arbeitsgeberverbande und der ORL-Osterreicheschen Gesellschaft zur Reinhaltung der Luft, 1010 Wein, Austria. 60 pp.
9. Kaforov, V.V. 1985. Wasteless Chemical Processing. Mir Publishers, Moscow, USSR. 251 pp.

10. Lindroos, Veikko, et al. 1986. Environmental High Technology from Finland. Kirjapaino R. Lunkka Ky, Finland. 186 pp.
11. Muir, W.R. 1987. Promoting Hazardous Waste Reduction: Six Steps States Can Take. INFORM Inc., 381 Park Ave. South, New York, NY 10016. 19 pp.
12. Nijpels, E. May 1987. "Prevention is better than cure. . . . We need fast decisions." Phoenix International. p. 18-21.
13. Oldenburg, K.U., and J.S. Hirschhorn. "Waste Reduction: From Policy to Commitment," in Hazardous Waste & Hazardous Materials. Vol. 4, Number 1, p. 1-8.
14. Overcash, M.R. 1986. Techniques for Industrial Pollution Prevention. Lewis Publishers, Inc. 121 South Main St., P.O. Drawer 519, Chelsea, MI 48118. 203 pp.
15. Piasecki, B.W., and G.A. Davis. 1987. America's Future in Toxic Waste Management: Lessons from Europe. Quorum Books, New York, NY. 325 pp.
16. Royston, M.G. 1979. Pollution Prevention Pays. Pergamon Press, New York, NY. 197 pp.
17. Schecter, R.N. 1987. Pollution Prevention Bibliography. The Pollution Prevention Pays Program, Division of Natural Resources and Community Development, P.O. Box 27687, Raleigh, NC 26711-7687. 162 pp.
18. Tavlarides, L.L. 1985. Industrial Waste Management, Process Modifications for Industrial Pollution Source Reduction. Lewis Publishers, Inc., 121 South Main St., P.O. Drawer 519, Chelsea, MI 48118. 150 pp.
19. Thompson, F.M., and C.A. McComas. 1987. "Technical Assistance for Hazardous Waste Reduction," in Environmental Science and Technology. Vol. 21 No. 12. p. 1154-1158.
20. Van den Akker, F. 1986. Low Pollution Process Technology in the Netherlands. The Ministry of Housing, Physical Planning & Environment, P.O. Box 450, 2260 MB, Leidschendam, The Netherlands. 350 pp.

Management of Hazardous Materials and Wastes: Treatment, Minimization and Environmental Impacts. Edited by S. K. Majumdar, E. W. Miller and R. F. Schmalz.

Chapter Eleven

REVIEW AND EVALUATION OF RCRA, HSWA, CERCLA AND SARA

JOHN W. UBINGER, JR.

Attorney at Law
Eckert Seamans Cherin & Mellot
600 Grant Street
Pittsburgh, PA 15219

INTRODUCTION

Subtitle C of the Resource Conservation and Recovery Act of 1976[1] ("RCRA") as amended most notably by the Hazardous and Solid Waste Amendments of 1984[2] ("HSWA"), and the Comprehensive Environmental Response Compensation and Liability Act of 1980[3] ("CERCLA" or "Superfund") as amended by the Superfund Amendments and Reauthorization Act of 1986[4] ("SARA") provide the current legal framework for regulating the management of hazardous wastes and the cleanup of hazardous substances released to the environment.[5] Subtitle C of RCRA and HSWA specify the parameters of a comprehensive management program to regulate hazardous wastes from "cradle to grave." CERCLA as amended by SARA and the "corrective action" provisions of HSWA reach beyond the grave and require the exhumation of hazardous wastes which have been disposed of or abandoned in an environmentally unsound manner. To the manager of hazardous substances or hazardous wastes the regulatory web of the RCRA and CERCLA programs poses formidable and often frustrating requirements that demand astute and diligent attention. To the uninitiated or careless, RCRA and CERCLA present a minefield where inadvertance or negligence can detonate significant noncompliance penalties and, when environmental contamination is involved, inordinate remedial or corrective action costs.

It is not possible in the relatively limited space allotted to each of the topics in this book to completely review and evaluate these complex and, in some respects, highly technical statutes. Rather, it is the purpose of this chapter to identify and briefly discuss selected features of the regulatory programs engendered by these laws that are particularly relevant to the manager of significant quantities of hazardous wastes.[6]

THE RCRA HAZARDOUS WASTE MANAGEMENT PROGRAM

The so-called "cradle-to-grave" RCRA Subtitle C hazardous waste management program includes four major categories of provisions: (1) technical specifications and regulatory conventions for identifying "hazardous wastes" which are subject to the RCRA program; (2) the elements of the cradle-to-grave tracking system, i.e., the hazardous waste generator notice and identification number requirements and the manifest and manifest reporting requirements imposed upon generators, transporters, and treatment, storage and disposal facility operators; (3) performance standards that govern parties that engage in hazardous waste management activities; and (4) behavior modification provisions which limit or ban certain previously prevalent hazardous waste management practices such as land disposal, and require new efforts to minimize the generation of hazardous wastes.

Identification of Hazardous Wastes

The bounds of the RCRA Subtitle C hazardous waste management program are established by EPA's regulations relating to the identification of "hazardous wastes." As required by Section 3001 of RCRA,[7] EPA has specified four characteristics for the purpose of identifying hazardous wastes: ignitability, corrosivity, reactivity and EP Toxicity.[8] If a solid waste exhibits one or more of the EPA-specified characteristics and is not otherwise excluded from coverage,[9] it is a hazardous waste for purposes of the RCRA Subtitle C hazardous waste management program.

The ultimate responsibility for determining whether a particular solid waste or class of solid wastes is a RCRA-regulated hazardous waste lies with the party that creates or generates the solid waste. In order to make that determination the "generator" should first consult the applicable list(s) of hazardous wastes[10] that have been compiled through the application of criteria established by EPA for purposes of identifying hazardous wastes on a generic basis.[11] If a potentially hazardous waste is not listed, the generator is required to make a waste-specific determination based upon either the performances of EPA-specified hazard characteristic test methods[12] or specific knowledge concerning the characteristics of the waste in question.[13]

Important regulatory implications flow from the distinction between hazardous wastes which are generically listed by EPA and hazardous wastes which are found on a waste-specific basis to exhibit one or more of the four hazardous waste characteristics. For example, a nonlisted hazardous waste ceases to be a hazardous waste as soon as it no longer exhibits its hazardous characteristic(s). On the other hand, a so-called listed hazardous waste remains a hazardous waste unless and until it is delisted by EPA. In addition, it is EPA's position that residues derived from treating, storing or disposing of *listed* hazardous waste and waste mixtures which include one or more *listed* hazardous wastes in any concentration are hazardous wastes until delisted by EPA,[14] even if such residues or mixed wastes do not exhibit any of the four hazardous waste characteristics.

Because of distinctions such as these it is important for the manager of potentially hazardous solid wastes to determine whether or not the waste is hazardous and to understand the regulatory basis for its hazard designation.

Thus far, the discussion has focused on the question of whether a *solid waste* is hazardous and therefore regulated by the RCRA Subtitle C hazardous waste management program. An even more fundamental question is whether a particular substance is a "solid waste" for purposes of RCRA.

"Solid Waste" for purposes of RCRA is:

> ". . . any garbage, refuse, sludge from a waste treatment plant, water supply treatment plant, or air pollution control facility and other discarded materials, including solid, liquid, semisolid, or contained gaseous material resulting from industrial, commercial, mining, and agricultural operations, and from community activities, but does not include solid or dissolved material in domestic sewage, or solid or dissolved materials in irrigation return flows or industrial discharges which are point sources subject to permits under (The National Pollutant Discharge Elimination System program of the federal Clean Water Act), or source, special nuclear, or byproduct material as defined by the Atomic Energy Act of 1954, as amended . . . "

(RCRA Section 1004(27) 42 U.S.C.A. §6903(27)(1983)).

In many instances the question of whether a given substance is a "solid waste" for RCRA purposes is not controversial. That is not the case where the substance in question can be used, reused, recycled, reclaimed or burned as fuel.

In establishing the initial RCRA hazardous waste management program EPA took the position that generated substances that could either discarded or used, reused, recycled, reclaimed or burned as fuel were "solid wastes" subject to regulation under RCRA, even when they were not discarded. However, after reviewing the substantial number of comments that were filed in response to that position, both pro and con, EPA opted to defer to a later date the formulation of rules governing the use and reuse of hazardous wastes and hazardous waste recycling and reclamation activities.[15] Consequently, EPA's initial RCRA rulemaking in this area was limited to the regulation of transportation and storage of discardable hazardous substances prior to use, reuse, recycling and reclamation.

EPA's subsequent rulemaking with respect to use, reuse, recycling, reclamation and burning of discardable hazardous substances was published in January

1985 as an amendment to the definition of "solid waste."[16] In the preamble to that rule, EPA reaffirmed its previous position that it had the authority under RCRA to regulate use, reuse, recycling, reclamation and fuel burning activities. It then went on to create what EPA itself characterized as a "complicated regulatory scheme" to govern some but not all activities involving the use, reuse, recycling, reclamation and burning of "recycled materials."[17]

Obviously, it is incumbent upon the manager of any activity that involves the generation, handling or disposition of a substance that could possibly be classified as a "hazardous solid waste" for RCRA purposes to understand these complex identification rules in order to avoid inadvertent noncompliance. On a more positive note, an understanding of the proper application of these rules, and an awareness of the various waste-specific variance, exclusion and delisting provisions that are available[18] may provide a means of lessening the regulatory burden on a specific activity.

The Hazardous Waste Manifest System

The keystone of the RCRA "cradle-to-grave" hazardous waste management program is the facility identification and manifest system. Under Section 3010 of RCRA[19] a party may not generate, transport, treat, store or dispose of a RCRA-regulated hazardous waste unless it has submitted facility-specific notice(s) of such activity on an EPA-prescribed form. When a party submits the requisite notice(s) it receives a distinct EPA identification for each facility that engages in a hazardous waste management activity.

A party that handles RCRA-regulated hazardous waste may not offer such waste to or accept it from any other party that does not have the requisite EPA identification number.[20]

The hazardous waste manifest is a federally-prescribed form[21] that is issued by a generator for each shipment of hazardous waste that leaves its site. In completing the form the generator specifically identifies the hazardous waste which is the subject of the manifest using certain prescribed codes and also specifically identifies each transporter and treatment, storage or disposal facility operator that will handle the waste prior to its ultimate disposition. The manifest accompanies the shipment of hazardous waste which it identifies and each of the specified parties that handles the waste through to its ultimate disposition must sign the manifest and retain a copy for its files. Once the hazardous waste is ultimately disposed of, the last party in the chain signs the hazardous waste manifest and returns a copy to the generator to close the loop. When the manifest is returned to the generator it is obligated to retain the form for a specified period of time. Information from the returned manifests is summarized in a report that the generator is required to submit periodically to either EPA or a RCRA authorized state.[22] If a manifest for any shipment of hazardous waste is not returned within a specified period of time, the generator is obligated to file an "exception" report with either EPA or a RCRA authorized state to alert the appropriate agency to the missing manifest and the potentially misdirected shipment of hazardous waste.[23]

While certain situations raise questions with respect to the completion of a manifest, the manifest system is relatively straight forward and uncomplicated when compared to other portions of the RCRA hazardous waste management program. Nonetheless, it is one of the most important components of the cradle-to-grave management concept. Consequently, with the assistance of computers, EPA and RCRA authorized states monitor the manifest system very closely and routinely penalize parties for failure to comply with manifest requirements. It is not unusual for the inadvertent omission of required information from an otherwise properly filled out manifest to draw the attention of either EPA or a RCRA authorized state and result in the imposition of a civil penalty for noncompliance. Therefore, it is incumbent upon the manager of hazardous waste to institute an effective program to insure that manifests are properly completed, properly executed and properly filed in an organized way to assure compliance with the manifest rules.

Performance Standards

The third category of RCRA hazardous waste management provisions relevant to the manager of hazardous wastes are the performance standards established for generators and transporters of hazardous wastes and the owners and operators of hazardous waste treatment, storage or disposal facilities. In addition to prescribing the use of manifests for the cradle-to-grave tracking of hazardous wastes and mandatory recordkeeping and reporting requirements applicable to generators, transporters and the owners and operators of hazardous waste treatment, storage and disposal facilities, these standards establish the rules which govern the day-to-day handling of hazardous wastes. In the case of treatment, storage and disposal facilities, the performance standards provide the bases for the issuance of hazardous waste management facility permits.

The performance standards applicable to generators relate primarily to the management of hazardous wastes prior to transport for treatment, storage or disposal. The most significant of these standards relates to the method in which the generator holds hazardous wastes prior to treatment, storage or disposal. Generally speaking, a generator may "accumulate" hazardous wastes prior to treatment, storage or disposal if the wastes are accumulated in either "containers" or "tanks," as those terms are defined in the RCRA rules[24] and the accumulation period for any specific unit of hazardous waste does not exceed ninety days.[25] If the generator accumulates hazardous waste in this method, it is not required to obtain a permit. However, it is important to note that the performance standards for the *storage* of hazardous wastes, an activity which does require a permit, are incorporated by reference in the rules governing *ac-*

cumulation. Therefore, while a generator may accumulate hazardous wastes for up to ninety days without a permit, in order to do so, it must meet many of the performance standards that apply to a permitted storage facility. It is also important to note that a generator is the only party that may "accumulate" hazardous wastes without a permit. A nongenerator that holds hazardous wastes that are not in transit for any period of time is required to obtain a storage permit.

In addition to the specific performance standards governing the accumulation of hazardous wastes prior to treatment, storage and disposal, the performance standards applicable to generators also expressly impose responsibility for packaging, labeling, marking and placarding of all shipments of hazardous wastes for off-site treatment, storage or disposal upon the generator. Therefore, even though a generator may contract with a hazardous waste transporter to perform that service, it is the generator that is responsible for assuring that hazardous wastes are packaged, labeled, and marked in accordance with federal Department of Transportation regulations and that the transport vehicle is placarded in accordance with DOT regulations.[26]

The EPA performance standards applicable to transporters of hazardous wastes are few in number and relatively straight forward. As is the case with all parties that handle hazardous wastes, the transporter is required to comply with the rules pertaining to the manifest system.[27] The only other specific performance standards imposed by the federal RCRA rules pertain to the action required in the event of an accidental discharge while the hazardous wastes are in transport.[28] As indicated previously, the transporter bears no express responsibility under the RCRA performance standards for packaging, labeling, marking and placarding; those obligations are imposed upon the generator.

The federal RCRA rules do not require hazardous waste transporters to be licensed. However, many states, including Pennsylvania, require transporters to obtain a state license.

The most complex and technically detailed performance standards in the RCRA rules are those that apply to treatment, storage and disposal facilities. The so-called "TSD" performance standards include generally applicable standards relating to such matters as general facility standards,[29] preparedness and prevention standards to minimize the possibility of fire, explosion or any unplanned sudden or nonsudden release of hazardous waste or hazardous waste constituents to any of the environmental media,[30] contingency planning and emergency procedure formulation,[31] manifest system recordkeeping and reporting,[32] monitoring specifications to detect releases from solid waste management units at treatment, storage and disposal facilities and specifications for "corrective action" in the event of releases,[33] standards for the closure and post-closure care of treatment, storage and disposal facilities[34] and financial requirements relating to liability insurance and financial assurance for the cost of closure and post-closure care.[35]

In addition to those generally applicable performance standards there are facility-specific performance standards governing the use and management of containers and the design, construction and operation of tank systems, surface impoundments, waste piles, land treatment, landfills, and incinerators.[36] It is far beyond the intended scope of this chapter to review or evaluate these lengthy and detailed performance standards. Suffice to say that they are, in many instances, complex and difficult to apply to specific situations. Consequently, the permit application process is long and tedious. If a party intends to construct a facility which requires a RCRA permit, a long lead time for the permit application process should be factored into the planning for the facility.

Hazardous Waste Management Modification Provisions

The primary purpose of the RCRA Subtitle C program, as initially enacted by Congress, was to provide a comprehensive regulatory program for the management of traditional hazardous waste generation, transport, treatment, storage and disposal activities. When Congress enacted HSWA in 1984, it not only sought to amend the program for regulating existing hazardous waste management practices, it set out to radically alter conventional thinking about the generation and disposition of hazardous waste. Thus, it is now the goal of RCRA to first reduce or eliminate to the greatest extent practicable the generation of hazardous waste and then to prohibit or strictly limit the land disposal of the reduced volume of hazardous waste that is generated.

To implement this new "national policy," HSWA requires generators of hazardous waste to expressly certify in writing that:

> "(1) The generator of the hazardous waste has a program in place to reduce the volume or quantity and toxicity of such waste to the degree determined by the generator to be economically practicable; and
> (2) The proposed method of treatment, storage or disposal is that practicable method currently available to the generator which minimizes the present and future threat to human health and the environment."
> (42 U.S.C.A. §§ 6922(b) and 6925(h) (West Supp. 1988)).

Of even more significance, HSWA sets in motion a schedule for the banning of land disposal of listed hazardous wastes—historically the most common method of disposal—

> "...unless [EPA] determines the prohibition on one or more methods of land disposal of such waste is not required in order to protect human health and the environment for as long as the waste remains hazardous, taking into account— (A) the long term uncertainties associated with land disposal,
>
> (B) the goal of managing hazardous waste in an appropriate manner in the first instance, and

(C) the persistence, toxicity, mobility and propensity to bio-accumulate of such hazardous wastes and their hazardous constituents." (42 U.S.C.A. § 6924(d) (West Supp. 1988)).

The so-called land ban is scheduled to be phased in over a period of sixty-six months beginning on November 8, 1984, the date of enactment of HSWA.

Quite obviously, the combination of the waste minimization certification requirements and the phased implementation of the land disposal prohibition will require the regulated community to fundamentally rethink the traditional methods by which it produces goods that generate hazardous wastes and the methods for ultimately disposing of the hazardous waste generated.

APPLICABILITY OF FEDERAL RCRA RULES IN PENNSYLVANIA

Section 3006(a) of RCRA provides that any state wishing to administer and enforce the federal RCRA hazardous waste management program could do so upon the grant of authorization by EPA.[37] In May 1981, Pennsylvania obtained interim authorization to administer and enforce a portion of the RCRA hazardous waste management program and on October 16, 1985, Pennsylvania formally applied to obtain final RCRA authorization.[38] EPA granted that final authorization effective January 30, 1986.[39] Consequently, as of January 30, 1986, Pennsylvania assumed the responsibility for administering and enforcing the basic RCRA hazardous waste management program, including responsibility for issuing permits to hazardous waste treatment, storage and disposal facilities based upon performance standards set forth in Pennsylvania's hazardous waste management rules.[40]

Because Pennsylvania has obtained final authorization to administer and enforce the federal RCRA program, its rules apply in lieu of any EPA RCRA rule that was in effect as of October 16, 1984.[41] In addition, Pennsylvania's EPA-approved RCRA rules apply in lieu of any amendment to a federal rule promulgated pursuant to *RCRA* pending the timely satisfaction of Pennsylvania's obligation to amend its rules to conform to the amended federal rule.[42] However, if EPA promulgates a hazardous waste management rule pursuant to *HSWA,* as distinguished from RCRA, the federal HSWA rule becomes immediately effective.[43] In that event, there is a dual federal and state regulatory program in effect until Pennsylvania promulgates new rules to conform to the HSWA-required rule promulgated by EPA.[44]

The rules relating to the enforceability of various types of amendments to the federal hazardous waste in RCRA-authorized states can be potentially confusing to the regulated community. It should be noted that each Federal Register Notice which publishes a new federal hazardous waste management rule specifies whether the rule is a requirement or prohibition prescribed by HSWA and thereby reserved to EPA for implementation and enforcement. It is incumbent upon managers of hazardous waste to understand the distinction between EPA implemented and enforced provisions and Pennsylvania implemented and enforced provisions, and to be aware of new rulemaking at both the federal and state levels that affect the federal-state relationship with respect to these areas. It is also important for parties that deal with other states to know whether the state in question has authorization to implement and enforce the RCRA program in lieu of EPA.

COMPREHENSIVE ENVIRONMENTAL RESPONSE COMPENSATION AND LIABILITY ACT AND THE SUPERFUND AMENDMENTS AND REAUTHORIZATION ACT

The comprehensive cradle-to-grave hazardous waste management program established by RCRA and HSWA is largely prospective in nature. With the exception of the corrective action requirements established by HSWA[45] neither RCRA nor HSWA deals with releases or threatened releases of hazardous substances from past disposal activities or sites which are not regulated by RCRA.

In December 1980, Congress enacted the Comprehensive Environmental Response Compensation and Liability Act to deal with releases or threatened releases from abandoned disposal sites and other sites not governed by RCRA, or some other federal environmental regulatory program. In order to accomplish its objective, CERCLA first required a specifically defined set of parties who would likely have knowledge of past or existing disposal activities at uncontrolled sites to give notice of the existence of those sites to federal EPA by June 1981.[46] CERCLA also required the development of a National Priorities List of sites determined by EPA to be prime candidates for remedial clean up actions under CERCLA.[47] CERCLA then provided the President of the United States with broad legal authority and where necessary federal funds to study uncontrolled sites at which there may be a release or threatened release of hazardous substances, to evaluate and select response actions for those sites and to implement the selected response action.[48] The President has delegated this CERCLA authority to EPA.

While CERCLA provides the federal government with the legal authority and funding to respond to releases or threatened releases of hazardous wastes at CERCLA sites, its primary goal is to impose the financial responsibility for such response actions upon private parties who are either directly or vicariously responsible for the presence of the hazardous substances which necessitate the response action. To that end, CERCLA, as amended by SARA, includes among those who are "potentially responsible parties" any person who owned or operated a facility at the time the hazardous substances in question were disposed of; the current owner or operator of the facility, regardless of whether that party caused or contributed to the release or threatened release which is the subject of the CERCLA response action; any person who by contract agreement or otherwise arranged for the transport of the hazardous substances to the facility or arranged for the disposal or treatment of the substances at the facility and the transporter of any hazardous substances in question.[49] Under the express provisions of CERCLA and SARA, each of these parties can be held strictly liable. More importantly, as the result of judicial opinions interpreting the liability provisions of CERCLA, it has been established that these parties can also be held jointly and severally liable. In other words a responsible party or subset of responsible parties can be required to bear a proportion of the response action costs which exceeds their pro rata contribution of hazardous substances to the Superfund site.

The original authorization of CERCLA expired in 1985. To that point both the pace of accomplishments under CERCLA and the results of some of the response actions undertaken by EPA and/or private parties pursuant to settlement agreements with EPA were unacceptable as far as Congress was concerned. Consequently, when Congress reauthorized CERCLA through the enactment of SARA in 1986, it established mandatory site identification and clean up timetables for EPA and mandatory minimum clean up standards which include compliance with "applicable or relevant and appropriate requirements" from other federal and state environmental laws. In addition, SARA increases the participatory role of both state regulatory agencies and the public. Finally, while Congress included provisions intended to induce and facilitate voluntary response actions by private parties, it also substantially increased EPA's leverage to impose its will in the negotiation of settlements through the virtual elimination of preenforcement of review of EPA decisions and by increasing the size of the "Superfund" by 8.5 billion dollars to provide EPA with more resources to undertake response actions in the absence of an agreement with potentially responsible parties.

There are numerous aspects of CERCLA and SARA which deserve discussion; however, such a discussion is beyond the intended scope of this chapter. For present purposes it is sufficient to say that:

> "For the practitioner, SARA is likely to make CERCLA, which was already a complex law, with extensive administrative and judicial elaboration, into even more of a specialized area of practice. Indeed, SARA, rather than directly addressing a central issue such as the strict, joint and several standard of liability in statutory language, makes the judicial opinions on this subject part of its legislative history. SARA also makes knowledge of other federal and state environmental laws essential. For example, the practitioner interpreting SARA needs to be familiar with RCRA to determine what groundwater cleanup and off-site disposal rules will be applied under

the SARA amendments, what substances will be included in the definition of hazardous substances, and what prospect there is that a cleanup or citizen suit can utilize RCRA rather than CERCLA procedures. In addition to RCRA, the new SARA cleanup provisions borrow standards and criteria from the FWPCA, SDWA, the Toxic Substances Control Act, and the Clean Air Act, as well as more stringent state rules. State siting, permit, and environmental impact statement requirements must also be considered. To these environmental laws, SARA adds references to the law of bankruptcy, insurance, indemnity and real estate."[50]

CONCLUSION

The legal framework for the regulation of the management of hazardous wastes and the clean up of hazardous waste disposal sites is a complex structure of highly technical federal and state statutory provisions and administrative rules. In the limited space devoted to this subject in this chapter, it is obviously not possible to do more than merely touch upon a select few of the many important concepts that a manager of hazardous wastes needs to know and understand to avoid noncompliance with the applicable legal requirements and to avoid disproportionate financial responsibility for the clean-up of hazardous waste disposal sites. The single most important thing that can be said in the context of this chapter is that diligent compliance with the applicable hazardous waste management rules and the exercise of due caution in dealing with hazardous waste management contractors, which in some instances may require taking steps which exceed the minimum requirements of the applicable rules, is absolutely essential. The management of hazardous wastes is a full-time undertaking and those managers who are unable to devote their full attention to keeping abreast of the ever-evolving hazardous waste management rules should seek the assistance of both technical consultants and legal counsel who routinely practice in this increasingly specialized area.

REFERENCES AND FOOTNOTES

1. The Resource Conservation and Recovery Act of 1976, Pub. L. No. 94-580, 90 Stat. 2795 (codified as amended at 42 U.S.C.A. § 6921 *et seq.* (1983 & West Supp. 1988).
2. The Hazardous and Solid Waste Amendments of 1984, Pub. L. No. 98-616, 98 Stat. 3221 (codified at 42 U.S.C.A. § 6921 *et seq.* (West Supp. 1988).
3. The Comprehensive Environmental Response Compensation and Liability Act of 1980, Pub. L. No. 96-510, 94 Stat. 2767 (codified as amended at 42 U.S.C.A. § 9601 *et seq.* (1983 & West Supp. 1988).

4. The Superfund Amendments and Reauthorization Act of 1986, Pub. L. No. 990646, 100 Stat. 1613 (codified at 42 U.S.C.A. § 9601 *et seq.* (West Supp. 1988).
5. It should be noted that The Toxic Substances Control Act, Pub. L. No. 94-469, 90 Stat. 2003, (codified as amended at 15 U.S.C.A. § 2601 *et seq.* (1982 & West Supp. 1988); regulates the manufacture, processing, distribution in commerce, use and *disposal* of one hazardous substance, polychlorinated biphenyls ("PCBs") and may be employed by the United States Environmental Protection Agency to regulate the management and disposal of other "toxic substances."
6. The RCRA Subtitle C hazardous waste management rules distinguish "small quantity generators" (i.e., parties that generate between 100 and 1,000 kilograms of hazardous wastes per month) from parties that generate or manage larger volumes of hazardous wastes. The so-called small quantity generator rules are a very significant part of the RCRA hazardous waste management program and have wide-spread impact on the regulated community. However, due to the focus of this chapter, the small quantity generator rules are not included in this discussion.
7. 42 U.S.C.A. § 6921 (1983).
8. 40 C.F.R. §§ 261.20 - 261.24 (1988) (specifying the characteristics of hazardous wastes).
9. *See,* 40 C.F.R. § 261.4 (1988) (relating to exclusions from the definition of hazardous waste).
10. *See,* 40 C.F.R. §§ 261.30 - 261.33 (1988) (identifying the listed hazardous wastes).
11. *See,* 40 C.F.R. § 261.11 (1988) (specifying the criteria for listing hazardous wastes).
12. *See,* 40 C.F.R. §§ 261.21 - 261.24 (1988) (specifying the characteristics of hazardous wastes).
13. *See,* 40 C.F.R. § 262.12 (1988) (relating to hazardous waste determinations by generators).
14. *See, e.g.* 53 Fed. Reg. 31,148 (1988) (relating to the so-called "derived-from" rule) and 45 Fed. Reg. 33,095 (relating to the so-called "mixture" rule).
15. *See,* 45 Fed. Reg. 33,092 (1980) (relating to EPA's initial position on the use, reuse, recycling or reclamation of hazardous wastes)
16. *See,* 50 Fed. Reg. 614 (1985) (relating to the definition of "solid waste").
17. *See,* 40 C.F.R. § 261.2 (1988) (relating to the definition of "solid waste").
18. *See,* 40 C.F.R. §§ 260.20 - 260.41 (1988) (relating to RCRA rulemaking petitions).
19. 42 U.S.C.A. § 6930 (1983).
20. *See, e.g.* 40 C.F.R. §§ 262.12(c) (relating to generators) and 263.20(a) (relating to transporters) (1988).
21. *See,* 40 C.F.R. Part 262 Appendix (1988) (relating to the uniform hazardous

waste manifest).

22. *See, e.g.* 40 C.F.R. §§ 262.40 and 262.41 (1988) (relating to manifest recordkeeping and reporting).
23. *See,* 40 C.F.R. § 262.42 (1988) (relating to missing manifest "exception" reporting).
24. *See,* 40 C.F.R. § 260.10 (1988) (relating to the definitions of "containers" and "tanks").
25. *See,* 40 C.F.R. § 262.34 (1988) (relating to the "accumulation" of hazardous wastes by a generator).
26. *See,* 40 C.F.R. §§ 262.30 - 262.33 (1988) (relating to packaging, labeling, marking and placarding).
27. *See,* 40 C.F.R. §§ 263.20 - 263.22 (1988) (relating to manifests and recordkeeping).
28. *See,* 40 C.F.R. §§ 263.30 - 263.31 (1988) (relating to hazardous waste discharges during transport).
29. *See,* 40 C.F.R. § 264, Subpart B (1988) (relating to general facility standards).
30. *See,* 40 C.F.R. § 264, Subpart C (1988) (relating to preparedness and prevention).
31. *See,* 40 C.F.R. § 264, Subpart D (1988) (relating to contingency plans and emergency procedures).
32. *See,* 40 C.F.R. § 264, Subpart E (1988) (relating to the manifest system, recordkeeping and reporting).
33. *See,* 40 C.F.R. § 264, Subpart F (1988) (relating to releases from solid waste management units).
34. *See,* 40 C.F.R. § 264, Subpart G (1988) (relating to facility closure and post closure care).
35. *See,* 40 C.F.R. § 264, Subpart H (1988) (relating to TSD facility financial requirements).
36. *See,* 40 C.F.R. § 264, Subpart I (1988) (relating to containers), Subpart J (relating to tanks), Subpart K (relating to surface impoundments), Subpart L (relating to waste piles), Subpart M (relating to land treatment), Subpart N (relating to landfills), and Subpart O (relating to incinerators).
37. 42 U.S.C.A. § 6926(a) (1983).
38. *See,* 50 Fed. Reg. 46,733 (1985) (relating to Pennsylvania's application for final RCRA authorization).
39. *See,* 51 Fed. Reg. 1,791 (1986) (relating to grant of final RCRA authorization to Pennsylvania).
40. *See,* 25 Pa. Code § 75.260 *et seq.* (1988) (relating to Pennsylvania's hazardous waste management rules).
41. *See,* 51 Fed. Reg. 1,792 (1986) (relating to the grant of final RCRA authorization to Pennsylvania).
42. *Id. See also,* 40 C.F.R. § 271.21(e) (1988) (relating to amendments to authorized state RCRA programs to incorporate amendments to the federal RCRA

program).
43. *See,* 42 U.S.C.A. § 6926(b) (West Supp. 1988) (relating to effect of HSWA-required amendments to the federal hazardous waste management program in RCRA-authorized states).
44. *See,* 51 Fed. Reg. 1,793 (1986) (relating to effect of HSWA-required amendments to federal hazardous waste management program in RCRA-authorized states).
45. Under the provisions of Sections 206 and 207 of HSWA (42 U.S.C.A. § 6924(u)&(v) (West Supp. 1988) and EPA's regulations thereunder, (40 C.F.R. §§ 264.100 and 264.101) the owner or operator of a hazardous waste management facility may be required to clean up previously closed solid waste management units at the facility and in some cases beyond the boundaries of the facility.
46. *See,* 42 U.S.C.A. § 9603(e) (1983) (relating to existing facility notices).
47. *See,* 42 U.S.C.A. § 9605 (8) (1983) (relating to the development of the Superfund National Priorities List).
48. *See,* 42 U.S.C.A. § 9604 (1983) (relating to Superfund response action authority).
49. *See,* 42 U.S.C.A. § 9607 (1983 & West Supp. 1988) (relating to CERCLA liability).
50. Atkeson, T.B., S. Goldberg, F.E. Ellrod III, and S.L. Connors. 1986. An Annotated Legislative History of the Superfund Amendments and Reauthorization Act of 1986 (SARA), pp. 8-9. In *Superfund Deskbook,* Environmental Law Institute, Washington, D.C.

Key to Abbreviations in References
"C.F.R." — Code of Federal Regulations
"Fed. Reg." — Federal Register
"Stat." — United States Statutes at Large
"U.S.C.A." — United States Code Annotated

Management of Hazardous Materials and Wastes: Treatment, Minimization and Environmental Impacts. Edited by S. K. Majumdar, E. W. Miller and R. F. Schmalz.

Chapter Twelve

THE COMMUNITY RIGHT-TO-KNOW LAW—SARA TITLE III

DEBORAH LUKESH[1] and ANTHONY NEVILLE[2]

[1]U.S. Environmental Protection Agency
401 M Street, S.W.
Washington, D.C. 20460
and
[2]SRA Technology
Alexandria, VA

INTRODUCTION

On October 17, 1986, Congress enacted into law the Superfund Amendments and Reauthorization Act (SARA). A free-standing act in SARA is Title III—The Emergency Planning and Community Right-To-Know Act, commonly referred to as SARA Title III or EPCRA. The objective of this chapter is to address the history and impetus for Title III, the major provisions of the law, and how these provisions support the intent of the law.

BACKGROUND

The Bhopal, India disaster in 1984, increased the general level of awareness and sensitivity to chemical accidents at all levels of government, in the regulated community, and in the private sector. The problems associated with the accidental release of acutely toxic chemicals into the air became highly apparent. As a result of this incident, the U.S. Environmental Protection Agency (EPA) established the Chemical Emergency Preparedness and Prevention (CEPP) program. The program emphasized the need for increased public awareness of chemical hazards in the community and comprehensive emergency planning as an effective mechanism for protecting public health and the environment in the event of an accidental chemical release.

State and local governments and industry also initiated new programs following the Bhopal incident. The Chemical Manufacturer's Association (CMA), for instance, established a similar program to CEPP, Community Awareness and Emergency Response (CAER). The CAER program emphasized the integration of facility emergency plans with community emergency response plans.

Congress recognized the merits of all these programs and provided statutory authority for CEPP through SARA Title III. Title III expands upon CEPP by increasing public knowledge of and access to information on the presence and releases of hazardous chemicals in their communities, by requiring the development of comprehensive emergency response plans, and by encouraging prevention of accidental releases.

SARA Title III establishes requirements for federal, state and local governments and industry regarding:

—Emergency planning & release notification
—Community right-to-know
—Hazardous & toxic chemical reporting

Other provisions provide for trade secrets, enforcement, and citizen suits.

Overall, Title III is intended to help communities, state and local governments, industry, and emergency response personnel be better prepared for responding to a chemical emergency accident.

SARA TITLE III PROVISIONS

Title III consists of the following three major provisions:

- Subtitle A—Emergency Planning and Notification
- Subtitle B—Reporting Requirements
- Subtitle C—General Provisions

The requirements under each subtitle are addressed in this section along with respective enforcement provisions, as provided in Section 325 of Title III.

SUBTITLE A

Subtitle A is designed to develop state and local emergency response and preparedness capabilities through better coordination and planning. It is divided into the following five (5) sections:

Section 301—Government Planning Structure
Section 302—Emergency Planning Notification
Section 303—Comprehensive Emergency Response Plans
Section 304—Emergency Release Notification
Section 305—Emergency Training and Review of Emergency Systems

Each of these sections identifies specific responsibilities for industry and state and local governments in implementing the emergency planning and notification provisions of Title III.

Government planning structure

Section 301 establishes a framework at the state and local government levels for implementing Title III. The Governor of each State is required to designate a state emergency response commission (SERC) which consists of members with technical expertise in emergency response, environmental and natural resources, public health, occupational safety, media, and transportation. All of the states have established a SERC as of April 17, 1987 as required by Section 301(a). The commission is responsible for establishing local emergency planning districts, appointing and overseeing local emergency planning committees (LEPC) and establishing procedures for handling public requests for information [Section 301(a) and (b)].

Under Section 301(c), a LEPC must include, at a minimum, representatives from the following groups: state and local officials, law enforcement, civil defense, fire fighting, environmental, hospital, media, public health, transportation, community, and facility owners and operators subject to the emergency planning requirements. The LEPCs are primarily responsible for preparing a comprehensive emergency response plan for their district by October, 1988 and for making information on hazardous chemicals, which is submitted under Title III, available to the public [Sections 301(c) and 303(a)].

Emergency planning notification

Section 302 requires EPA to publish a list of extremely hazardous substances (EHS) within 30 days of SARA enactment. This list consists of those substances identified under CEPP as posing a potential hazard to a community upon release into the air environment.

Under Section 302, any facility which uses, stores, or produces any of these 366 EHSs in quantities equal to or greater than their respective threshold planning quantity (TPQ) must notify the SERC by May 17, 1987. If a facility becomes subject to the requirements of this section in the future, notification must be made within 60 days to the commission, as well as to the local committee [Section 302(c).]

Representative businesses covered by Section 302 include not only major chemical manufacturing facilities but also a wide variety of chemical users/producers such as farmers, dry cleaners, and other service-related businesses. Exemptions under this section apply only to vessels (ships/boats), federal facilities, and transportation. Storage incidental to transportation is exempt provided that the materials are still moving under active shipping papers and have not reached the final consignee.

Comprehensive emergency response plans

Section 303 defines the requirements for preparing a comprehensive emergency response plan by the LEPC [Section 303(a)]. As part of the planning process, the LEPC must evaluate available resources for developing, implementing, and exercising the plan [Section 303(b)]. The plan must include the following:

—Identification of facilities subject to Subtitle A
—Identification of transportation routes for EHSs
—Identification of risk related facilities
—Methods and procedures for response
—Designated community and facility coordinators
—Procedures for public notification
—Methods for determining release occurrence and area affected
—Description of emergency equipment and facilities and those responsible
—Evacuation plans and training programs

To ensure the participation of those facilities which notified under Section 302, each facility must have designated a facility emergency coordinator by September 17, 1987 [Section 303(d)]. The designation of the facility coordinator must be made to the LEPC or the SERC, if there is no LEPC. This facility representative serves to provide the LEPC with any information on the facility which the committee needs to develop and implement its plan. The coordinator must also keep the LEPC up-to-date on any relevant changes which occur at the facility to maintain plan integrity. The Governor or SERC can designate additional facilities to be subject to the requirements of Subtitle A [Section 302(b)(2)]. Designation can only be made after public notice and opportunity for comment.

Assistance in plan preparation and review is provided under Section 303(e). The National Response Team (NRT), composed of fourteen (14) Federal agencies with emergency response capabilities, is required to prepare guidance on emergency response planning. This document, entitled the *Hazardous Materials Emergency Planning Guide,* was published by the NRT in March 1987 [Section 303(f)].

Although not required under Title III, EPA, in conjunction with the Federal Emergency Management Agency (FEMA) and the Department of Transportation (DOT), published a supplement to the *Hazardous Materials Emergency Planning Guide* in December 1987. This document, entitled *Technical Guidance for Hazards Analysis,* provides technical assistance to LEPCs in assessing the lethal hazards associated with potential airborne releases of EHSs. Additional supplements are planned to be published in 1988 for assessing flammable, corrosive, and explosive hazards.

Under Section 303(e), the completed emergency response plan must be reviewed by the SERC and annually by the LEPC. Regional Response Teams, composed of regional Federal agencies and state representatives, can review plans and provide technical assistance upon request of the LEPC [Section 303(g)].

Emergency release notification

Section 304 addresses the emergency release notification procedures for the regulated community. Any facility which uses, stores, or produces hazardous chemicals must notify the SERC and the LEPC when a release occurs of an EHS or a CERCLA hazardous substance in quantities equal to or greater than the substance's reportable quantity (RQ). A release of a CERCLA hazardous substance also requires notification to the National Response Center (NRC) in Washington, D.C. (800-424-8802) [Section 304(a)]. Listings of the EHSs and the CERCLA hazardous substances are in the April 22, 1987 Federal Register (40 CFR Parts 300 & 355) and in the September 29, 1986 FR (40 CFR Parts 117 & 302), respectively.

Notification of a release must be made immediately by the facility owner or operator via telephone, radio, or in person to the SERC, to the community emergency coordinator for the LEPC, and to the NRC, if a CERCLA hazardous substance [Section 304(b)(1)]. A transportation-related release notification can be established by dialing 911 or the operator. The release notification must include the following information, if available [Section 304(b)(2)]:

—Chemical identify
—Indication if EHS
—Amount released
—Duration of release
—Released to what media (air, water or land)
—Known or anticipated acute or chronic health risk and appropriate source of medical information, if applicable
—Precautions to be taken and evacuation information
—Name and telephone number of person to be contacted for further information

Section 304(c) requires a written followup report, as soon as practical, which provides more specific data on the release, the response, and the effects. The SERC and the LEPC are the recipients of these written reports. Specifically, the following information must be addressed in the report:
—Response and containment procedures
—Any known or anticipated acute or chronic health risk
—Where appropriate, advice regarding medical attention necessary for the exposed individuals

The following are important exemptions under the emergency release notification provisions:
—Federally permitted releases as defined under CERCLA
—Releases which result in exposure to persons solely within the facility boundaries

—Releases exempt from CERCLA reporting
—Releases which occur at a facility which does not store, produce, or use hazardous chemicals
—Emissions from engine exhaust of motor vehicles, rolling stock, aircraft, or pipeline pumping stations
—Application of a Federal Insecticide, Fungicide, and Rodenticide Act (FIFRA) registered pesticide as defined under CERCLA in accordance with its intended purpose

Section 304 is the only Title III provision which does not exempt hazardous chemicals which are being transported or stored incidental to transportation under active shipping papers.

Emergency training and review of emergency systems

Section 305(a) authorizes the Federal Emergency Management Agency (FEMA) to provide $5 million for each of fiscal years 1987 through 1990 to support and enhance state and local hazardous materials training programs. These funds are intended to improve emergency planning, preparedness, mitigation, response, and recovery capabilities in support of Title III.

Section 305(b) requires EPA to conduct a "review of emergency systems for monitoring, detecting, and preventing releases of extremely hazardous substances at representative domestic facilities that produce, use, or store" these substances. The final report is to be submitted to Congress in June/July 1988.

The findings and recommendations of the report address the status of devices to monitor, detect, and prevent releases, in addition to the status of devices, systems and procedures for providing timely and effective public warning of accidental releases. The review serves to acknowledge the necessity for integrating current technologies for preventing releases and alerting the public with effective management practices.

Enforcement—Subtitle A

Federal enforcement provisions for violations of Sections 302, 303, and 304 include:

—Any facility owner or operator who fails to comply with Sections 302(c) and 303(d) can be assessed civil penalties of up to $25,000 for each day of violation [Section 325(b)].
—Facilities in violation of Section 304 can result in fines of up to $25,000 per violation per day. Subsequent violations can be assessed fines of up to $75,000 per violation per day [Section 325(b)].
—Criminal penalties can be assessed on individuals who knowingly and willfully fail to provide notice under Section 304 of not more than $25,000 or imprisoned for not more than two years. Subsequent or second con-

victions can be assessed fines of up to $50,000 or five years imprisonment [Section 325(b)].

State enforcement provisions are also applicable where states have incorporated Title III provisions into state law.

SUBTITLE B

Subtitle B establishes the mechanism for assuring the awareness of chemical hazards within the community. Key elements of Subtitle B are:

Section 311—Material Safety Data Sheets
Section 312—Emergency and Hazardous Chemical Inventory Forms
Section 313—Toxic Chemical Release Forms

Material safety data sheets (MSDS) and Emergency and hazardous chemical inventory forms

Any facility owner or operator who is required to prepare or have available a MSDS for a hazardous chemical as required under the Occupational Safety and Health Act of 1970 (OSHA) is potentially subject to Sections 311 and 312. Under Section 311, the facility owner or operator must submit a MSDS for each hazardous chemical or a list of such chemicals by October 17, 1987 or three (3) months after the facility is required to prepare or have available a MSDS for a hazardous chemical under OSHA regulations. If a facility submits a list, the hazardous chemicals must be grouped by the following five (5) health and physical hazard categories [Section 311(a)]:

—Immediate (acute) health hazard
—Delayed (chronic) health hazard
—Fire hazard
—Sudden release of pressure hazard
—Reactive hazard

Under Section 312, a facility owner or operator who must comply with Section 311 is required to submit additional information on the hazardous chemicals present at the facility. Beginning March 1, 1988 and annually thereafter, the facility must submit an emergency and hazardous chemical inventory form which supplies information on an estimate of the maximum amount of hazardous chemicals present at the facility during the preceding calendar year, an estimate of the average daily amount, and the location of these chemicals at the facility [Section 312(a)].

There are two reporting tiers of information under Section 312. The Tier I form contains the required information on amount and location of hazardous chemicals by grouping the data into the five previously identified hazard categories. The Tier II form contains the same information, but on the individual hazardous chemical. The Tier I form must be submitted annually, and the

Tier II form is only submitted upon request. A state, however, can request a Tier II form in lieu of a Tier I form. The October 15, 1987 Federal Register (40 CFR Part 370) contains a uniform format for the Tier I and II forms.

The MSDSs or lists and the inventory forms must be submitted to the SERC, LEPC and local fire department. In order to ease the burden of data management for the receiving groups and to provide facility owners or operators the opportunity to learn about their reporting responsibilities, EPA instituted a phase-in reporting schedule for Sections 311 and 312. MSDS and inventory reporting is based upon a threshold level of 10,000 pounds for 1987 and 1988, 0 pounds in 1989 and thereafter. This 0 pound threshold will be finalized, pending further study by EPA. The threshold quantity for EHSs, however, is 500 pounds or the Threshold Planning Quantity (TPQ) whichever is lower.

It should be noted that OSHA is expanding to include not only manufacturers and importers in Standard Industrial Classification (SIC) codes 20-39, but all non-manufacturers (i.e., SIC codes 1-89). At the time this expansion is finalized (projected Summer/Fall 1988), newly covered facilities will have 90 days to come into compliance with Section 311. Section 312 compliance will be required for the next annual submission. Reporting requirements will be based upon the previously described threshold schedule.

The following hazardous chemical exemptions apply to Sections 311 and 312 [Section 311(e)]:

—Any food, food additive, color additive, drug, or cosmetic regulated by the Food and Drug Administration

—Any substance present as a solid in any manufactured item to the extent exposure to the substance does not occur under normal conditions of use.

—Any substance to the extent it is used for personal, family, or household purposes; or is present in the same form and concentration as a product packaged for distribution and use by the general public.

—Any substance to the extent it is used in a research laboratory or a hospital or other medical facility under the direct supervision of a technically qualified individual. This exemption does not apply if small quantities are produced for distribution, such as by a pilot testing lab.

—Any substance to the extent it is used in routine agricultural operations or is a fertilizer held for sale by a retailer to the ultimate customer. This applies only to the user of the chemical.

Toxic chemical release reporting

Under Section 313, EPA is mandated to obtain information regarding the routine and accidental releases of certain toxic chemicals and to establish a computerized database. The database will provide information to federal, state and local governments and the public about releases of toxic chemicals in the environment. The information will also assist in research and the development of regulations, guidelines, and standards.

Facility owners or operators who are subject to this section must complete a toxic chemical release form for specified chemicals. This form must be submitted on an annual basis beginning July 1, 1988.

Reporting requirements apply to facilities under the following conditions [Section 313(b)]:

- —Ten (10) or more full-time equivalent employees
- —In primary SIC codes 20-39
- —Manufacture, process, or otherwise use a listed toxic chemical in excess of the following threshold quantities:
  - —Otherwise use over 10,000 lbs/yr
  - —Manufacturing/processing of quantities
    - —Over 75,000 lbs in 1987
    - —Over 50,000 lbs in 1988
    - —Over 25,000 lbs thereafter

The list of toxic chemicals consists of 309 chemicals and 20 chemical categories and are published in the February 16, 1988 Federal Register (40 CFR Part 372). This federal register also contains the format for the reporting form. Any person can petition for changes to the list. EPA can add or delete chemicals from the list based upon evidence regarding adverse health and environmental effects [Section 313(e)].

Enforcement—Subtitle B

Federal enforcement provisions for violations of Sections 311, 312, and 313 include:

- —Violations of reporting requirements under Section 311 include civil penalties not to exceed $10,000 for each violation [Section 325(c)].
- —Violations of reporting requirements under Section 312 include civil penalties of up to $25,000 per violation per day [Section 325(c)].
- —Violators can be assessed civil penalties not to exceed $25,000 for each violation under Section 313 [Section 315(c)].

State enforcement provisions apply for those states which have incorporated provisions of Title III into state law.

SUBTITLE C

The General Provisions Subtitle contains the following sections:

Section 321—Relationship to other Laws

Section 322—Trade Secrets

Section 323—Provisions of Information to Health Professionals, Doctors, and Nurses

Section 324—Public Availability of Plans, Data Sheets, Forms, and Followup Notices
Section 325—Enforcement
Section 326—Civil Actions

Relationship to other laws

Section 321 states that SARA Title III does not preempt any state or local law.

Trade secrets and Provisions of information to health professionals, doctors, and nurses

Section 322 allows the withholding of the specific chemical identity of an EHS, a hazardous chemical, or a toxic chemical by any facility owner or operator subject to reporting under emergency planning [Sections 303(d)(2) and (d)(3)], community right-to-know reporting (Sections 311 and 312), and toxic chemical release reporting (Section 313). If the chemical identity is withheld, the generic class or category of the chemical must be submitted [Section 322(a)(1)]. Proposed regulations for trade secret claims are in the October 15, 1987 Federal Register (40 CFR Part 370).

Section 323 provides for the disclosure of chemical identity to health professionals for diagnostic purposes and to local health officials for prevention and treatment measures during non-emergency situations. The information can only be provided during a non-emergency, if the health professionals sign a written confidentiality agreement and provide a written statement of need [Section 323(a) and (c)]. During an emergency, a facility owner or operator must provide chemical identify information to physicians or nurses. Following stabilization of the emergency, a facility may require a written confidentiality agreement and statement of need [Section 323(b)].

Enforcement provisions for violations of Sections 322 and 323 include the following:

—Civil and administrative penalties of $25,000 per claim for trade secret claimants filing frivolous claims [Section 325(d)].

—Civil and administrative penalties of up to $10,000 per day against facilities who fail to provide trade secret information in medical emergencies [Section 325(c)].

—Civil and administrative penalties of up to $10,000 per day against facilities that fail to provide required information to EPA when making trade secret claims [Section 325(d)].

Public availability of plans, data sheets, forms, and followup notices

Section 324 provides the public with access to emergency response plans,

MSDSs or lists, inventory forms, toxic chemical release forms, and followup emergency notices. This information must be made available to the public during normal working hours at a location designated by the SERC or LEPC, as appropriate [Section 324(a)]. In addition, the LEPC must annually publish a notice in local newspapers indicating that this information is available at a specified time and location [Section 324(b)].

Enforcement (refer Subtitles)

Civil actions

Section 326 provides authorities for citizens and states or local governments to bring civil actions regarding Title III.

Any person may commence a civil action on his/her own behalf against:

- —A facility owner or operator for failure to do any of the following [Section 326(a)(1)(A)]:
  - —Submit a followup emergency notice under Section 304(c)
  - —Submit a MSDS or a list under Section 311(a)
  - —Complete and submit an inventory form containing Tier I information under Section 312(a)
  - —Complete and submit a toxic chemical release form under Section 313(a)
- —The EPA Administrator for failure to [Section 326(a)(1)(B)]:
  - —Publish inventory forms under Section 312(g)
  - —Respond to a petition to add or delete a chemical under Section 313(e)(1) within 180 days after receipt of the petition
  - —Publish a toxic chemical release form under Section 313(g)
  - —Establish a computer database for toxic chemical release reporting under Section 313(j)
  - —Promulgate trade secret regulations under Section 322(c)
  - —Render a decision in response to a petition under Section 322(d) regarding trade secret provisions within 9 months after receipt of the petition
- —A State Governor, a SERC, or EPA for failure to [Section 326(a)(1)(C)]:
  - —Provide a mechanism for public availability of information as described under Section 324(a)
- —A State Governor or SERC for failure to [Section 326(a)(1)(D)]:
  - —Respond to a Tier II form request under Section 312(e)(3) within 120 days after the request

Any state or local government may commence a civil action against a facility owner or operator for failure to [Section 326(a)(2)(A)]:

- —Provide notification to the SERC under Section 302(c)
- —Submit a MSDS or list under Section 311(a)

—Make available information requested under Section 311(c)
—Complete and submit a Tier I inventory form under Section 312(a)

A SERC or LEPC may commence a civil action against a facility owner or operator for failure to [Section 326(a)(2)(B)]:

—Submit Tier II information under Section 312(e)(1)
—Provide information under Section 303(d)

A state may commence a civil action against EPA for failure to [Section 326(a)(2)(C)]:

—Provide information to states regarding trade secrets [Section 322(g)]

DISCUSSION/CONCLUSION

SARA Title III identifies responsibilities for federal, state, and local governments and the regulated community. The primary objective of the law is to develop local community preparedness and response capabilities for hazardous chemical accidents by requiring the preparation of plans, industry participation, and public availability of chemical information. The focus of this summary discussion is to present how the emergency planning and community right-to-know provisions support this community goal of emergency preparedness, response, and prevention.

Subtitle A establishes a relationship between all levels of government, the private sector, and the general public for the proper management of hazardous chemicals in the community. Broad-based participation in the development of emergency plans is promoted by the designation of the SERCs and LEPCs and the participation of chemical users, producers, and storers. The list of EHSs and TPQs places initial focus on the chemicals, facilities, and transportation routes of most immediate concern for community planning and response. Attention is centered at the local level where chemical accidents occur. This framework incorporates the technical expertise of organizations which can effectively plan for and respond to such an incident.

Title III data mandates the availability of a full range of chemical data to improve the effectiveness of emergency response. Emergency release notification requires that state commissions and local committees be informed of accidental releases of EHSs and CERCLA hazardous substances. The submission of MSDSs and inventory forms to the SERCs, LEPCs, and local fire departments ensures the availability of necessary information for effective response to a chemical accident. This data, along with the toxic chemical release information, represent additional sources for identifying other areas of chemical risk in the community.

EPA's study of emergency systems and procedures identifies gaps in technologies and practices for alerting the public and for detecting, monitoring, and preventing releases of chemicals. It stresses the need for management commitment to chemical safety within the facility itself and the community at large. EPA is undertaking several initiatives to improve safety. They include coordinating with organizations having technical expertise in systems design, assisting in program development for identifying facilities at risk, and for developing techniques for communicating chemical risk.

SARA Title III identifies that everyone has a role in protecting human health and the environment. The structure of the law creates a framework in which the concept of partnership between industries and their communities in preventing chemical emergencies starts at the local level. This local partnership is strengthened by private, state and federal technical assistance. This partnership between the government and the regulated community is a vital ingredient of an effective local emergency response program.

REFERENCES

1. Superfund Amendments and Reauthorization Act of 1986 (SARA), Public Law 99-499, U.S. Congress, October 17, 1986.
2. *Chemical Emergency Preparedness Program Interim Guidance,* Washington, D.C.: U.S. Environmental Protection Agency, November 1985.
3. *Community Awareness and Emergency Response (CAER) Handbook,* Washington, D.C.: Chemical Manufacturer's Association, April 1985.
4. "Environmental Protection Agency: Extremely Hazardous Substances List and Threshold Planning Quantities; Emergency Planning and Release Notification Requirements; Final Rule;" 40 CFR Parts 300 and 355; Federal Register: Vol. 52, No. 77, Wednesday, April 22, 1987.
5. "Environmental Protection Agency: Superfund Programs; Reportable Quantity Adjustments; Final Rule;" 40 CFR Parts 117 and 302; Federal Register: Vol. 51, No. 188, Monday, September 29, 1986.
6. *Hazardous Materials Emergency Planning Guide,* Washington, D.C.: National Response Team, March 1987.
7. *Technical Guidance for Hazards Analysis,* Washington, D.C.: U.S. Environmental Protection Agency, Federal Emergency Management Agency and U.S. Department of Transportation, December 1987.
8. "Environmental Protection Agency: Extremely Hazardous Substances List; Final Rule;" 40 CFR Part 355; Federal Register: Vol. 52, No. 242, Thursday, December 17, 1987.
9. "Environmental Protection Agency: Extremely Hazardous Substances List; Final Rule;" 40 CFR Part 355; Federal Register: Vol. 53, No. 37, Thursday, February 25, 1988.
10. Comprehensive Environmental Response, Compensation, and Liability Act of 1980 (CERCLA), Public Law 96-510, U.S. Congress, December 11, 1980.
11. "Environmental Protection Agency: Emergency and Hazardous Chemical

Inventory Forms and Community Right-to-Know Reporting Requirements; Final Rule;" 40 CFR Part 370; Federal Register, Vol. 52, No. 199, Thursday, October 15, 1987.

12. "Department of Labor, Occupational Safety and Health Administration, Hazard Communication Standard;" 29 CFR Part 1910.1200.
13. "Department of Labor, Occupational Safety and Health Administration, Hazard Communication; Final Rule;" 29 CFR Parts 1910, 1915, 1917, 1918, 1926, and 1928; Federal Register: Vol. 52, No. 163, Monday, August 24, 1987.
14. "Environmental Protection Agency: Toxic Chemical Release Reporting Form; Community Right-to-Know; Final Rule;" 40 CFR Part 372; Federal Register: Vol. 53, No. 30, Thursday, February 16, 1988.
15. "Environmental Protection Agency: Trade Secret Claims for Emergency Planning and Community Right-to-Know Information; and Trade Secret Disclosures to Health Professionals; Proposed Rule;" 40 CFR Part 350; Federal Register, Vol. 52, No. 199, Thursday, October 15, 1987.

Management of Hazardous Materials and Wastes: Treatment, Minimization and Environmental Impacts. Edited by S. K. Majumdar, E. W. Miller and R. F. Schmalz.

Chapter Thirteen

CHEMICAL INVENTORY AND COMMUNITY RIGHT TO KNOW IN PENNSYLVANIA

JAMES H. TINNEY

Director
Bureau of Worker and Community Right to Know
Pennsylvania Department of Labor and Industry
Harrisburg, PA 17120

The last decade has witnessed the dawning of a new era of public awareness concerning the presence of hazardous chemicals in our lives.

From the advent of the federal efforts to address the "super" challenge of cleaning up chemical waste dumps and the tragic loss of life due to an accidental chemical release in Bhopal, India, to the all too frequent headlines about neighborhood evacuations caused by truck and railroad tank car leaks and industrial spills, the American public has seen the potential for hazardous chemical damage to our environment and to ourselves. As more becomes known about the health effects of the approximately half-a-million chemicals already in use in American industry, and as new chemicals are developed and introduced into use annually, the potential for harmful chemical exposure has been correctly recognized as both real and growing.

Government has been called upon by labor, environmental and public interest groups, and emergency responders to provide some form of regulation to protect citizens from hazardous chemical dangers.

The principal response of federal, state, and local governments to the hazardous chemicals issue has been to fashion chemical awareness laws which provide chemical information to workers, citizens, and emergency response personnel who have a "right to know" about the presence and potential dangers of exposure to chemicals found where they live and work. The rationale underlying these laws is that a citizenry with rights of access to information about the presence and dangers of hazardous chemicals will be better prepared for chemical emergencies, better able to make informed decisions about the location of chemical facilities in the community, and better versed in the precautions individuals should take in handling hazardous chemicals in the workplace.

In the Commonwealth, the Pennsylvania Worker and Community Right to Know Act of 1984, Title III of the federal Superfund Amendments and Reauthorization Act of 1986, and the federal Occupational Safety and Health Administration (OSHA) Hazard Communication Standard, as expanded in 1987, constitute a joint governmental effort mandating a system for workplace hazardous chemical inventory and protecting the community's right to know about both the presence of such chemicals and about the methods of their release into the environment.

Though separate, these three laws interact at the state and local levels to provide an expanding base of chemical information. In turn, these three laws establish a framework through which government facilitates citizens' access to this base of information.

PENNSYLVANIANS' RIGHT TO KNOW

Signed into law on October 5, 1984, the Pennsylvania Worker and Community Right to Know Act was just that, a law providing for workers' and community members' right to know about hazardous chemicals. In addition, the law provided for local emergency response agency direct access to information about the hazardous chemicals in workplaces in their areas, thus giving them the means to better plan for response to incidents involving these chemicals.

While employees working in manufacturing facilities were not covered by the law in their roles as workers—because the federal OSHA Hazard Communication Standard already regulated employees access to such information inside manufacturing workplaces—they were covered by the state law in their roles as community members.

As the brief Right to Know Chronology at the end of this chapter indicates, the initial, relatively clear distinction between state Right to Know jurisdiction and federal OSHA jurisdiction was to become blurred by court action that was not resolved until October 1987, three years after the law was signed, and almost a full two years after the Right to Know regulations were printed as final in the August 2, 1986, *Pennsylvania Bulletin.*

The court's conclusion: While OSHA may regulate inside Pennsylvania's manufacturing workplaces, the Commonwealth, in its historic role of regulating community health and safety, could require manufacturers to comply with additional state requirements for completion of workplace hazardous chemical inventories and identification of chemical packages sent from workplace to workplace.

The other major impact on state Right to Know jurisdiction came as a result of OSHA's expansion of its Hazard Communication Standard on August 24, 1987. Effectively, OSHA expansion preempted state Right to Know regulation inside almost all private sector nonmanufacturing workplaces. Pennsylvania's private sector manufacturing workers *and* nonmanufacturing workers, in their roles as employees, would now proceed by OSHA rules to obtain information about chemicals inside their workplaces. Public sector workers would continue under the state Right to Know rules to obtain information about chemicals in their workplaces. In addition, all persons living or working in the Commonwealth would continue to have rights of access to any workplace's hazardous chemical information under the community access provisions of the state law. Local emergency response agencies' state-based rights of direct access to all workplaces' hazardous chemical information remained unchanged by either the court ruling or federal OSHA expansion.

The effect of this "shaking out" of federal and state jurisdiction over regulating access to hazardous chemical information has been to confirm the applicability of Pennsylvania's requirements for annual completion of a workplace Hazardous Substance Survey Form (chemical inventory) by *all* workplaces and to verify that all persons who are members of the state's "community" do indeed have some recourse in gaining hazardous chemical information under the state Right to Know law.

SARA TITLE III IN PENNSYLVANIA

On October 17, 1986, the President signed the Superfund Amendments and Reauthorization Act (SARA) into law. One component of SARA is Title III, the Emergency Planning and Community Right-to-Know Act of 1986. SARA Title III mandated requirements under which the federal Environmental Protection Agency (EPA) and the individual states and their local communities could obtain and provide for citizen access inventories of workplaces' hazardous chemicals and information on the means of hazardous chemical release into the environment.

In terms of the community's right to know, SARA Title III established a mechanism complementing the state's Worker and Community Right to Know Act provisions which citizens and local emergency responders could use as a second avenue to information about the location and dangers of workplace chemical hazards.

SARA Title III required that the individual states establish a system of community-based Local Emergency Planning Committees (LEPCs) responsible for coordinating collection of chemical hazard information from employers and emergency response planning based on this information.

In the Commonwealth, Governor Robert P. Casey's Executive Order of April 20, 1987, established the Pennsylvania Emergency Response Commission (PERC). The Commission was charged with the "primary responsibility for overall policy and direction of statewide emergency planning and notification activities and organization consistent with the provisions of" SARA.

Lieutenant Governor Mark S. Singel was named Commission chairman and the Pennsylvania Emergency Management Agency (PEMA) was assigned lead state agency responsibility for coordinating SARA Title III program development and execution. The Department of Labor and Industry was assigned a support role in collecting hazardous chemical reports filed under SARA Title III, with the Secretary of the Department, Harris Wofford, assigned a seat on the Commission.

On the local level, the PERC assigned LEPC coordination responsibilities in each of the state's 67 counties to PEMA's individual County Emergency Management Coordinators. As mandated by SARA, the membership of the county-based LEPCs is made up of elected state and local officials and representatives from the local media, industry, emergency response agencies, environmental groups, and others with interests and expertise in establishing plans for local chemical emergency response.

For the purposes of this chapter, the two key points to be made about SARA Title III compliance concern the employers' requirements to do SARA reporting and the system through which such employers' SARA chemical inventories and other reports will be provided to federal and state governments and local agencies.

All employers covered by the federal OSHA Hazard Communication Standard, including those private sector nonmanufacturing employers affected by the 1987 OSHA expansion, are subject to SARA Title III reporting requirements. The brief SARA Title III Chronology at the end of this chapter indicates how OSHA regulated nonmanufacturing employers are gradually brought under the SARA reporting requirements along with manufacturing employers.

As the chronology also shows, covered employers provide SARA Title III-regulated chemicals' Material Safety Data Sheet (MSDS) information, Chemical Inventory Forms, and Toxic Release Forms to various federal, state, and local entities. MSDSs go to the state Right to Know office as Pennsylvania's collection point for SARA chemical information, while lists of the MSDSs go to the LEPC and local fire departments. Chemical Inventory Forms listing specific information on the quantity and location of SARA-regulated hazardous chemicals in a workplace are provided to the state Right to Know office, the LEPC, and local fire departments. Toxic Release Forms providing information on the quantity and means of release of hazardous chemicals from a workplace are given to the state Right to Know office and the federal Environmental Protection Agency.

In its current support role, the state Right to Know office is proceeding with the processing of submitted SARA Title III materials for the purpose of community access. At the time of this writing, November 1988, SARA MSDSs (approximately 85,000) are being processed into a computer base. The office is also compiling and indexing Chemical Inventory Forms (approximately 1,500) and Toxic Release Forms (approximately 3,000).

To facilitate public access to this information, a SARA Title III Request for Information Form has been developed. In addition, a walk-in reading room for Toxic Release Forms has been set up in the Right to Know office, with a similar facility for Chemical Inventory Forms to be established shortly. To aid employers in understanding and complying with the SARA Title III requirements, printed Right to Know materials such as the annual Hazardous Substance Survey Form mailing and the program newsletter carry notices and updates about SARA compliance rules and the program's Electronic Bulletin Board regularly offers SARA-related bulletins.

THE OUT YEARS

As we in Pennsylvania, the nation, and, indeed, the world come more and more to realize the formidable tasks involved in handling hazardous substances, we can hope that programs such as the chemical inventory and community right to know initiatives noted above will provide the footing and the sound base of information we can use to accurately answer the question, "Where are we now?", in terms of hazardous chemical inventory, control, and emergency response.

The answer to that question and the public debate that accompanies it will be two of the gauges by which we and future generations can assess the success or failure of our efforts in managing hazardous materials.

PENNSYLVANIA'S RIGHT TO KNOW CHRONOLOGY

October 5, 1984—The Worker and Community Right to Know Act is signed into law, P.L. 734, No. 159.

August 17, 1985—The list of Hazardous Substances to be regulated under the Right to Know Act is published in the Pennsylvania Bulletin.

August 2, 1986—Regulations implementing the worker and community access provisions of the Act are published as final in the Pennsylvania Bulletin. Most of the regulations are scheduled to go into effect on December 1, 1986.

The published regulations take into account ongoing court action affecting manufacturer compliance with the community access provisions of the Right to Know program. The regulations exempt manufacturers from employee-access provisions because such access is already regulated by the federal Occupational Safety and Health Administration (OSHA).

September 12, 1986—The U.S. Court of Appeals for the Third Circuit rules that manufacturers must comply with the community access provisions of the

Right to Know regulations. The decision is subsequently appealed to the U.S. Supreme Court.

August 3, 1987—As scheduled in the regulations, the rules for hazardous chemical labeling and sharing of other information about hazardous chemicals go into effect.

August 24, 1987—Under court order, OSHA expands its regulation of employee access to the nonmanufacturing sector. This action immediately preempts the State from regulating employee access to workplace chemical information in almost all private sector workplaces.

September 19, 1987—The Department issues notice in the Pennsylvania Bulletin about the impact of OSHA expansion. The State will continue to regulate public sector employee access and all community access provisions of the Right to Know regulations.

October 5, 1987—The U.S. Supreme Court denies a hearing of the manufacturers' appeal. The lower court's ruling requiring manufacturing compliance with community access regulations stands.

March 26, 1988—Amendments to the Right to Know regulations are published in the Pennsyvlania Bulletin and go into effect immediately. The amended regulations require manufacturer compliance with community access rules and exempt nearly all private sector employers from having to comply with employee access rules.

August 2, 1988—As scheduled in the regulations, rules for labeling nonhazardous chemicals go into effect.

SARA TITLE III CHRONOLOGY

October 17, 1986—The federal Superfund Amendments and Reauthorization Act of 1986 (SARA) is signed into law.

April 20, 1987—By Executive Order 1987-8, the Governor establishes the Pennsylvania Emergency Response Commission.

July 17, 1987—Deadline for the Pennsylvania Emergency Response Commission to designate emergency planning districts. In Pennsylvania, the districts are the state's 67 counties.

October 17, 1987—Manufacturers make initial submission of Material Safety Data Sheets for SARA Title III chemicals to the state (Right to Know office) and lists of these chemicals to the Local Emergency Planning Committee and local fire departments.

March 1, 1988—Manufacturers make initial submission of Chemical Inventory Forms for SARA Title III chemicals to the state (Right to Know office), the Local Emergency Planning Committee, and local fire departments.

July 1, 1988—Manufacturers make initial submission of Toxic Release Forms listing toxic chemical emissions for the prior calendar year to the state (Right to Know office) and to the federal Environmental Protection Agency.

September 24, 1988—OSHA regulated nonmanufacturers make initial submission of lists of Material Safety Data Sheets for SARA Title III chemicals to the state (Right to Know office), the Local Emergency Planning Committee, and local fire departments.

October 17, 1988—Local Emergency Planning Committees complete their initial local emergency response plans.

March 1, 1989—OSHA regulated nonmanufacturers make initial submission of Chemical Inventory Forms for SARA Title III chemicals to the state (Right to Know office), the Local Emergency Planning Committee, and local fire departments.

Management of Hazardous Materials and Wastes: Treatment, Minimization and Environmental Impacts. Edited by S. K. Majumdar, E. W. Miller and R. F. Schmalz. © 1989, The Pennsylvania Academy of Science.

Chapter Fourteen

PENNSYLVANIA'S CHALLENGE TO IMPLEMENT HAZARDOUS MATERIALS SAFETY AND TITLE III OF THE SUPERFUND AMENDMENTS REAUTHORIZATION ACT

JOSEPH L. LAFLEUR and SANDERS A. CORTNER

Pennsylvania Emergency Management Agency
P.O. Box 3321
Harrisburg, PA 17105

The heavily industrialized Commonwealth of Pennsylvania has long been a producer and user of hazardous chemicals and is a key cog in the development of chemical based products used throughout the world. The capability to respond safely and effectively in the event of a spill of a hazardous material has been a concern for many years. Pennsylvanians are concerned about clean water, clean air and a healthy environment. Governor Robert Casey in his campaign made the environment one of his "3 E's" as a policy centerpiece: Education; environment; economy.

On October 17, 1986, the President signed the "Superfund Amendments and Reauthorization Act of 1986" (SARA) into law. One part of the new SARA provisions is Title III: the "Emergency Planning and Community Right-to-Know Act of 1986." These Title III amendments are a direct result of Congress's effort to reduce the potential for an incident like the Union Carbide release which took place in Bhopal, India that killed several thousand nearby residents and injured untold others. Title III establishes requirements for Federal, State, and local governments, and industry regarding emergency planning and community right-to-know reporting on hazardous chemicals. This legislation builds upon the Environmental Protection Agency's (EPA's) Chemical Emergency Preparedness Program (CEPP) and numerous State and local programs aimed at helping communities to meet their responsibilities in regard to potential chemical emergencies. The passage of this Act has increased the attention to and the pace of chemical emergency preparedness and response activities, long needed, in Pennsylvania.

Title III has four major sections: emergency planning (Section 301-303), emergency notification (Section 304), community right-to-know reporting requirements (Section 311, 312), and toxic chemical release reporting—emissions inventory (Section 313). The sections are interrelated in a way that unifies the emergency planning and community right-to-know provisions of Title III. While many consider "Superfund" as just waste site cleanup, these added sections identified new emergency response requirements that focus on the source of extremely toxic substances and other hazardous materials. The emergency planning sections are designed to develop State and local government emergency preparedness and response capabilities through better coordination and planning, especially at the local level. More and better information will enhance plans. Responders, as well as the public living near these locations, will be more sensitive to their own safety.

Nationally, awareness of spilled chemicals and the threat they pose to both citizen and responder safety has not had the attention of policy makers and budget formulators. The passage of SARA, Title III should help gain their appropriate attention. Unfortunately, the public and these decision makers are beginning to accept as normal the current rate of hazmat incidents. This is similar to the historical patterning of fire incident and death rates in the United States which have leveled off but still remain among the highest in the industrialized world. The current number of chemical spills that pollute our environment and threaten our citizens is unacceptable and must be reduced, not accepted. We must battle against this spill rate and not succumb to the currently fatiguing spate of "mistakes" which is causing spills. Over time, the spill rate will be reduced, as storers and users seek to control and reduce the burden the regulations place on them by reducing the number of toxic chemicals they possess.

During the 1976-84 period, the Congressional Office of Technology Assessment reported that Pennsylvania had 6,718 transportation incidents that spilled hazardous chemicals. The Campaign Against Toxic Waste staff reported that Pennsylvania had the most reported spills in 1987 with 540, Illinois was next with 481, followed by Ohio with 441, Texas with 423, and California with 372. SARA is a fixed facility program; however, the two situations are tied together. Reports of transportation and fixed facility spills are increasing based on improved awareness. We are now in a situation in the United States where people go home at the end of the day, turn on their television and watch the news, weather, sports and the spill of the day. Through improved reporting, the Pennsylvania Emergency Management Agency (PEMA) will provide the Pennsylvania Department of Environmental Resources (DER) information on the spillers for follow-up regulatory action and source point mitigation. PEMA is working with county and local government, volunteer groups and the private sector to improve planning, training and response to future incidents.

IMPLEMENTING TITLE III

An initial analysis of the requirements of SARA, Title III would indicate that DER might have been the appropriate state agency to handle the implementation of the law. However, further consideration of the emergency preparedness aspects of the law in conjunction with PEMA's broad mission areas of emergency planning and coordination of emergency responses clearly established PEMA as the state implementing agency for SARA, Title III. An additional factor in favor of PEMA's lead involvement was the fact that PEMA had an organizational infrastructure with three area offices and the 67 county and more than 2,600 municipal government emergency coordinators throughout the Commonwealth. Combined with this infrastructure is the legislative mandate P.L. 1332, "Emergency Services Act of 1978", as amended, that requires each county and local government to have an emergency operations plan, which could serve as a base document for the community emergency plans required by SARA. In view of all the factors described above, it was decided by the Governor that PEMA would be the lead state agency for implementing SARA, Title III.

PEMA was created, (Section 711) of the ACT, for the purpose:

> To assure prompt, proper and effective discharge of basic Commonwealth responsibilities relating to civil defense and disaster preparedness, operations and recovery, . . .

In compliance with SARA, Title III, Governor Robert P. Casey signed an executive order on April 20, 1987, creating the State Emergency Response Commission (SERC). The Commission membership is composed of the existing Pennsylvania Emergency Management Council (key members of the Governor's cabinet, the leadership of the State Senate and House) plus the Secretary of Labor and Industry and members from the public and private sector. The Council is the formally recognized body under the ACT, for emergency management policy. Lieutenant Governor Mark S. Singel was appointed SERC chairman by the Governor coincident with his Council chairmanship. On July 13, 1988, Governor Casey signed Senate Bill 1167 expanding membership on the Council which will lead to the rescinding of the original executive order and once again unify all emergency management policy making and implementation of SARA through the Council. The functions of emergency management often overlap from hazard to hazard and the same capabilities are required for program management and response (evacuation, shelter, warning, direction and control, etc.). Hence, maintaining one policy body clarifies program channels for local emergency managers.

The original executive order also designated PEMA as the lead state agency responsible for coordinating the development, execution, and management of programs and functions of state agencies and emergency planning districts (each of the 67 counties was designated as an emergency planning district in the order). The planning districts were directed to establish a local emergency planning committee (LEPC) in each county. The county emergency management coordinator is a key advisor to the committee. These planning committees are to prepare the emergency off-site plan for each facility that has a SARA extremely hazardous substance. The SARA, Title III list stands at 366 extremely hazardous substances at this writing, down from 406. The reduction is due to various groups petitioning removal of some chemicals from the list or EPA removing chemicals as new studies bring to light more information.

In 1986, the Director of PEMA and the Director of Response and Recovery made several inquiries concerning this forthcoming law to identify chemicals and to control chemical storage and transportation in the Commonwealth. After several discussions with facilities, federal officials, and the National Governors' Association, they determined that a law could be forthcoming within the next year. Based on this, the Director of PEMA directed the Response and Recovery staff to begin implementing an automated chemical mapping and information storage program. PEMA requested EPA to provide a list of fixed facilities which used, stored, or shipped chemicals in what was then to be the 406 chemical listing. PEMA contracted with Research Alternatives to originate an automated, geographic, mapping program to include geographical locations, Material Safety Data Sheets, a list of chemicals, a chemical dispersion model, population densities and key facilities (schools, health care facilities, jails, etc.) of the local communities surrounding the chemical facility locations. EPA initially provided PEMA with a list of 500 facilities in Pennsylvania and Research Alternatives programmed these facilities as described above.

SARA TRAINING

The only funding that has been made available to aid implementation by the federal government has been for training. The federal law states that $5 million will be made available for training only for each of four years, beginning in 1987-1988. Pennsylvania was allocated $134,000 for training the first year and $138,000 for the second year. The lack of funds for the county/state level has been a political and operational sore point in federal/state relations and state/county relations. However, the Congress sees chemical preparedness as an industry-state-LEPC responsibility.

PEMA's training plan, using the Federal funds, was designed in a multi-year approach. First year funds were allocated to multiple delivery of three basic awareness level training courses approved by the National Fire Academy using only Federally or state approved instructors. These courses are: The Pesticide Challenge, Hazardous Materials Incident Analysis, and Recognition and Identification of Hazardous Materials. These courses were delivered throughout the state, for all necessary interested parties during business hours. Fifteen one-day SARA orientation seminars were offered for public officials also during business hours. These courses discussed LEPC formulation, Title III requirements, and facility plan development. This course delivery process was essentially a saturation run with goals of (1) raising SARA awareness of public officials, (2) increasing attention to basic responder safety, (3) stimulation of future requests for courses, (4) adding media attention to fuel the interest/concern, (5) begin momentum toward the huge planning task, and (6) create government-facility linkage. All of these goals were reached through the delivery of the first 42 courses. Through this process, an expanded core group of participants beyond county emergency managers and some fire chiefs was formed. They did as we hoped, stimulate interest "back home." The expanded group was from law enforcement, public works and the emergency medical community. The courses also unleashed a flood of information requests and speech requests on PEMA and the Department of Labor and Industry as agents for the SERC.

PEMA originally planned to conduct 42 hazardous materials training seminars statewide in the fall of 1987 (the first grant). Approximately 1,900 members of local emergency services organizations and local and county officials were to attend these seminars. The funding for these seminars came from EPA through the Federal Emergency Management Agency (FEMA). A second round of 22 courses was projected for the Spring of 1988, as need was high and could be accommodated because courses were running under the original budget projections. However, demand far outstripped projections and the PEMA staff rose to the occasion. Working closely with county emergency managers, fire, police and elected officials, 104 activities were conducted for 4,077 participants, all within budget.

Some of the SARA funds were directed to the State Fire Academy (SFA), for direct enhancement of teaching aids for present and future hazardous materials team training. SFA instructors have been attending courses at national training sites to gain additional information and attend train-the-trainer courses. The investment of time and funds in staff development is a prerequisite to delivery of the one week and two week hazmat team courses scheduled to begin in year two of the program.

Through prudent management of the initial SARA funds, additional offerings of the three FEMA approved awareness courses are being delivered in 1988. The delivery strategy can and has changed to include county by county delivery rather than regionalized delivery. Courses are offered evenings and weekends. Reimbursement of student expenses for mileage, meals, rooms and materials, where appropriate, continues.

EPA's use of the FEMA delivery system was a smart move by the Federal agencies. EPA does not have the same legislative latitude to reimburse students for expenses that FEMA, through the Emergency Management Institute (EMI) field program has. It was a pleasure to see the two agencies work together for the benefit of the students at the volunteer grass roots level. Many volunteer firefighters, Red Cross, Emergency Medical Service (EMS) and other students were reimbursed for their personal expenses or those of their agencies which are usually gained through fund raising. This was one way the "feds" eased SARA implementation and reduced the direct burden at the local and state government, individual and volunteer agency levels.

PEMA's intent in the second year of the training program is to follow federal guidance in a three-fold fashion. First, it will continue to press for plan development. It will do this by stressing attendance at planning related courses, both at the federal and state level. These are three-to-five day courses and include the aspects of technical assessment. Second, some (but fewer than the first year) awareness level courses will be delivered. Third, an emerging emphasis area of training for hazmat teams consisting of tactics, radiological monitoring/safety, and chemistry will begin. These specialized, expense reimbursable, one and two week courses will allow attendance after awareness level prerequisites. They will be available to team members of officially authorized teams who meet administrative training and equipment availability criteria. This program mix, coupled with increasing use of hazardous materials events in the on-going PEMA-FEMA exercise program will continue into the third and forth years. Federal training funds are vital to the states so that momentum can be gained for proper implementation. Training in the various aspects of Title III is the necessary agent that will bring the program into reality most quickly. However, long term preparedness maintenance/stability must/will be predicated on providing an administrative support (staff and equipment) structure. Government at all levels must realize and avoid this potential shortfall.

PLANNING AND PROGRAM ADMINISTRATION

PEMA has drafted a model off-site emergency plan, which meets all of the EPA requirements of SARA to help LEPC's over the initial hurdles of complimentary fit to the existing county emergency operations plan and the site specific requirements. Copies of this plan were given to all local emergency planning committees and to the participants at the training seminars with detailed guidance on how to use the model in the Fall of 1987. Additionally, PEMA is providing a diskette with the model plan on it for use by those LEPC's that have a computer data management capability. SARA plans for the estimated 2,000 plus facilities were due to LEPC's on October 17, 1988. The training program, as noted above, has been the catalyst to move the planning program forward. The need to manipulate the data and begin program development was made to assist the counties. PEMA spent $40,000 of its operating funds to purchase chemical preparedness software for the counties from its 1987-88 budget. A major expenditure of travel funds for staff to go "on-site" to provide technical assistance was reprogrammed. Extensive staff time was allocated to LEPC assistance from other duties. Postage and duplicating costs soared as request for SARA information outstripped PEMA's budget.

Lieutenant Governor Singel, as the Governor's appointed chairman of SERC, approved a PEMA 1988-89 budget submission request for 17 SARA staff. To date, five positions are approved for funding in 1989. The total PEMA staff, due to this and other proposed augmentation, has risen to the point that four administrative support personnel are also required. However, they are outside the direct program staff request. The 17 staff were for support to local government, as shown below:

1 Position to SFA - Faculty Hazmat Specialist
9 Staff to PEMA Areas/HQ for Planning/Training/Response/Data Entry
7 Staff 24-hour Emergency Operations Center

In the future, EPA indicates the tightening and adding of planning and reporting requirements in chemical emergency preparedness. Additional staff may be required. However, adequate staffing now, to catch up to the fast growing power curve, may negate some of the future need. If the state and counties get too far behind, staff levels above the state's 17 and currently unknown projected county needs will be higher. The unfortunate possibility exists that industry could be fined for non-compliance due to the state/county inability to be ready to receive the documents and participate in plan development and annual exercises. This would be unfair to industry and poor public sector management/regulation of the private sector. Ultimately, this lack of programmatic readiness becomes a negative tax revenue impact and generates a larger need for additional fee collection. Most states are considering, and some have enacted, legislation assessing a fee for SARA work on the chemical industry (Wisconsin, Florida, Kansas). This may also happen in Pennsylvania. Legislation is currently drafted to help Pennsylvania implement SARA, Title III. It will be reviewed with the chemical industry and the counties prior to submission. The proposed program is funded in a pattern after the Federal law using industry fees and state and county tax revenues.

Planning is the real shirt sleeve work on the front burner. Through the use of the planning outline PEMA developed the on-site safety plan or spill prevention and control plan, of which 50-75% of the working document is already in "draft." The remaining 25-50% revolves around the hazard assessment and the complexity of the corresponding response, evacuation, and recovery considerations. EPA has developed a technical "workbook" for determining

PENNSYLVANIA EMERGENCY MANAGEMENT AGENCY

COMMUNITY-BASED EMERGENCY RESPONSE SYSTEM

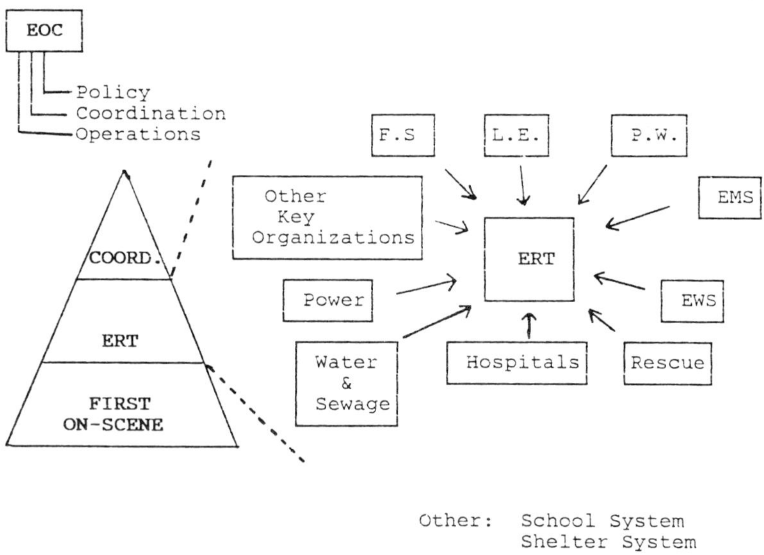

PENNSYLVANIA EMERGENCY MANAGEMENT AGENCY

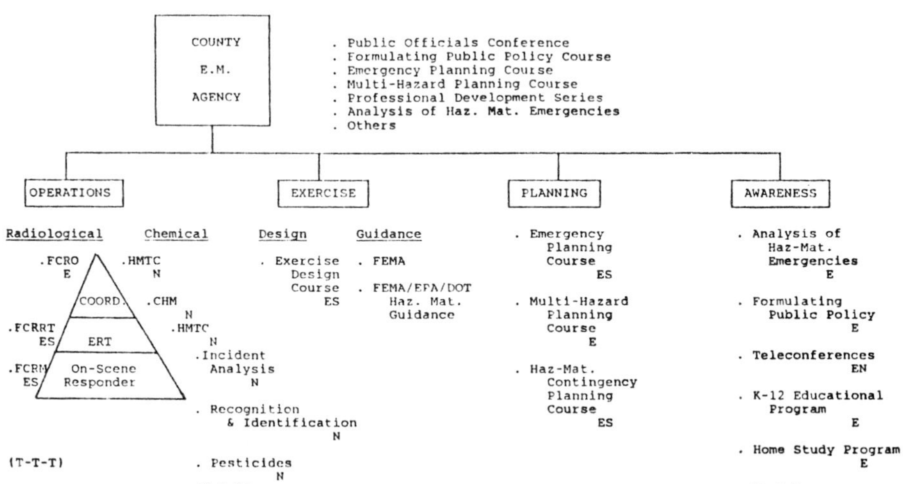

FCRO - Fundamentals Course Radiological Officer
FCRRT - Fundamentals Course Radiological Response Team
FCRM - Fundamentals Course Radiological Monitor
HMTC - Hazmat Tactics Course
CHM - Chemistry Hazmat
E - Emergency Management Institute
N - National Fire Academy
S - State Delivery
TTT - Train-the-Trainer

evacuation distances. The Federal Department of Transportation (DOT) *Emergency Response Guidebook for Hazardous Materials* is also a helpful fixed site and transportation planning and response document. FEMA took the lead and coordinated the development of the March 1987 National Response Team *Hazardous Materials Emergency Planning Guide.* FEMA also developed a training catalog of federal hazardous materials courses (radiological and chemical). FEMA and DOT co-authored a Report to Congress on *Hazardous Materials Training, Planning and Preparedness,* July 1986. In addition, teleconferences and studies have assisted state and local implementation.

We have found no chemical facilities at this time which have refused to cooperate once they learn the requirements. However, many remain ignorant of the law, particularly the smaller companies who do not have the staff to readily follow regulatory changes. Many are willing to hire consultants, provide typists, or donate services to complete the work. Local planning committees are expanding to meet the workload demand or developing a sub-committee structure, or both. A few counties are behind the time line due to lack of attention at the elected official level. Obviously, both industry and EPA will be concerned, as well as the state, if deadlines are missed. The Community Awareness Emergency Response (CAER) program was industry initiated to complement and supplement EPA's Chemical Emergency Preparedness Program (CEPP). This voluntary industry program put participating companies on record in their efforts to visibly improve safety.

The structural inadequacy of the law (depending on an unfunded committee to do this hands-on portion of the emergency preparedness work) has been overcome to some degree by the national structure of emergency preparedness as promoted/promulgated loosely by FEMA. The state and local emergency management structure may yet accomplish this worthy goal of EPA and the Congress. Funds should have been provided to FEMA for an expanded emergency management assistance program and for personnel and administrative expenses. EPA's initial thoughts of setting up a totally separate chemical emergency preparedness structure would have failed for several bureaucratic and financial reasons. Fortunately, this option was discarded early on by the senior executive cadre at FEMA and EPA and voices more experienced in the art of the possible prevailed. Although a big elephant to eat, bite by bite, SARA will be accomplished by the nation's state and county emergency managers. The long term quality of the system, as stated earlier, depend on staff and program management funds, as well as training funds.

The state has used an existing organization that is set up in the Department of Labor and Industry (L&I) to handle the reports required under Sections (311) Material Safety Data Sheets (MSDS), (312) Chemical Inventory reports and (313) Toxic Chemical Release reports. This has been extremely helpful because as of Nov. 1, 1988, the Department had received over 95,480 multi-page MSDS and 1,587 Tier I and II forms. 3,149 toxic chemical release inventories were

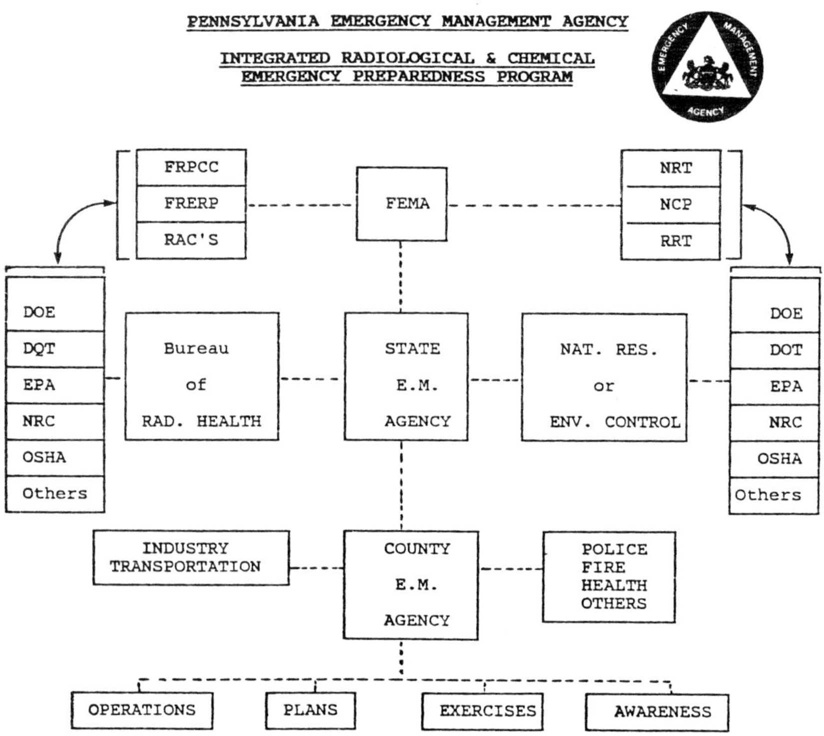

FRPCC - Federal Radiological Preparedness Coordinating Committee
FRERP - Federal Radiological Emergency Response Plan
RAC - Regional Assistance Committee
NRT - National Response Team
NCP - National Contingency Plan
RRT - Regional Response Team

registered as of Nov. 1, 1988.

The volume of paperwork involved with these reports is tremendous and getting larger. The backlog in processing MSDS is currently 1065 inches or 8 months, 3 weeks and 2½ days. The state is considering a program using a computerized optic disk image management system which will help in the storage and rapid retrieval of essential information for use by emergency responders. The system will cost about $800,000. Annual operating expenses must also be appropriated. A terminal to access the system would be placed in the state operations center for 24 hour access in emergencies, exercises and for planning.

As an additional outreach program, PEMA requested that L & I include information concerning SARA in their quarterly newsletter authorized by the state right-to-know law. This newsletter was mailed on March 16, 1988, and went to 220,000 employers throughout the state. Subsequent L&I,

PEMA, EPA, and industry mailings have also provided information. Based on this out-reach program, we believe that with rare exception everyone who should know about SARA does so now. We fully expect to have over 2,000 facilities with one or more extremely hazardous substances present and, therefore, are subject to the requirement for an off-site emergency plan. PEMA and L&I are developing a compliance manual for SARA to aid implementation.

HAZARDOUS MATERIALS INCIDENT RESPONSE TEAM

A team of multi-disciplinary responders is required if we in emergency management are to efficiently address hazardous materials incidents which do take place in the future. Since incidents will take place, everyone needs to get their teams trained, equipped, and administratively ready, now, to meet the complexities of the ever growing chemical and radiological inventory.

The chemical development boom took place post World War II as scientists turned their forced and rapid wartime development of chemicals to peaceful achievement and corporate profit. Fixed site manufacturing and use was the norm for chemicals and chemical based products until the 1960's. Spurred by the space race, chemicals grew more complex as more additives took chemicals beyond petroleum brands and basic plastics to multiple component mixtures that bubble and boil and have more exotic names that a witch could say over the caldron as she made her strange brew. As chemicals grew more sophisticated, they improved and replaced existing metals and other natural components. The better they proved to be; the greater the demand. The greater the demand; the more profit potential. Many chemical producing and using sites developed. Greater transportation requirements also evolved for in motion storage time, traveling distance stability, and cargo transfer containerization. During the 1970's, fixed site use and production and transportation spillage also reached a crescendo that caused the Congress to pass the Hazardous Materials Transportation Act (HMTA) of 1974.

1974 was an epic year in the passage of emergency management legislation as the Disaster Assistance Act (PL 93-288) was also passed. Public Law 93-288 was an initiative in response to Hurricane Agnes which caused about $2.3 billion in damage in Pennsylvania alone. We should not lose site of the fact that nearly every time there is a significant natural disaster, there is a fixed site or transportation chemical spill that takes place and further pollutes our environment, because response is often difficult and delayed.

History of hazardous materials (hazmat) response is heavily vested in the fire service. Fire companies were the initially called units because they could "wash away" the mistakes of others. As the spills became more difficult to cope with and improved tank specifications were still slow in developing, the fire company with the necessary characteristics, such as a disciplined source of manpower, available at no or little cost, with water, bringing their own protective gear, and available in a moment's notice, became the first knight on-scene in shining armor. However, as the weeks changed to years of increasingly more complex chemicals, radiological components and transuranics, the fire company was no longer able to universally cope in the hazmat world.

Enter the 1980's. Enter more specialists on the hazmat scene - heavy equipment operators, toxicologists, paramedics, biologists, physics professors, doctors, emergency managers, attorneys and environmental engineers. The team concept is in full development throughout the country.

Research had shown that firefighters were, on an ever increasing basis, being subjected to more dangerous and long term life threatening situations. To their credit, fire chiefs began reaching out to other disciplines to protect their firefighters. They told their plight to elected and appointed officials. They sought multi-agency and inter-governmental coordination to help them respond. They told coordinating, planning and recovery agencies like the Federal Emergency Management Agency (FEMA) and state and local emergency management agencies that they, too, belonged in this vast ocean of hazmat; that we are all in this together; and that it was time to start pulling on the oars to set standards, acquire funding, do evacuation and shelter plans and, in general, take these increasing and improperly placed burdens off the fire service. The first objective of the fire service, they reminded, was fire suppression, not necessarily emergency medicine or hazmat response. They agreed, however, that they were in it for the long haul.

Obviously, the size and complexity of any given incident determines how many different "types" of responders and what the magnitude of the response should be. Sometimes the team stands and "waits out" the incident, something not readily viewed as part of the psyche of emergency responders. Today's hazmat team is like a surgical hospital. Depending upon what special chemical has been brought out to be operated on, we choose which specialists on our staff list are best able to cope with, operate on, and advise as to how the culprit should be handled. It's one from column A, two from column B, three from column C, and one from column D operations response for selecting any given days responder "team".

The hazmat team should be administratively approved by a governing body. In Pennsylvania, we lean toward county supported and approved teams, supported for insurance purposes, contributed to for equipment acquisition, and administratively supported by the emergency management office for mapping, evacuation zone estimates, and local emergency planning committee integration on the annual SARA exercise requirement.

The "hazmat team(s)" need not respond to all hazmat spills/incidents. Often, regular fire, police and public works staff can dike, contain, wash down or pick up the materials in question. But, for example, sophisticated mitigation of contents, under pressure, with evacuation imminent under life threatening conditions requires the higher level team.

Training to nationally recommended and state approved standards will shortly be upon us for hazmat teams. The number of hours, types of courses, and testing procedures are in the system now. The time consuming process of bringing together the Federal Department of Transportation, FEMA, Nuclear Regulatory Commission (NRC), EPA, and the Occupational Safety and Health Administration (OSHA) is on the move through federal interagency mechanisms called the National Response Team (NRT) and the Federal Radiological Preparedness Coordinating Committee. They are using the National Fire Protection Association (NFPA) for assistance in standards development. Training and equipment standards are now in development. Emergency response coordination, planning, enforcement, tactics, chemistry, training, exercise design, cleanup, radiological training, first responder awareness and analysis courses are readily available in train-the-trainer or advanced level from FEMA's Emergency Management Institute (EMI) and National Fire Academy (NFA) in Emmitsburg, Maryland, as well as DOT's Oklahoma City training center, EPA centers in Cincinnati, Ohio, and Edison, New Jersey, and OSHA's Chicago school. Many state, local and county counterpart agencies offer similar courses of lesser difficulty, home-study, fundamental and field versions funded with their own and other governmental level funds or private sector assistance.

The private sector has, of recent times, realized their vulnerability to civil suits through liability cases. Therefore, the railroad industry makes cars with DOT Federal Railroad Administration (FRA) prompting with inner and outer skins, valve protectors, and recessed connection mechanisms. All chemicals today are required to be placarded with the standard United Nations numbering and color coding system. Manifests must be in the possession of truck drivers and conductors. Material safety data sheets (MSDS) are necessary to be filed in local and state repositories for all to see and use for responding and planning. The program is coming together.

In 1987, a PEMA conducted study found that:

1. Thirty-three (33) counties were within five (5) miles of known, high level radioactive waste spent fuel routes.
2. Ten (10) counties indicated that they have an established county controlled hazmat team. (Bucks, Crawford, Cumberland, Erie, Fayette, Lebanon, Philadelphia, Washington, Westmoreland, York).
3. Thirteen (13) counties indicated that they have established municipal hazmat teams. (Allegheny (Pittsburgh), Berks (4), Blair (2), Bucks (3), Chester (1), Lancaster (2), Mercer (2), Mifflin (1), Montgomery (3), Somerset (1), Venango (2), Warren (18), Westmoreland (1)).
4. Fifteen (15) counties indicated that they have personnel trained under Act 147 radiological safety (225 personnel total).
5. Forty (40) counties indicated that they have hazmat trained personnel. (1951 personnel hazmat trained - various degrees of training).
6. Twelve (12) counties indicated that they have mutual aid agreements for

support from industry; most are verbal agreements.

We in Pennsylvania, feel that SARA, Title III is a good law and, when funded and fully implemented, will provide for greatly improved capabilities for all responders to hazardous materials accidents. Like no other law before, SARA, Title III calls on industry to air publicly much of its chemical processes and safety procedures. The law was enacted as part of legislation to revise and extend the federal Superfund program and, as such, opens up to scrutiny nearly every chemical action that happens behind plant walls. Additionally, our attention is also on train, plane and roadway spills as the statistics demonstrate. No more does the "lone responder", as the masked man of TV fame, ride in the save the day. The "team" is the answer and mitigation and preparedness are the long term cures. We need to become proficient and coordinated in multiagency intergovernmental response to meet today's challenges. Yesterday's "Lone Responder" needs many trusty companions to keep the chemicals moving safely and the environment clean.

BIBLIOGRAPHY

Public Law 99-499, Superfund Amendments and Reauthorization Act of 1986 (SARA Title III).

Emergency Management Services Code, As Amended (35 Pa. C.S.A. Section 7101 *et seq.*) (P.L. 1332).

Governor"s Executive Order 1987-8, Pennsylvania Emergency Response Commission, April 20, 1987.

40 CFR 350, Trade Secrecy Claim for Emergency Planning and Community Right-To-Know Information and Trade Secret Disclosures To Health Professionals.

40 CFR 355, Emergency Planning and Notification (Sections 302, 303 & 304).

40 CFR 370, Hazardous Chemical Reporting: Community Right-To-Know (Sections 311, 312 & 313).

29 CFR 1910.1200, Occupational Safety and Health Administration"s Hazard Communication Standard.

Emergency Response Guidebook, 1987, U.S. Department of Transportation, DOT P.5800.4.

Report to Congress on Hazardous Materials Training, Planning and Preparedness, July 1986. Prepared by Federal Emergency Management Agency and U.S. Department of Transportation.

Hazardous Materials Emergency Planning Guide, March 1987, NRT-1, Prepared by the National Response Team.

Technical Guidance for Hazardous Analysis, Dec. 1987, Prepared by U.S. Environmental Protection Agency, Federal Emergency Management Agency, and U.S. Department of Transportation.

Hazardous Materials Transportation Act of 1974.

Public Law 93-288, Disaster Relief Act of 1974, As Amended.

Management of Hazardous Materials and Wastes: Treatment, Minimization and Environmental Impacts. Edited by S. K. Majumdar, E. W. Miller and R. F. Schmalz. © 1989, The Pennsylvania Academy of Science.

Chapter Fifteen

HAZARDOUS WASTE MANAGEMENT IN HOSPITALS

MARTHA A. ANDERSON

Director, Environmental Health and Safety
Jefferson Medical College
Thomas Jefferson University
Philadelphia, Pennsylvania, 19107

Health care facilities, including primarily hospitals, but also long-term care institutions, clinics, physicians and dental offices, veterinary clinics, pharmaceutical companies, commercial laboratories, academic or industrial research laboratories and blood banks, are recognized as generators of hazardous and infectious wastes. Even homes in which health care is being administered produce small amounts of these wastes which if not properly handled and disposed could be potentially harmful to others who come in contact with them. The Centers for Disease Control (CDC) estimates that the typical hospital manages more than 10 pounds of potentially hazardous wastes per patient per day. Indeed this amount may be greater, even approaching 15-20 pounds per day per patient since the implementation of "universal precautions" which health care facilities have had to more closely enforce for employee protection since December 1987. This increase would not necessarily be from an actual increase in the amount of infectious waste realized from each patient but rather from the more stringent precautions of treating each patient's wastes as if they are being generated by someone with a potentially infectious condition. This is resulting in an overall increase in disposable employee protective equipment like gloves, gowns, masks, safety glasses etc. as well as additional amounts from patient areas of solid wastes like dressings, tissues, paper towels etc. which have previously been disposed as regular trash.

To fully understand the complexities of handling and disposing of hazardous and infectious wastes, it is important to acknowledge the various entities and agencies which have some control, whether voluntary or mandatory, over hospital operations. Who are they and what are there individual or collective roles?

—American Association of Blood Banks

This agency sets criteria for handling and disposal of blood and blood products.

—American Hospital Association (AHA)

This voluntary association sets guidelines for handling of waste in hospitals.

—Association of Operating Room Nurses (AORN)

This professional organization published guidelines on disposal of surgical waste and isolation case waste as well as equipment management.

—College of American Pathologists Laboratory Accreditation Program (CAP)

This accrediting group sets criteria and evaluates clinical laboratories for compliance with standards for safe handling and disposal of hazardous materials and wastes. These inspections must meet HCFA approval.

—Department of Transportation (DOT)

This federal agency regulates handling of hazardous materials and wastes across state lines including shipping, handling, labeling and marking and packaging.

—Environmental Protection Agency (EPA)

This federal agency is responsible for ensuring a clean environment and has the most effect on waste management programs through its own or state authorized departments which regulate and enforce solid waste disposal regulations. EPA regulations which directly or indirectly apply to hospital settings include the Clean Air Act, the Clean Water Act, the Safe Drinking Water Act, the Toxic Substances Control Act (TSCA), the Comprehensive Environmental Response, Compensation and Liability Act (CERCLA or Superfund), the Superfund Amendments and Reauthorization Act of 1986 (SARA) and the Resource Conservation and Recovery Act (RCRA) and Hazardous and Solid Waste Amendments (HSWA).

—Food and Drug Administration (FDA)

This federal agency conducts research and sets standards for the composition, quality and safety of foods, drugs, medical devices and cosmetics. The National Center for Toxicological Research is a division of the FDA which researches biological effects of toxic substances on the environment, and evaluates the effects on humans of toxic chemicals and environmental pollutants.

—Health Care Financing Administration (HCFA)

This entity approves facilities for Medicare and Medicaid reimbursement.

—Joint Commission on Accreditation of Healthcare Organizations (JCAHO)

This organization sets the standards for operations and safety of health care facilities, surveys them for compliance and certifies the institutions.

—National Fire Protection Association (NFPA)

This is a voluntary, nongovernmental organization that sets standards for fire and electrical safety concerns. Specific NFPA codes address hospital operations including the Life Safety Code (NFPA 101), Flammable and Combustible Liquids Code (NFPA 30) and Health Care Facilities (NFPA 99).

—Nuclear Regulatory Commission (NRC)

This federal agency (formerly the Atomic Energy Commission) is responsible for the use, handling and disposal of radioactive materials which requires a licensing agreement with the agency or its state-authorized counterpart.

—Occupational Safety and Health Administration (OSHA)

This branch of the United States Department of Labor is responsible for regulating the management of hazards in the workplace as they relate to worker safety.

—Pennsylvania (or state) and Philadelphia (or local) agencies (PADER;DOH)

State and local governmental agencies also have jurisdiction over health care facilities for safety and health standards. In some cases authority for compliance and enforcement is delegated to the state or local agencies but more often a health care facility can expect separate compliance surveys from federal, state and local governmental entities in addition to inspections from accrediting organizations.

How does a health care facility translate all of these applicable standards and regulations into workable and compliant programs?

An effective and total waste management program for health care facilities can be developed using a 4D approach:

DETERMINE identify all departments which have hazardous or infectious materials and produce potentially hazardous wastes from their activities through surveys and internal audits as well as from reviews of pertinent literature or of successful programs from similar institutions.

DEVELOP develop and implement a program addressing the particular needs as ascertained from the DETERMINE step.

DISPOSE adequately treat on-site or arrange for proper off-site disposal of all wastes generated by all operations of the facility.

DOCUMENT keep accurate records of all activities relating to the facility's waste management programs including but not necessarily limited to information from DETERMINE and DEVELOP steps including surveys, policies, procedures (for handling and disposal), training, annual audits, manifests, hauler and disposal contracts etc.

The 4D approach to health care waste management programs can be followed for each of four major areas of concern:

chemical waste
radioactive waste
infectious waste
chemotherapeutic waste.

Each major area is discussed in the 4D management approach is employed for each area in the following text.

CHEMICAL WASTE MANAGEMENT

The EPA formulated specific regulations for small quantity generators which became effective in September 1986. Since that time, all health care facilities have been covered for chemical waste management and disposal in one of the four generator-quantity classes: 1) up to 100 kg (220 pounds) of non-acutely hazardous waste per month, 2) 100 kg (220 pounds) to 1000 kg (2200 pounds) of non-acutely hazardous waste per month, 3) more than 1000 kg (2200 pounds) of hazardous waste per month and 4) more than 1 kg (2.2 pounds) of acutely hazardous waste per month. Such wastes would include sodium azide, osmium tetroxide, arsenic and cyanide - containing compounds etc. Those facilities which are included in classification 1) are considered "conditionally exempt"; those in classification 2) are referred to as small quantity generators and those in classification 3) are large quantity generators by definition. Each classification has specific applicable requirements. While most health care facilities are "conditionally exempt" small quantity generators, any one could become subject to the requirements for large quantity generators for any month in which it generates more than 1 kg acutely hazardous waste.

Therefore, a health care facility must first determine which generator classification applies to its activities. Then, the administration should designate or hire a hazardous waste coordinator whose first program responsibilities should include defining the institution's status. This can effectively be done through surveys of each department to ascertain which chemicals are purchased and how they are currently used and disposed. The survey results should be compiled and used to determine by EPA definitions if any of the chemicals are hazardous and in what amounts they are being generated as wastes. Many survey forms are available upon request from large hospitals that have already completed this process; a sample one from the Tucson Medical Center in Tuscon, Arizona is included in this text. The designated Coordinator should also perform internal audits to verify information provided by the various departments. For example, the engineering or maintenance department should have listed potential exposure and wastes from asbestos, caustics, mercury, flammable gases, paints and solvents, PCB's, gasoline, oils and lubricants, freons, pesticides etc. Housekeeping should have reported chlorine, iodine, corrosives, glutaraldehyde, phenol and quatenary ammonium; the morgue should have reported formaldehyde etc. The Coordinator should review all surveys for accuracy and completeness before formally defining the institution's chemical waste stream. A review of current literature and applicable regulations will be useful. Hazardous chemicals are found throughout most facilities and will include known and suspected carcinogens, and highly toxic materials such as asbestos, mercury, PCB's, ethylene oxide, formaldehyde, anesthetic agents, benzene etc. Some chemicals commonly found in clinical laboratories and morgues include acetic acid, acetonitrile, ammonium hydroxide, chloroform, dioxane, ethyl acetate,

TMC HAZARDOUS MATERIAL USE FORM

ITEM		TOXICS/ HAZARDS	USE				MANUFACTURER		
Brand Name (1)	Application (2)	Chemical Components (3)	How Disposed (4)	Packaging (5)	Qty Used/ Period (6)	Annual Est Use (7)	Name (8)	Address (9)	Phone (10)

DEPARTMENT ____________ DATE ____________ PAGE ____ OF ____

TMC HAZARDOUS MATERIAL USE FORM
INSTRUCTIONS

ITEM:

Column 1: Enter the primary name of the item, product, chemical or material, and in ( ) any additional trade names by which it or its contents are known.

Column 2: Enter the primary application for which this item, product, chemical or material has been purchased and in ( ) any additional uses.

TOXICS/HAZARDS:

Column 3: Enter the principal chemical constituents.

USE:

Column 4: Indicate the primary disposal technique for the item, product, chemical or material. Examples would be regular trash, protective care, sewer, incineration, completely used up, etc. Disposal techniques for containers holding hazardous/toxic chemicals should be noted in this column as well.

Column 5: Enter the weight of each separately packaged unit of the item, product, chemical or material; for example, 42 lbs./ bottle or 50 grams/vial.

Column 6: Enter the quantity of the item, product, chemical or material used in the most recent month with (mo). If not used in the most recent month, indicate most recent quarterly usage with (qtrly). Example: 5 bottles/(qtrly) or 300 vials/(mo).

Column 7: Enter your best estimate of the quantity of the item, product, chemical or material used during the past year or for a recent 12 month period.

MANUFACTURER

Column 8: Enter name of manufacturer or distributor of the item, product, chemical or material.

Column 9: Enter address of manufacturer or distributor.

Column 10: Enter phone number of manufacturer or distributor. Please indicate "800" number available.

Should you have any questions with filling out this form, please contact the Hazardous Waste Taskforce member assigned to your department.

Note: Can use each survey sheet for multiple entries.

ethyl ether, ethylene dichloride, ethylene oxide, formaldehyde, hydrochloric acid, hydroquinone, mercuric chloride, methyl alcohol, methylene chloride, nitric acid, sodium hydroxide, sodium nitrite, sodium periodate, sulfuric acid, tetrahydrofuran, toluene and xylene.

Next, with the support and assistance of all hospital personnel and participation of the safety committee, the Coordinator must develop policies and procedures to manage, contain, segregate, collect and store the wastes prior to disposal. The policy should be set by the chief executive officer, hospital director or administrator to ensure that all employees, medical staff and volunteers understand and appreciate the importance of the program to the institution. Once chemical inventories and uses are determined and a policy is established, specific procedures for packaging, collection and disposal can be instituted by the Coodinator whose authority is designated in the policy. These procedures should include information on proper packaging and identification requirements, as well as directions for secondary containment precautions such as on-site transport carts and cardboard boxes or plastic or steel drums for off-site disposal requirements. Since it is not practical to arrange for daily disposal of the chemical wastes generated, some area in or near the facility must be designated for storage of these materials prior to disposal. Storage does not require a permit unless large quantity generators hold wastes for longer than 90 days or small quantity generators who ship wastes farther than 200 miles hold wastes for longer than 270 days. Regardless of the holding time, the storage area, at minimum, should have secure and limited access, be safe according to local and other applicable fire codes, and equipped with emergency and personal protective equipment in case of a spill or any other problem which might occur. The Coordinator should ascertain storage requirements applicable to the specific health care facility location. Throughout the packaging, collection and disposal operations, attention must be given to compatabilities of the chemicals; this will ensure a safe handling of the waste within the facilities and more expedient operation when the contractor arrives to collect wastes for off-site disposal.

If possible, the Coordinator should consider all options of disposal available in order to reduce the quantities which must be sent off-site for incineration or landfill burial. Some local sanitary sewer systems will accept small quantities of chemicals like acids, alkalais, alcohols and formaldehyde which are commonly used. Often, contracts can be established with companies which reclaim solvents (some even redistill and sell the distillates back to the generator at reduced costs) like xylene and toluene. Other companies reclaim mercury (collected from broken thermometers and blood pressure cuffs) or reclaim silver from spent photographic fixers and developers commonly used in X-ray departments. For those wastes which cannot be sent to the sewers or reclaimed, the Coodinator must arrange for off-site disposal. Although the facility may have an incinerator on-site for infectious and general trash, no chemical waste should be incinerated on-site. Pathological waste incinerators acre not really designed to reach the high temperatures and retention times which are required for complete combustion of chemical wastes, and even if the unit can perform to these standards, the EPA requires a special permit which is in itself costly and time consuming to obtain. The volumes of wastes generated by even large health care facilities do not warrant this effort. Therefore, the Coordinator should have contracts with reputable firms which can package and transport and arrange for or provide direct disposal to an approved facility, preferably an incinerator. All contractors must be required to show proof of insurance, should be provided with the facility's EPA "generator identification number", should assist in completing the proper paperwork including the manifest for each shipment (of which a complete copy should be received before final payment is made), and should not object to being monitored and visited occasionally.

All of these procedures and operations must be thoroughly documented for eventual scrutiny by facility inspectors, governmental agencies and even insurance carriers. Therefore, the Coodinator must keep records of inventories, collections, storage, packaging, transportation and disposal of all chemicals and the resulting wastes. Records must also be maintained of employee training for waste handling, emergency and spill responses and any other activity which relates to this program. In addition, the Coordinator, with the assistance of other key personnel, must develop an emergency and contingency response plan which addresses the items included in 40 CFR Part 265, Subparts C and D. The plan must include actions which would be taken in case of fire, explosion etc., personnel and equipment assignments, procedures for spill response etc. The committee should review all applicable regulatory requirements to make the final plan as complete as possible; periodically, the plan should be tested with a mock chemical emergency to ensure that it is not only complete but that it is also effective when implemented.

While it is not a simple task to develop an effective chemical waste management scheme for a health care facility, it is also not an impossible assignment. Cooperation of all personnel is as critical an element in the development of the program as is the need to give the designated coordinator the time and resources to make it happen.

RADIOACTIVE WASTE MANAGEMENT

In health care facilities, radioactive wastes are produced by the clinical laboratories, nuclear medicine and radiology departments and any research laboratories which may be associated with the facility. The most common radioisotopes used in health care activities include hydrogen (H-3), carbon (C-14), phosphorous (P-32), cobalt (Co-57, Co-58, Co-60), iodine (I-123, I-125, I-131); other less common ones include sulphur, chlorine, iron, xenon, iridium, thallium, mercury etc. The departments which use these materials produce solid and liquid wastes and some "sealed sources," all of which must be properly handled and disposed. The requirements for a radioactive waste management program are very specific and enforced through a license which each health care facility must obtain from the Nuclear Regulatory Commission (NRC) or from a state agency if they are located in an agreement state (with the NRC).

In order to adequately protect employees, the health care facility must ensure that radioactive materials are properly identified, shielded and contained. The need for shielding, unless obvious by the declared activity of the source material, can be ascertained by instrument surveys of the area performed by the designated radiation safety officer. Equal care must be taken in 1) using, 2) storing and 3) disposing of the radioisotopes.

The primary concern of the program coordinator should be to minimize the amounts and types of isotopes on-site. Once he has determined facility users and radioactive isotopes in stock, he should develop a program which includes recommendations for using isotopes which can be stored for decay or disposed to the sewer system. If the half-life is 65 days or less, the health care facility can easily store the waste materials until they "decay" to levels that are no longer considered radioactive. This certainly reduces the amounts of wastes which must be disposed off-site, thus also reducing the significant associated costs. Other isotopes can be disposed of to the sewer system because they are also present in our natural environment. Wastes which cannot be disposed with these accepted procedures must be segregated, collected, packaged and shipped to an approved disposal site. Contaminated plastic and glass items, liquid scintillation vials and paper wastes are collected in yellow colored, thick, plastic bags labeled with yellow and black tape which is marked with the radiation symbol and "CAUTION-RADIOACTIVE MATERIAL". The original bags or containers must also have a label which indicates isotope, activity, date, and generator's name. Then, these are transported by trained, authorized personnel to a designated handling and storage area where the original containers are placed in boxes or other stronger containers while determinations for proper disposal are made. The same procedures must be followed for radioactive contaminated biological wastes including animal carcasses. It is interesting to note that a waste product may be considered an infectious or a chemical waste if the infectious properties or the toxic properties of the waste are determined to be more hazardous than its radioactive content. Once, however, a waste is determined radioactive, it must be stored properly in an area that is secured, signed and shielded as well as adequately ventilated and constructed to meet local fire safety codes. A log must be kept of all wastes by type, activity, date and location in the storage area; the log can also be used to indicate the method and date of disposal.

Disposal of some radioactive liquid wastes to the sewer system can be carried out in specially designated and labeled sinks if the materials are dispersed or become soluble in water and the concentrations do not exceed the daily, monthly or annual concentrations specified by the NRC. Scintillation fluids cannot be sewer disposed. As previously stated, some radioisotope wastes can be stored until they reach an acceptable concentration which has been set at 10 half-lives after which the waste can be handled as regular trash. Monitoring must be performed to prove minimum concentration and dates must document adequate decay times. All labels should be removed before these materials are disposed as trash.

In March, 1981 the NRC set "de minimis" radioactive concentrations in which tritium or carbon-14 contaminated scintillation liquids, items contaminated with these fluids, scintillation vials and absorbents, animal bodies and organs can be sent to a pathological incinerator or sanitary landfill. For wastes which meet the requirements set in 10 CFR 20.306, no NRC approval is necessary, but records must indicate that the materials were disposed. Also, the program Coodinator must then determine whether these wastes are included in the EPA regulations and require disposal as chemical wastes. The facility can seek approval to incinerate on-site tritium and carbon-14 wastes with activities greater than those in 10 CFR 20.306 by submitting specific information to NRC. All radioactive wastes which cannot be disposed as discussed above must be shipped to an approved site for off-site burial. Even these wastes can often be reduced in amount by solidifying the liquids and compacting the solids in specially designed units prior to shipment. In addition, vendors will often accept the return of sealed radiation sources which they recover or dispose as part of their contract.

INFECTIOUS WASTE MANAGEMENT

Infectious wastes from patients in health care facilities may be found in the forms of stool and urine, blood, exudates, secretions, and solid wastes such as disposable instruments, linens, paper and plastic products. The most common methods of disposal for these materials are incineration, sewer disposal and removal by contractors for disposal off-site.

An article in *Laboratory Management,* 16(6): 37-44, 1978 listed the following as common contaminated wastes from medical laboratories: culture dishes, pipettes, syringes and other sharps, tissue culture bottles and flasks, membrane filters in plastic dishes, collection bottles, cups, and tubes from specimens of blood, urine, feces, saliva, exudates, or secretions, microtiter plates used for hemagglutination testing, slides and cover slips from blood specimens or tissue or colony picking, disposable rubber gloves, lab coats, and aprons, swabs, capillary tubes, and spreaders used to take or transfer samples containing pathogens, and centrifuge tubes. These waste products are obviously often generated by other than medical and clinical laboratories; therefore, recent regulations apply to any laboratories or facilities which produce contaminated wastes.

It is easily noted that each agency which regulates or addresses infectious wastes has a similar yet somewhat different definition of what constitutes infectious. Perhaps the most important definition a health care facility needs to follow is the one provided by the governmental agency which has jurisdiction over the collection and disposal of its wastes. The Environmental Protection Agency is being pressured to develop specific regulations for infectious wastes even though the agency already has published the 1986 *EPA Guide for Infectious Waste Management* which was intended as a standard for the development of infectious waste programs. The EPA also intended, and is yet supporting its position, that regulation of infectious wastes should be performed at the state level with county and city government involvement; local control is expected to better address specific local issues and will likely lend to more locally sited incinerators than a larger regional approach. Therefore, in Pennsylvania, for instance, the definitions of infectious wastes included in the *Pennsylvania Bulletin* Volume 18, Number 15, April 9, 1988 Part II Municipal Waste Management Regulations are the most important as they define which and how health care wastes are to be handled thus pre-empting any other previous definitions for facilities within the state. Infectious waste is as stated in Section 271.1 "Municipal waste which, is or may be contaminated by a disease-producing microorganism or material, or may harm or threaten human health. The term includes the following wastes unless they are generated by individual residences: (i) Wastes generated by hospitalized patients who are isolated, or on blood and body fluid precautions, in order to protect others from their severe and communicable disease. (ii) Cultures and stocks of etiologic agents. (iii) Animal waste blood and animal blood products which are known or are suspected to contain contagious zoonotic pathogens, and human waste blood and blood products. (iv) Tissues, organs, body parts, blood and body fluids that are removed during surgery and autopsy. (v) Wastes generated by surgery or autopsy of septic cases or patients with infectious diseases. (vi) Wastes that were in contact with pathogens in any type of laboratory work, including collection containers, culture dishes, slides, plates and assemblies for diagnostic tests; and devices used to transfer, inoculate and mix cultures. (vii) Sharps. (viii) Wastes that were in contact with the blood of patients undergoing hemodialysis at hospitals or independent treatment centers. (ix) Carcasses and body parts of animals exposed to contagious zoonotic pathogens. (x) Animal bedding and other wastes that were in contact with animals suffering from contagious zoonotic diseases due to natural infection or laboratory research, and their excretions, secretions, carcasses or body parts. (xi) Waste biologicals—for example, vaccines—produced by pharmaceutical companies for human or veterinary use. (xii) Food and other products that are discarded because of contamination with etiologic agents. (xiii) Equipment and equipment parts contaminated with etiologic agents".

To best comply with these regulations it is necessary to understand the definition of zoonotic. The 1982 *EPA Draft Manual for Infectious Waste Management* noted that nearly 200 zoonoses have been identified and that animals or animal tissues are probably involved in 30-40% of all laboratory acquired infections. Examples of the more common zoonotic diseases (those which can be transmitted from animals to man) are Newcastle disease (poultry), lymphocytic choriomeningitis (rodents), tularemia (rabbits), rabies (opossum, skunk, fox, dogs), toxoplasmosis (cats), Q-fever, anthrax, tuberculosis, cow pox (cattle, sheep, goats and pigs), tuberculosis, hepatitis, *Herpes simiae,* malaria, SV40 and rabies (nonhuman primates).

Pennsylvania health care facilities are also subject to the new OSHA requirements for protection of health care employees from HIV and AIDS infections which include the precaution that each patient's wastes are to be treated as if they are infectious. Therefore, it should be obvious that a health care facility must first determine which wastes being generated from its operations are infectious by regulatory definitions. Once this is accomplished, the person designated to manage the waste flow (usually the housekeeping director) should establish some procedures to segregate infectious waste from non-infectious waste which will reduce the handling time and disposal costs. Infectious wastes should be separated at the source where trained employees like the nursing staff can best differentiate; these wastes should then be placed in red or red-orange colored containers like thick plastic bags which are tied or closed off and labeled with the universal biohazard symbol. Heavier containers like cardboard boxes should be used at the point of generation for sharps; these containers must also be labeled. Then, housekeepers or other employees trained to handle infectious wastes should collect the containerized, labeled materials and transport them in separate waste carts to a central collection point or loading dock where the wastes should all be placed in cardboard containers which minimize the possibility of exposure from bag tears and "spills". Since infectious wastes must be treated or disposed on the same day they are generated (longer on-site storage requires wastes be refrigerated), an effective infectious waste management program must provide for on-site treatment through autoclaving or incineration, or off-site treatment and disposal.

Autoclaving or steam sterilization is dependent on both steam pressure and temperature and the time allowed the process. The effectiveness of the technique is determined by the minimum time required to kill a control organism, *Bacillus stearothermophillus,* in steam at a specific temperature. Loosely packed, infectious wastes of plastic and paper are easily treated by autoclaving; denser wastes and liquid wastes are sometimes not completely decontaminated with this process unless excessive times are allotted. In all cases of steam sterilization, monitoring is required with the indicator set in a position which will ensure that the entire load is thoroughly treated. Once the decontamination is complete, the autoclaved wastes can be considered noninfectious.

Many health care facilities operate pathological incinerators which ordinarily operate at temperatures around 1400°F; these units have been effective for the infectious wastes from body fluids and parts. They have not been efficient for burning the disposable plastic wastes as the units cannot reach the higher temperatures and do not usually have secondary chambers in which the gaseous residues can be destroyed. Therefore, many health care facilities have been forced by local air pollution agencies to improve or close existing incinerator installations. Until recently, it was an easy decision to cease operating an on-site incinerator because costs of disposal were competitive and relatively inexpensive. Within the past six months, bans or limitations on the uses of landfills have forced many waste generators and haulers to turn to incineration for disposal, leading to a very competitive market because there are not yet as many commercial incinerators as there were landfills available and to higher prices per pound than disposal of infectious wastes has ever cost before. Now, the trend is back to on-site incineration but not without some difficulties from more stringent permitting requirements which have been promulgated in the interim and from increased public interest in locations and operations of new units. The new requirements include units designed with two chambers which will handle higher temperatures and with pollution control equipment which can continuously monitor for combustion emissions like CO, CO_2, O_2 and opacity; permits must be obtained from air and solid waste management authorities and facility information must include descriptions of wastes generated, training, emergency and contingency plans as well as incinerator and equipment descriptions and specifications. Even though the procurement and permitting processes have become more stringent, it is still prudent for a health care facility to consider installing its own incinerator. Off-site disposal costs are not likely to decrease or even stabilize in the near future and as with chemical wastes, the generator always remains principally responsible for the wastes regardless of any improper disposal operations or methods used by the contractors who haul and dispose of the wastes. Indeed some states like Pennsylvania are now requiring tests to be performed on incinerator ash to establish that the process did kill all the microorganisms present in the original infectious wastes.

In view of the publicity and criticism directed toward a health care facility which does not handle its wastes properly, it is imperative to not only establish an effective infectious waste management program but also to document all activities associated with the program. Records must be kept for all collections, shipments, methods of disposal whether on or off site, costs, contracts, training, emergency and contingency plans and drills, permit documents and approvals etc. If the program is well managed and documented, there is less possibility that the health care facility will become better known for its wastes than for its patient care.

CHEMOTHERAPEUTIC WASTE MANAGEMENT

Treatments for cancers often include chemotherapy which involves the use of cytotoxic drugs or antineoplastic agents which inhibit the growth of cancer cells through nutritional interference, hormonal imbalance, or by inhibiting DNA replication or protein synthesis. Special care must be taken when working with these chemotherapeutic drugs or agents because of their toxic acute and chronic effects on animals and the environment. Wastes from the drugs listed by EPA in 40 CFR 261.33 are regulated as hazardous chemical wastes while wastes from the other chemotherapeutic drugs or agents are either regulated in state programs (as in Pennsylvania) or not regulated at all. Control of these agents in a health care facility is not difficult as they are usually purchased, stored and distributed from the pharmacy directly to nursing or oncology from which they are administered to patients. According to the American Society for Hospital Engineering, there are about 30 drugs commonly used as chemotherapeutic agents or drugs including alkylating agents like cisplatin, nitrogen mustard, chlorambucil and streptozocin; investigationals like azacytidine and melphalan; antimetabolites like fluorouracil and mercaptopurine; antibiotics like dactinomycin, mitomycin C and daunomycin; mitotic inhibitors like vinblastine and two miscellaneous, L-asparaginase and dacarbazine. The pharmacy also produces most of the chemotherapeutic wastes which are usually leftover or incorrectly mixed liquids. Other wastes like contaminated syringes, gloves, paper etc. are generated from nursing and patient areas as well as the pharmacy. All wastes must be properly packaged and labeled. To comply with the new Pennsylvania regulations, a health care facility must use at least (2) 3 mil thick, leakproof, yellow, plastic bags for disposable wastes and yellow, leakproof, puncture-resistant containers with lids for contaminated sharps. Each type of container must be labeled as either "chemotherapeutic waste" or as "biohazard" with the universal biohazard symbol. It is interesting to note that use of designations specified in the Pennsylvania regulations will likely lend to more confusion within the health care facility as yellow has already been used for radioactive wastes and the term "biohazard" is most often used to denote an infectious material rather than a chemical exposure.

Next, these wastes should be collected by personnel who have been trained and outfitted with personal protective equipment. They should transport the containers to a central storage and handling area where the the bags should be placed in boxes or other stronger containers to minimize the possibilities for exposures and spills. In fact, although the EPA regulates the disposal of only seven of these agents, it is considered prudent practice to treat all chemotherapeutic wastes as chemical wastes for disposal purposes. However, some chemotherapeutic wastes can still be disposed to the sewer and sanitary landfills if they are in the unregulated category and are not in great quantities. Also, containers or vials with less than 3% by volume of all agents but the ones designated by EPA for chemical disposal are considered to be empty and can be discarded legally to a sanitary landfill. Off-site incineration is the best disposal method as temperatures of 1800°F are necessary to destruct the liquid chemotherapeutics. As with all health care waste programs, and especially for chemotherapeutics wastes, as they are yet the least regulated, it is prudent for facility personnel to review the regulations of all agencies which may have jurisdiction to ensure legal compliance. It is strongly recommended that health care facilities follow the most prudent operating procedures to protect employees and patients from potential exposures and to document their compliance not only with legal requirements but also with public perception of care.

CONCLUSIONS AND RECOMMENDATIONS

Health care facilities must address the issues relevant to managing, generating and disposing of chemical, radioactive, infectious and chemotherapeutic materials and their resultant wastes. This can be accomplished by determining the amounts and locations of the materials, developing programs which control uses and waste handling, arranging for proper on-site or off-site disposal of the wastes and documenting all activities associated with the programs. The governmental control of waste disposal is a relatively recent process which is continuously changing and therefore must be continually monitored to ensure that the facility is indeed following all of the rules and regulations. While hazardous waste disposal is certainly not as important to a health care facility as providing good patient care, it is, nonetheless, a critical management responsibility and must be addressed adequately and soon.

REFERENCES

1. *Environmental Management News,* Volume 3, No. 3, April, 1988, pp 1-3.
2. Environmental Protection Agency, Hazardous Waste Management Regulations, USEPA, 40 CFR, Government Printing Office, Washington, DC.
3. *EPA Draft Manual for Infectious Waste Management,* SW-957, September, 1982, USEPA, Washington, DC, pages 2-12.
4. *EPA Guide for Infectious Waste Management,* SW-86-014, 1986, USEPA, Washington, DC.
5. *Federal Register,* Volume 51, #56, "Hazardous Waste: Generators of Hazardous Waste, On-Site Storage . . . Final Rule," March, 1986.
6. *Hazardous and Infectious Waste Management for Health Care Facilities,* Lawrence C. Wylie, Fred S. James and Co., Inc., 1984, 263 pages.
7. *Hazardous Materials Management,* A Guide for Hospitals, ECRI, Plymouth Meeting, Pennsylvania, 1987, 50 pages.

8. *Hazardous Waste Management Strategies for Health Care Facilities,* Nelson S. Slavik, Ph.D., American Society for Hospital Engineering of the American Hospital Association, Chicago, Illinois, 59 pages, 1987.
9. *Laboratory Management,* Volume 16, Number 6, 1978, pages 37-44.
10. Nuclear Regulatory Commission, Standards for Protection Against Radiation, USNRC, 10 CFR, Government Printing Office, Washington, DC.
11. *Occupational Health and Safety News Digest,* Volume 3, No. 4, April, 1987, pages 1-4 and 22.
12. Occupational Safety and Health Administration, General Duty Clause OSH Act Section 5 (a) (1) and 29 CFR 1910.132,29 CFR 1910.22 (a) (1), 29 CFR 1910.141 (a) (4) (ii), and 29 CFR 1910.145, Government Printing Office, Washington, DC.
13. *Pennsylvania Bulletin*, Volume 18, Number 15, April 9, 1988.
14. "Recommendations for Prevention of HIV Transmission in Health-Care Settings," *Morbidity and Mortality Weekly Report,* Volume 36, No.2S, August, 1987, pages 3S-18S.

Management of Hazardous Materials and Wastes: Treatment, Minimization and Environmental Impacts. Edited by S. K. Majumdar, E. W. Miller and R. F. Schmalz.

Chapter Sixteen

LOW-LEVEL RADIOACTIVE WASTE MANAGEMENT OPTIONS

ROBERT F. SCHMALZ

Department of Geosciences
The Pennsylvania State University
University Park, Pennsylvania, 16802

In 1980, the federal government delegated responsibility for disposing of commercially-generated low level radioactive waste to the states (Low Level Waste Policy Act, P.L. 96-573). This action stimulated prolonged public debate and controversy which appears quite disproportionate to the volume and toxicity of low level radioactive waste relative to other forms of hazardous waste. Indeed, many believe that social and political obstacles will prove far more intractable than any technical problems associated with the secure disposal of low level radioactive waste. These non-technical problems are discussed elsewhere in this volume and by Bord (1985), Friedman (1985) and Witzig *et al* (1983) as They relate to radioactive waste management in particular. Worldwide, we have nearly thirty years' experience managing non-military low- and medium-level wastes. It is the purpose of this paper to review that experience and to consider the merits of the management options available.

CHARACTERISTICS OF LOW LEVEL RADIOACTIVE WASTE

In the United States, commercially generated radioactive waste is divided into two categories by the Low Level Waste Policy Act (P.L. 96-573) and the Nuclear Waste Policy Act of 1982 (P.L. 97-425): (1) "high level radioactive waste", comprising uranium and thorium mine and mill tailings, spent reactor fuel and alpha-emitting trans-uranic elements (TRU's) and (2) "low level waste" comprising all radioactive waste not specifically identified as high level waste. The same legislation assigns responsibility for the disposal of all high level radioactive waste to the federal government.

The four major commercial sources of low level radio-active waste are nuclear power plants, industry, medical diagnosis and treatment, and research laboratories. Recent estimates of the volume and activity of wastes generated by each of these sources in Pennsylvania and in the United States have been summerized by Dornsife (1985) and Singh (1987), (Table 1). Much of the volume of low level radioactive waste is in the form of housekeeping trash (protective clothing, clean-up materials, etc.) disposable paper- and plastic-ware, and contaminated liquids. The bulk of such wastes can be significantly reduced by compacting, evaporation or incineration. The extent to which on-site volume reduction can facilitate low level radioactive waste management is evident from the data in Table 2. The residual volume of waste will be characterized by the presence of radionuclides which may be divided into three broad classes:

i) Volatile constituents.
   Generally gases or nuclides which form gaseous or volatile compounds under near-surface conditions (eg: tritium, carbon-14).

ii) Mobile constituents.
   Soluble elements or nuclides which form water-soluble compounds

TABLE 1

Quantities of Low Level Radioactive Waste Shipped for Disposal in the U.S. and Pennsylvania, by Source

Source	Pennsylvania (1986) volume (ft^3)	activity (Ci)	U.S. (1982 83) volume (ft^3)	activity (Ci)
Power reactors	134,371	22,602	1,780,260	289,563
Medical	8,176	11		
Industrial	47,797	1,029	1,045,602	170,055
Research	2,634	10		

(Sources: Dornsife, 1985; Singh, 1987)

TABLE 2

*On-site Reduction of LLRW Volume and Activity, Pennsylvania, 1982.**

Source	Volume (cubic feet) Generated	Shipped	Activity (Curies) Generated	Shipped
Reactors	206,160	185,496	167,7000	6,607
Medicine	34,012	17,244	12,620	15
Industry	250,245	6,506	11,490	8,378
Research	10,389	2,640	22	5
TOTAL	500,806	211,886	191,832	15,005

Volume reduction: 58%
Activity reduction: 92%

*Exclusive of quantities released to air, water or sanitary sewer.
(Pollog, 1983; Tables 3, 18.)

in near-surface environments (eg: cesium-134, cesium-137, strontium-90).

iii) Immobile constituents.
Metals or other radionuclides which are generally insoluble and/or chemically inert in near-surface environments.

Obviously, volatile nuclides will be confined in a disposal facility only with great difficulty. Tritiated water (HTO) either as a liquid or vapor, or tritiated methane (CTH_3) will be virtually impossible to retain for a long period in any disposal site. The same will be true of ^{14}C-bearing carbon dioxide or methane. There is strong evidence that gas leakage constitutes the only significant path by which radio-nuclides have escaped from the shallow land burial facility at West Valley, New York (Matuszek & Robinson, 1983). It is probably an important factor elsewhere as well.

The mobile constituents, particularly those which are water soluble, will also present serious problems of confinement. These problems, however, can be reduced to manageable proportions by (i) restricting surface- and ground-water access to the waste, (ii) minimizing the movement of contaminated leachate away from the site, and (iii) selecting a site where geological conditions will favor the 'fixation' of the more soluble isotopes (Farrett, 1983; Kohl, 1986). Particularly soluble waste material may require stabilization in specially formulated cement or packaging in high-integrity water-proof containers.

The immobile nuclides will generally present no special disposal problems, although metals may require isolation from organic chelating agents found in some soils.

Obviously, if it is to be successful, management of low level radioactive wastes must combine on-site waste volume reduction and geologically-based site selection, with well-planned waste packaging and emplacement, an effective plan for surface- and ground-water management, and adequate long-term maintenance. Responsibility for the first of these requirements rests with the generator; the last three requirements can be combined conveniently into the design and operation of the facility.

FACILITY DESIGN OPTIONS

There are four basic designs of low level radioactive waste management facility: shallow land burial, earth-mounded bunkers, surface storage, and deep geological emplacement. Practical experience with each type reveals strengths and weaknesses, although all have proven effective. In the discussion which follows each design is described in a simple form but it should be understood that the basic design can be modified by engineered features which may enhance its performance, raise public confidence, or both, usually at a substantially increased cost.

SHALLOW LAND BURIAL

All of the commercial low level radioactive waste disposal facilities in the United States are of this type. Technically, shallow land burial facilities (SLB) are simple extensions of familiar sanitary landfills. Disposal "cells" twenty-five to fifty feet deep and up to six hundred feet in length are excavated at the site. A French drain or sump is cut along one side of the excavation, and a drainage gradient is provided across the trench to the drain, as well as longitudinally. Monitoring wells are located at regular intervals along the drain, and permeable sand or gravel floor is placed in the trench to permit movement of water beneath the waste to the drain. Packaged waste is stacked as compactly as practicable, the interstices filled with dry sand, and the entire mass then covered with a cap designed to divert surface water away from the trench and to prevent (insofar as possible) the penetration of water into the waste beneath.

Shallow land burial facilities are currently operating in the United States at Beatty, Nevada, Richland, Washington and Barnwell, South Carolina. Environmental contamination has not been a significant problem at any of these facilities, although similar facilities at Sheffield, Illinois, Maxey Flat, Kentucky and West Valley, New York, experienced operational problems prior to (or shortly after) their closure (Clancy, *et al*, 1984). At all three facilities contaminated, leachate is believed to have escaped from the burial cells (Cartwright, 1982; Kelleher, *et al*, 1983; Grant, 1982).

Several factors appear to have contributed to the operational problems at these three sites. Firstly, original licensing regulations permitted disposal of contaminated liquids in shallow land burial sites; such liquids, usually packed in steel drums or glass containers, have been released as the containers break down. Secondly, the early regulations allowed a wide range of packaging materials in shallow land burial sites. These containers, including paper drums, wooden or pasteboard boxes and even plastic bags, have decayed or collapsed, releasing their contents. Thirdly, because disposal costs were low initially (less than one dollar per cubic foot) little effort was made to compress the waste, to fill containers completely, to utilize the available trench volume efficiently or to backfill interstices among waste packages. Consequently, as the waste settled and the containers decayed, the trench cap was left unsupported. The cap then collapsed, cracked and allowed surface water access to the waste (Lu, 1978; Prudic, 1980, 1986). Finally, jurisdictional disputes, particularly at West Valley, delayed timely corrective action when problems were first recognized (Schmalz, 1987).

Lessons learned in the operation of these three early "failures" have led to revised regulations and improved operating procedures that have contributed significantly to the success of the facilities now in operation. Free liquids may no longer be included in buried waste, rising costs (now exceeding $50/cubic foot in most cases) encourage on-site volume reduction, strict packaging standards ensure structural integrity of waste, careful backfilling procedures and improved site engineering and operational management all have contributed to a level of performance and environmental protection which exceeds mandated standards (Smith, 1984).

Successful as these facilities may be, many legislators and environmental groups have advocated further refinement of shallow land burial practices (eg: Sierra Club). It has been proposed that trenches be lined and covered with impermeable 'geofabric' or plastic membranes, or that the simple trench be replaced by a subterranean concrete vault. One form of below-grade vault is used for storage of some waste forms at Ontario Hydro's Bruce Nuclear Power Development near Kincardin, Ontario (Figure 1) (Carter & Rao, 1984). Further enhancement of ordinary shallow land burial practices can be achieved by the use of high-integrity plastic containers (HIC's) or rigid concrete cells like the SUREPAK units proposed by the Westinghouse Corporation's Waste Technology Services Division (Westinghouse Corp., 1987). Encapsulating waste in specially formulated grout or concrete, either inside containers, in trenches or vaults, or both, may provide additional confinement. Any of these enhancements will, of course, increase the cost of waste disposal, but the increase may be justified by improved site performance or increased public confidence or both.

Shallow land burial offers the advantages of short start-up time, proven effectiveness, low maintenance cost, efficient land use, accommodation of large waste volumes and the potential for alternative land use at the close of the mandated 100-300 year institutional control period. The primary disadvantages of shallow land burial appear to be a perceived difficulty in monitoring contaminant leakage, the high cost of corrective action in the event of a leak, and the difficulty of recovering the buried waste, should this become necessary at some future date.

EARTH-MOUNDED BUNKERS

Earth-mounded bunkers (EMB) are used for radioactive waste disposal in Russia and France. Their design combines sub-grade burial, usually reserved for intermediate activity waste, with above-grade disposal; the entire mound being covered and shielded by an engineered soil cover. A sectional view of a typical earth-mounded bunker used at the Centre de la Manche (near Cherbourg, France) is shown in figure 2.

Construction of an earth-mounded bunker begins with an excavation approximately 40 x 80 meters (130 x 260 feet) and 10 meters (33 feet) deep. A reinforced concrete base slab and internal drainage channels are then constructed and the bunker is ready to receive waste. The waste is packaged in 200 liter steel drums, grout-filled concrete cylinders ("blocs") approximately 4 feet high, and steel or concrete boxes ("caissons") with a capacity of 185 cubic feet. These are arranged on the base slab, enclosed in temporary concrete forms, and embed-

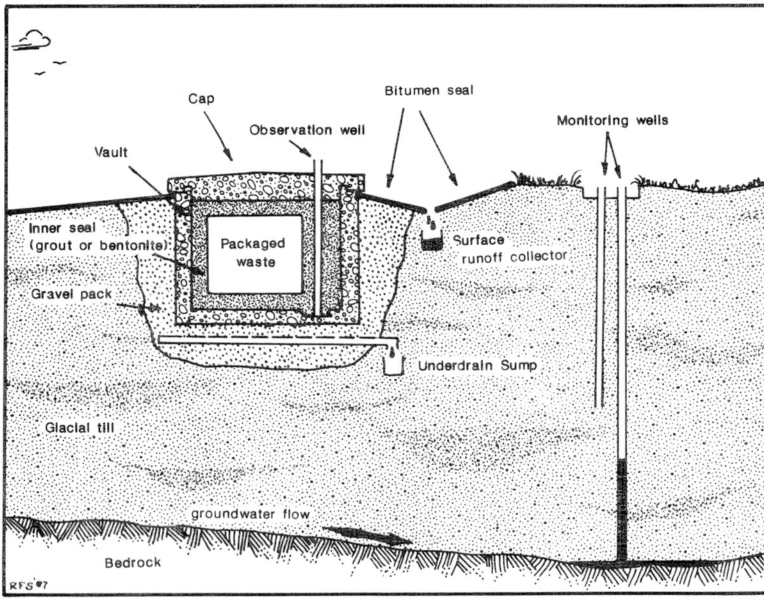

FIGURE 1. Shallow sub-surface storage vaults used at Bruce Nuclear Power Development (Ontario Hydro). Multiple envelopes provide effective containment of waste, and several levels of monitoring. (redrawn from Carter & Rao, 1982.)

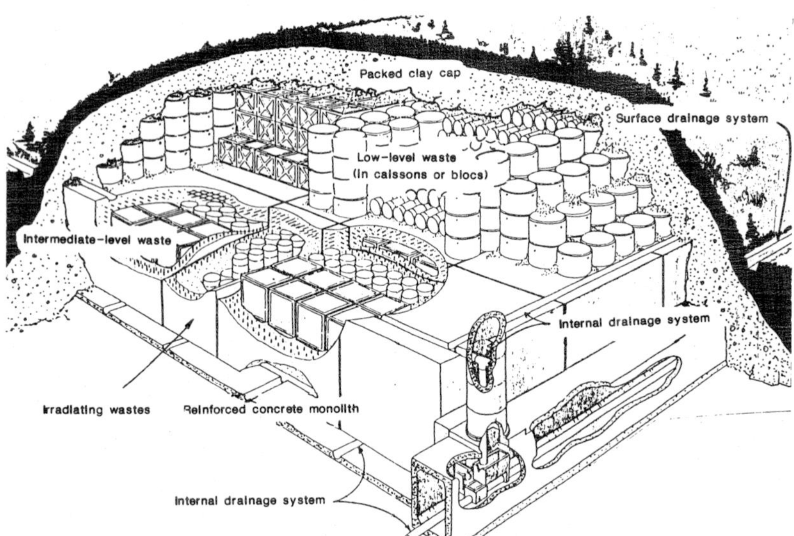

FIGURE 2. Sectional view of earth-mounded bunker or tumulus in use at the Centre de la Manche, France. In this facility provision is made to dispose of intermediate-level and irradiating wastes as well as low-level waste.

(from Bennett & McAneny, 1985.)

ded in concrete. At grade level, a second reinforced concrete slab with additional internal drains is poured on top of the embedded waste monolith, and low level waste packed in drums and blocs is stacked upon it to a height of 16-18 feet above grade. When the entire bunker is completed, the interstices are packed with sand and gravel, and a "temporary" impervious cover of packed clay with surface drains is built up in 50 centimeter (20 inch) lifts. The design of the final permanent cover has not yet been decided upon. Each such tumulus or mounded bunker will accommodate six to eight months' waste received at the site (Lavie & Barthoux, 1982).

In France, surface drainage (runoff from the tumulus cover) is discharged directly into a nearby stream; seepage (if any) from the internal drainage system is discharged into the liquid waste storage tanks of the COGEMA (Compagnie Generale des Matieres Nucleaires) fuel reprocessing plant nearby.

The site operator, ANDRA (Agence Nationale pour la Gestion des Dechets Radioactifs) reports no problems with the operation of the disposal facility, but is seeking a design for the permanent tumulus cover which will reduce permeability to one one-hundredth that of the temporary covers now in use. This implies that significant quantities of water are infiltrating the tumuli, but the sampling regimen is not adequate to determine whether leaching of radioactive materials is occurring (Bennett & McAneny, 1985).

To the extent that reliable information is available, the Russian experience with earth-mounded bunker disposal systems is similar to that described in France.

The earth-mounded bunker offers a comprise between the economic benefits and convenience of shallow land burial and the security of a monolithic, grouted, mined repository at shallow depth. Its disadvantages are much the same as those of shallow land burial - perceived problems of monitoring seepage and even greater difficulty correcting contamination if it occurs. The EMB affords good security against inadvertant intrusion and provides extra shielding for intermediate and higher level waste. It does not permit alternative land uses at the close of the institutional control period, and it makes retrieval of most of the waste virtually impossible.

MINED GEOLOGICAL REPOSITORIES

Mined geological repositories are used for disposal of low level radioactive waste in Sweden and Germany. Similar repositories are under consideration for high level waste in the United States, and their characteristics have been studied experimentally in the Climax Mine at the Nevada Test Site. The Forsmark repository for low-, intermediate- and high-level radioactive wastes near Osthammer, Sweden, is carved out of granite 50 meters beneath the floor of the Gulf of Bothnia. This is probably the most costly low-level waste disposal facility in the world (Anon., 1982). By contrast, German engineers have utilized the existing workings of former iron (Konrad Mine) and salt mines (Asse, Wintershall Mines), and thereby reduced disposal costs significantly (Anon. 1980; Salander, *et al*, 1980).

In both Swedish and German repositories, low- and medium-level wastes are packed in 200 liter steel drums or (in Sweden) concrete boxes of approximately 1.8 cubic meter capacity. At Forsmark, these waste packages are stacked in galleries and periodically backfilled with a sand-bentonite mixture and walled off. In galleries used for low-level waste, conventional warehouse equipment is used for waste handling. Intermediate- and high-level waste are handled by remote-controlled machinery. Large silos are reserved for disposal of high-level waste, which is packed into 2.5 meter square cells and grouted in place.

In the German mines, drums are stacked in the modified mine tunnels, backfilled with crushed iron ore (Konrad) or salt. Then sealed with shotcrete. Both Swedish and German plans are highly conservative, and both presume that once sealed, the waste will be essentially irrecoverable.

Where suitable existing underground mines are available for the purpose, mined geological repositories may offer the most secure and economically viable means of disposing of low-level radioactive waste. Unfortunately, in the United States, few such mines exist, and little effort has been devoted to the development of specialized waste disposal facilities of this type.

The cost of mined geological repositories ranges widely depending upon regional geology, availability of suitable mined-out caverns or galleries, their depth (hence the hoisting costs), the regional water table and other factors. Disposal in mined repositories combined with shotcrete or grouted rubble barriers probably affords maximum security against environmental contamination or inadvertant intrusion. The primary disadvantage is the cost, but the long start-up time, finite site capacity, operational complexity and scarcity of suitable sites all detract from mined repositories. Finally, once waste is emplaced and sealed in such a repository, corrective action or waste retrieval becomes virtually impossible.

Whereas the unique geological characteristics of Sweden make specially-constructed mined repositories the only practicable solution to the waste disposal problem this is not true of the United States. In this country, most people agree, equally secure alternatives are likely to be far more cost effective.

ABOVE-GROUND STORAGE

Although frequently referred to as a "disposal" alternative, above-ground repositories are, in fact, only temporary storage facilities. No large-scale above-ground facilities exist in the United States, but two forms of above-ground storage are in use in Canada at the Bruce Nuclear Power Development operated by Ontario Hydro near Kincardin, Ontario.

Ontario Hydro constructed a series of large, above ground structures called "Quadricells" primarily to store cylindrical shipping containers filled with reactor exchange resins. The structures are built of reinforced concrete, approximately cubical, six meters (20 feet) square and approximately six meters high. They are divided internally into four chambers (hence "Quadricells"). Each chamber has a separate reinforced concrete cover slab, and can be loaded with waste packaged in steel, concrete or high-integrity plastic containers of standard design. With the cover in place, each internal cell will accommodate approximately 3.5 tons of dewatered exchange resin with adequate shielding for near-permanent storage. The Quadricell structure requires regular maintenance, and the concrete shows signs of spauling after several years' exposure to Ontario winter weather. The limited capacity of the Quadricells, the large capital investment and high maintenance cost which they represent led Ontario Hydro to experiment with steel- and concrete-lined boreholes ("tile holes") of comparable capacity. These simple borings proved to be cheaper to construct, and easier to maintain than the Quadricells, and the tile holes provide more effective shielding at substantially lower cost. They have largely replaced the Quadricells for storage of medium-level waste at Bruce (Carter & Rao, 1984).

Ontario Hydro is also experimenting with large above-ground storage buildings (LLSB) for low-level waste. These structures are built of pre-cast concrete panels, 18 inches thick to provide shielding, erected on a reinforced concrete slab. Roofs are fabricated of somewhat thinner (6.4 inch) concrete slabs supported by reinforced concrete columns. A sub-grade drainage system with sumps and monitoring wells facilitates detection of leakage and may allow interception of contaminants. The low level storage buildings have a design capacity of nearly 250,000 cubic feet (each). The structures were designed specifically to provide relatively low cost storage for large volumes of waste with radiation fields of 1 Rem/hr or less. Intermediate level waste could be transferred into the LLSB from tile holes, Quadricells or vaults after they had decayed to this level, thus freeing the more costly facilities for incoming intermediate level wastes (Carter & Rao, 1984).

Although initial costs for the LLSB are low, and construction time is short, the structures do require continuing maintenance in the Ontario climate. The structures do not appear to afford special advantages in terms of leak detection or contaminant control relative to in-ground disposal facilities, and with the first unit in service only since 1985, it is too early to evaluate their long-term effectiveness. Neither is it possible to predict the problems they may present in the event of a severe storm (hurricane or tornado). However, one unanticipated problem has emerged: the radiation level inside the LLSB has risen unexpectedly rapidly, due perhaps to accumulating gases (HTO, TCH_3, or ^{14}C-dioxide), to spillage or to a combination of these. Whatever the cause, radiation exposure of wastehandlers working inside or near the building is high enough to be of concern, although the building is filled to less than one-half its design capacity. Work schedules must be curtailed and operational procedures must be revised if the full capacity of the LLSB is to be realized.

Above ground storage offers the advantages of simple engineering, low initial cost, short start-up time and easily expanded waste capacity. The disadvantages include high continuing maintenance costs, uncertain security in the event of severe storms, little protection against deliberate intrusion and possible radiation hazard to workmen in or near the storage structures. The structures may be unsightly, and they do not allow alternative land use at the close of the institutional control period. While it is widely believed that above ground storage affords better opportunity for leak detection and simpler and less costly corrective action in the event of a leak, there is no hard evidence to support this belief available at the present time. It is of interest, however, that as host state for the Appalachian Low Level Radioactive Waste Compact (serving Pennsylvania, Delaware, Maryland and West Virginia), Pennsylvania has enacted legislation which mandates above-ground "disposal" as well as a 300- to 500-year institutional control period.

SUMMARY

Experience has demonstrated that each of the four basic management options can provide effective isolation and confinement of radioactive wastes. Each option offers technical or economic advantages, and each appears to have drawbacks. All four options admit of engineered modifications to the facility or the waste package which may improve performance or raise public confidence. Such engineered "enhancements" entail increased costs, and may jeopardize the performance of the facility by destroying the integrity of the site. The choice among the four basic options should reflect the careful evaluation of the economic and environmental costs relative to the potential benefits of each. In the last analysis, the choice among them will probably rest with public perception of the environmental hazard each presents.

SITE SELECTION

Whatever waste management option is used, and however that disposal option is modified by enhanced packaging, engineered barriers, or improved operational practices, site selection will play a vital role in radioactive waste management as it does in the management of any hazardous waste. Because any man-made container, any engineered structure or barrier, any detection or control system *may* fail, the secure containment of the waste will depend ultimately on the geological characteristics of the site itself (Schmalz, 1983). The requisite site characteristics will be determined by the properties of the waste materials, the nature of the hazard they represent, and the manner in which they may be carried from the disposal facility to enter the biosphere. Site selection and evaluation procedures must be based on the premise that containment of the waste throughout its hazardous lifetime will be ensured by the geologic properties of the site and that man-made containers and engineered barriers will serve only as enhancements providing additional security.

Low level radioactive wastes present hazards of just two basic types: (i) direct radiation exposure, and (ii) chemical toxicity. Because chemically toxic materials make up a small part of low level radioactive wastes, and because chemical toxins must be controlled in essentially the same fashion as the soluble radionuclides in the waste, it is appropriate to concentrate on the problems of site selection as they relate to the radioactive contaminants.

Direct exposure to alpha, beta or gamma radiation is known to cause tissue damage, and may be carcinogenic, teratogenic or mutagenic (RERF, 1987; Sabo, 1985). Usually relatively high doses are necessary to cause the more serious effects, but ingestion of low level radioactive materials, even alpha-emitters, can lead to dangerous radiation doses to specific organs. Protection from direct radiation exposure is afforded by physical isolation, shielding and reducing exposure time. Shielding and controlled exposure time are the best means of reducing radioactive exposure for waste handlers; physical isolation and shielding afford the best protection for the general public. None of these measures will succeed, however, if radio-nuclides are not immobilized and confined to the controlled environment of the disposal site. Unfortunately, as noted earlier, there are several paths by which radioactive isotopes may escape from a disposal site and enter the biosphere.

Gases, particularly those of low molecular weight generated by chemical reactions or organic decay at the site, are extremely difficult to contain. Tritiated hydrogen gas, methane, or water vapor and ^{14}C-enriched methane or carbon dioxide have been detected escaping from low level disposal facilities in the United States and elsewhere. Such gases probably represent the principal form of radionuclide releases, in terms of both volume and curies, from disposal facilities (Matuszek & Robinson, 1982). Fortunately, the volume of escaping radioactive gas is small relative to the circulating atmosphere above, and natural dilution and dispersal of the gases in the atmosphere provides ample protection for the biosphere.

The second route by which radionuclides may escape from a disposal site and enter the biosphere is in solution in surface- or groundwater. To protect against solution transport, the disposal facility must be designed to preclude contact between the waste and natural waters. In the event that design proves less than completely effective, the time of contact between the water and the waste should be minimized. If successful, these two steps will prevent significant dissolution of hazardous constituents in the waste. Should contaminated leachate be generated despite all efforts to prevent it, movement of the leachate toward the biosphere must be blocked, diverted or delayed sufficiently to allow any contained radionuclides to decay to a level where they no longer represent a hazard. These goals can be achieved only if the facility is sited above the groundwater table in the region where soil (or regolith) is extremely impermeable, where fractures, caverns or solution passages are absent, and where the mineralogy and chemistry of the soil and rock will absorb or react with escaping radionuclides to "fix" them in the ground, preferably within the boundary of the disposal site. Performance objectives for disposal sites are detailed in Subpart C of 10 CFR 61; these regulations provide the basis for all site selection criteria. Geologic factors in site selections and evaluation are far too complex to discuss fully here; the site selection process has been treated in summary form by the Nuclear Regulatory Commission (NRC WM 408-4, 1987) and in greater detail by Jarrett (1983) and Schmalz (1983, 1984) among others.

CONCLUSION

Low level radioactive wastes can be managed effectively and at reasonable cost. Each of the four basic management options, shallow land burial, above ground storage, earth-mounded bunkers and mined geological repositories has been used successfully in the United States or elsewhere in the world. Each of these options offers economic or technical advantages; each has drawbacks. The choice of management options and or enhancements to increase the security of the site will depend upon the local conditions , applicable regulations, and a thorough cost-benefit analysis which includes consideration of the socio-political aspects of site development and operation. Finally, inasmuch as effective containment of the waste will depend ultimately upon the geological characteristics of the site, the principle criterion for site selection and facility design must be the geology of the specific site. Site selection and pre-development evaluation will probably play the most critical role in determining the technical success or failure of any hazardous waste disposal facility, just as socio-political factors will probably play the deciding role in its acceptability to the public.

REFERENCES

Anon. (1980). *R&D Programme for a Feasibility Study of the Konrad Mine for the Disposal of Radioactive Waste, Summary Status Report.* (GSF-T114. Gesellschaft fur Strahlen und Umweltforschung mbH, Munchen. (141 pp.)

Anon. (1982-a,-b). *Radioactive Waste Management Plan 82, Parts 1 and 2,* (TR82-09:1, TR82-9:2)), Svensk Karnbransleforsorjning AB, Stockholm.

Bennett, R.D., W.O. Miller, J.B. Warriner, P.G. Malone, and C.C. McAneny, (1984). Alternative Methods for Disposal of Low-Level Radioactive Wastes.

Task 1: Description of Methods and Assessment of Criteria. U.S. Nuclear Regulatory Commission, Washington, D.C.(NUREG/CR-3774) volume 1, 76 pp.

Bennett, R.D. and C.C. McAneny (1985). Inspection of Low- and Intermediate-Level Radioactive Waste Disposal Facilities and Research and Development Projects in France, Sweden and West Germany, 30 April-16 May, 1984. Geotechnical Laboratory, USAE Waterways Experiment Station, Vicksburg, MS, July, 1985. (143 pp. + 4 appendices.)

Bord, R.J. (1985). Problems in Siting Low Level Radioactive Wastes: A Focus on Public Participation, Chapter 14, in: Majumdar and Miller, *Management of Radioactive Materials and Wastes: Issues and Progress.* The Pennsylvania Academy of Science, PA. 1985, pp. 189-202.

Carter, T.J. and P.K.M. Rao (1984). *Fifteen Years of Radioactive Waste Management at Ontario Hydro.* Ontario Hydro, Toronto, Ontario, M5G 1X6. (Preprint).

Cartwright, K. (1982). A geological case history: Lessons learned at Sheffield, Illinois. in: Yalcintas, 1982, volume 2, pp. 67-82.

Clancy, J.J., D.F. Gray and O.I. Oztunali (1981). Data Base for Radioactive Waste Management: Volume l, Review of Low-Level Radioactive Waste Disposal History, U.S. Nuclear Regulatory Commission, Washington, D.C. (NUREG/CR-1759, vol. 1).

Dornsife, W.P. (1985). Classification of Radioactive Materials and Wastes. Chapter 1 in: Majumdar and Miller, *Management of Radioactive Materials and Wastes: Issues and Progress.* The Pennsylvania Academy of Science, PA, 1985, pp. 2-9.

Friedman, R.S. (1985). Political Considerations of Nuclear Waste Policy. Chapter 15 in: Majumdar and Miller, *Management of Radioactive Materials and Wastes: Issues and Progress.* The Pennsylvania Academy of Science, PA, 1985, pp. 203-215.

Grant, J.L. (1982). Geotechnical measurements at the Maxey Flats, Kentucky, low-level radioactive waste disposal site—lessons learned. in: Yalcintas, 1982, volume 2, pp. 151-166.

Jarrett, A.R. (1983). Site Selection, Chapter 5 in: Witzig, *et al,* 1983., volume 3. 29 pp.

Kelleher, W.J. (1979). Water problems at the West Valley burial site. in: M.W. Carter (Ed.), *Management of Low Level Radioactive Waste,* Pergamon Press, NY., 1979, pp. 843-850.

Kelleher, M, J. Lyons, K. Tong, W. Greenman & T. Waldman (1983). "Report: Investigation of a Low Level Waste Burial Site at the Western New York Nuclear Services Center." Power Authority of the State Of New York, Albany, New York.

Kohl, Dagmar (1985). *Hydrogeology and Radionuclide Adsorption as Factors in Low Level Radioactive Waste Disposal Site Selection.* B.S. Honors Thesis,

Pennsylvania State University, University Park, PA., 63 pp.
Lavie, J.M. & A. Barthoux (1982). The Management of Low and Medium Level Radioactive Waste in France, in: *Waste Management '82,* Proceedings of the Symposium on Waste Management, Tuscon, Arizona, March, 1982.
"Low-Level Waste Policy Act," P.L. 96-573, 94 Stat. 3347, December, 1980.
Lu, A.H. (1978). Modeling of Radionuclide Migration from a Low-Level Radioactive Waste Burial Site. Health Physics, vol. 34, pp. 39-44.
Majumdar, S.K. and E.W. Miller, Eds., (1984). *Hazardous and Toxic Wastes: Technology, Management and Health Effects.* Pennsylvania Academy of Science, Easton, PA, 1984. (405 pp. + appendix, index.)
Majumdar, S.K. and E.W. Miller, Eds., (1985). *Management of Radioactive Materials and Wastes: Issues and Progress.* Pennsylvania Academy of Science, Easton, PA, 1985. (355 pp. + 2 appendices, index.)
Matuszek, J.M. & L.W. Robinson (1983). Respiration of Gases from Near-Surface Radioactive Waste Burial Trenches. in: *Waste Management '83,* Proceedings of the Symposium on Waste Management, Tuscon, Arizona, February, 1983.
"Nuclear Waste Policy Act of 1982," P.L. 97-425, 96 Stat. 2201, December, 1982.
Pollog, T.E., (1984). Pennsylvania Low-Level Radioactive Waste Management Survey. Pennsylvania Department of Environmental Resources, Harrisburg, PA. (13 pp. + 2 appendices.)
Prudic, D.E. (1980). Permeability of Covers Over Low-Level Radioactive Waste Burial Trenches, West Valley, Cattaraugus County, New York. U.S. Geological Survey, Water Resources Investigations 80-55.
Prudic, D.E. (1986). Ground-Water Hydrology and Subsurface Migration of Radionuclides at a Commercial Radioactive-Waste Burial Site, West Valley, Cattaraugus County, New York. U.S. Geological Survey, Professional Paper 1325, 83 pp.
RERF (1987). *US-Japan Joint Reassessment of Atomic Bomb Radiation Dosimetry in Hiroshima and Nagasaki, Final Report.* Radiation Effects Research Foundation (RERF), Hiroshima, Japan, 1987. Volume 1, 430 pp.
Sabo, A.T. (1985). Radiation Protection Standards and Radiation Risks. Chapter 22 in: Majumdar and Miller, *Management of Radioactive Materials and Wastes: Issues and Progress.* The Pennsylvania Academy of Science, PA, 1985, pp 285-293.
Salander, C., R. Proske, and E. Albrecht. (1980). The Asse Salt Mine: The World's Only Test Facility for the Disposal of Radioactive Waste. Interdisciplinary Science Reviews, vol. 5, pp. 292-303.
Schmalz, R.F. (1983). Geology - Pennsylvania Specific. Chapter 6 in: Witzig *et al,* 1983, Volume 3, 49 pp.
Schmalz, R.F. (1984). Geological Factors in Disposal Site Selection. in: Majumdar and Miller, *Hazardous and Toxic Wastes: Technology, Management and Health Effects.* Pennsylvania Academy of Science, Easton, PA, 1984,

pp. 124-136.

Singh, K.N. (1987). *Pennsylvania and Maryland Low-Level Radioactive Waste Management Survey - 1986.* Commonwealth of Pennsylvania, Department of Environmental Resource, Harrisburg, PA. 1987. (17pp. + 3 appendices.)

Sierra Club, Radioactive Waste Campaign Fact Sheet (undated). *Insecure Landfills: The West Valley Experience.* Sierra Club Radioactive Waste Campaign, Buffalo, NY.

Sierra Club, Radioactive Waste Campaign Fact Sheet (undated). *West Valley: A Challenge for the 80's.* Sierra Club Radioactive Waste Campaign, Buffalo, NY.

Smith, R.L. (1984). Low Level Radioactive Waste Disposal at the Barnwell Waste Management Facility. in: *Low Level Waste Disposal Handbook,* Electric Power Research Institute (NP-2488-LD).

US NRC: 10 CFR 61, "Licensing Requirements for Land Disposal of Radioactive Waste." FEDERAL REGISTER, vol. 47, no. 248, December 27, 1982, pp. 57446-57482.

US NRC: WM 408-4 (1987). *Draft Regulatory Guide and Value/Impact Statement: Guidance for Selecting Sites for Near-Surface Disposal of Low-Level Radioactive Waste.* Nuclear Regulatory Commission, Office of Nuclear Regulatory Research, Washington, D.C. 17 pp.

Vince, M.A. (1985). Handling of Radioactive Materials and Waste in Hospitals. Chapter 12, in: Majumdar and Miller, *Management of Radioactive Materials and Wastes: Issues and Progress.* The Pennsylvania Academy of Science, PA, 1985, pp. 157-173.

Witzig, W.F., W.P. Dornsife and F.A. Clemente (Eds.). (1983) *Low Level Radioactive Waste Disposal Siting: A Social and Technical Plan for Pennsylvania.* Final Report for Subcontract No. C29-007909, EG&G, Idaho, Inc., Institute for Research on Land and Water Resources, The Pennsylvania State University, University Park, PA., 1983. (4 volumes, LW8303-1 through -4.)

Yalcintas, M.G., (Ed.), Proceedings of the Symposium of Low-Level Waste Disposal, Volume 2. Arlington, VA., June, 1982. Oak Ridge National Laboratory, Oak Ridge, TN. (NUREG/CP-0028 CONF-820674)

Westinghouse Corporation (1987). State of Illinois, Low-Level Radioactive Waste Disposal Site, License Application. Westinghouse Waste Technology Services Division, Waltz Mill, Madison, PA.

Management of Hazardous Materials and Wastes: Treatment, Minimization and Environmental Impacts. Edited by S. K. Majumdar, E. W. Miller and R. F. Schmalz.

Chapter Seventeen

PREREQUISITES FOR A MULTITECHNOLOGY WASTE MANAGEMENT SYSTEM

JERRY S. WALLS, AICP

Executive Director
Lycoming County Planning Commission
Williamsport, PA 17701

STRATEGIC PERSPECTIVE

History repeats itself! The problem is that when history repeats itself it becomes more expensive. Today we are witnessing a return to the waste incineration popularity which was widespread during the first half of this century. However, today purveyors of resource recovery systems (as they are now called) have embellished the "Plain Jane" incinerator with many technology options and protective devices because incinerators do have a potential for environmental and health impact. As a result, we must face a typical cost for even a small scale modern incineration system of $10 million and it is not uncommon for communities to pursue $150 to $200 million resource recovery facilities using incinerators as the base component.

The Lycoming County solid waste management philosophy is not anti-incineration! In fact, the long-range Solid Waste Management Plan of Lycoming County does incorporate selected waste incineration as one component. The current rush to develop large-scale resource recovery systems built primarily around incinerators and the respondent uprising by those who question the environmental soundness of such an approach serves to illustrate the importance and distinction of the conceptual underpinnings of the Lycoming County solid waste management philosophy. Strategic management requires as prerequisites long-range comprehensive planning and total technology integration. In America, given our fierce dedication to private enterprise, without the mechanism to assign comprehensive responsibility to private enterprise for all component waste management functions, we have historically limited ourselves to a single-technology narrow solid waste system.

The consequences of single technology management systems are many and costly. The single technology "low tech" landfill approach is the least expensive in capital cost but could well become the most costly in terms of the long-range environmental cleanup and risk of indirect health costs. Likewise, the limited technology approach can be extremely "costly" in terms of economic opportunities lost. Such "opportunities" could be provision of inexpensive waste management for existing industries in a community in order to retain jobs and keep those industries healthy, or to use waste management services to attract industry to the community. On the other hand, to undertake a high technology high cost resource recovery system, if not fully warranted, can impose on the community a restriction of choice because of the long-term locked-in commitments to that debt amortization which can prevent the community from shifting to less expensive waste technologies as they become available. In a technological environment which is so highly dynamic as is the waste management field, this locked in restriction cannot be fully estimated as to the true magnitude of costs to the community.

FUNCTIONAL PREREQUISITES

At a strategic level, from cost considerations alone, it is evident that a multi-technology management approach may have considerable advantages. To achieve a multi-technology system generally requires an organizational approach which is capable of encompassing the full range of jurisdictional, legal, technological, financial, operational, and systems planning elements. The magnitude of jurisdictional, legal, and organizational scope needed makes it nearly impossible to achieve a multi-technology totally integrated solid waste management system without substantial involvement by an appropriate level of government. Even with substantial governmental involvement, it will still be necessary to have meaningful private enterprise participation in various elements of the solid waste system.

In order to achieve a truly strategic level of solid waste management, it is also necessary to intervene at the earliest planning stage, which in this function entails the development of a policy commitment from both the public sector and private sector parties who determine how a specific waste is to be addressed, even before that waste is generated. Likewise, this policy commitment must follow completely through the chain of transactions all the way to the point of disposal and post-disposal perpetual management. Any approach short of this full range from conception to perpetuity program will most certainly lose potential opportunities for a more effective waste management system and potentially leave openings for major unmanaged liabilities.

Each segment of the total waste stream must be identified, fully understood, and the most appropriate technology selected for managing that waste stream segment. However, that is not enough. In order to approach a truly strategic concept of management, it is also necessary to overlay a hierarchy of waste management technology preferences in order to select the best approach in light of broad environmental goals. For example, while it may be very attractive to send the good quality office paper and newsprint to the resource recovery incinerator because of the energy value enhancement which those clean burnable materials will produce, the broader environmental goal of preserving diminishing natural resources dictates that we reuse the wood fiber content of those high quality papers in order to avoid depletion of that natural resource. Therefore, the higher order technology preference is to reuse a waste material rather than burn it, even for energy recovery. But there is a more fundamental preference than reuse. The most far-reaching solution to many of our waste management problems is the fundamental prevention of waste generation by addressing the throw-away practices of our society. Convenience in our daily lives has a high price. Most plastics do not biodegrade. The extra materials for one-person portion-packaging as opposed to bulk packaging creates a much higher volume of waste. Strategic waste management therefore demands strong initiatives at the federal and state levels to impose some reasonable limitations and incentives which can foster the minimization of waste at the point of commodity manufacturing.

Local and county governments do not have the jurisdictional reach to regulate manufacturing, therefore their waste management approach must pick up at the point of waste material generation. The Lycoming County concept for strategic solid waste management starts with the overarching jurisdictional program covering household (i.e. municipal) institutional, commercial and industrial waste types. Through a combination of contractual arrangements, pricing incentives, and operational practices, the Lycoming County Solid Waste System reaches into the factory, institutional complex, and major commercial installations to specify the source separation of those waste materials which are appropriate for special waste management.

THE LYCOMING COUNTY SOLID WASTE MANAGEMENT PLANNING APPROACH A CASE STUDY IN STRATEGIC SOLID WASTE MANAGEMENT

THE CONTEXT

To understand solid waste management planning in Pennsylvania, one must be aware of somewhat unique factors which dictate the planning approach. First, let me describe the circumstances existent in 1968-70 when Lycoming County prepared its first Solid Waste Management Plan. Eleven dumps were scattered about the County. The largest was operated by the City of Williamsport on leased land and received approximately 60% of all waste generated in Lycoming County. Two medium size dumps were operated by municipalities. The remainder were simply open dumps with infrequent attempts to cover the accumulated waste with soil or to "torch" the waste material in order to reduce the volume of trash. While most of the open unattended dumps were free with the cost of covering periodically borne by the municipality, or the Pennsylvania Bureau of Forestry, the prevailing rate for those which did charge was the equivalent of $.50 - $1.50 per ton.

The Constitution of Pennsylvania established a structure that gave individual townships, boroughs and cities sovereign jurisdiction over solid waste planning, collection, storage, transportation and disposal. Lycoming County had never been involved in the solid waste function. Lycoming County did not and still does not have a Public Works Department nor a County Engineer on staff.

Lycoming County is located in Northcentral Pennsylvania approximately two hours north of Harrisburg astraddle U.S. Route 15 and Interstate 180. The County contains 1,215 sq. miles (1 sq. mile larger than the State of Rhode Island) and 52 incorporated municipalities. While over 118,000 people now live in Lycoming County, 72% of that population resides in the Williamsport Urban Area which leaves vast segments of the county rural and undeveloped. It is a strongly conservative area with a very healthy John Birch Society.

THE PLANNING PROCESS

The process used by the Lycoming County Planning Commission to prepare the Solid Waste Management Plan has been somewhat skewed by the existence of 52 municipalities within the County which, under prior Pennsylvania Law, must consent to the transfer of functional planning responsibility to the County government. When the Pennsylvania Solid Waste Management Act was first passed in 1968, 51 of the 52 municipalities (through formal resolution or less formal written communication) requested the Lycoming County Planning Commission to undertake the solid waste planning required by law. The Solid Waste Plan was undertaken in 1969 and completed in 1970. That plan proposed a consolidated county-wide land disposal system with an unlined landfill site near Level Corners in the western portion of the County. When the Grange and farmers organizations reacted strongly to oppose that landfill site, the County Commissioners chose to pursue a site in the northern section of the County on Coal Mountain. This concept would have involved a transfer station with contract hauling by the strip mine operator and a contract landfill operation on the coal mine property owned privately and operated privately. The Consolidated Sportsmen and affiliated friendly groups rallied to pressure the Bureau of Mines of the Department of Environmental Resources into a veto of the landfill permit, even before hydrogeologic and engineering investigations could be completed. Then followed a desperate search for a landfill site in 1971 and early

1972 since the large landfill operated by the City of Williamsport was running out of space.

In early 1972, when it was learned that the United States Bureau of Prisons had engaged an engineering consultant to search for a replacement landfill site for their own purposes, negotiations were immediately instituted. Agreement was reached in the spring of 1972 that a Use Permit for the landfill would be issued by the United States Bureau of Prisons at no cost for use of the prison land if Lycoming County would design, construct, and operate a landfill facility which would provide free disposal of all wastes generated by the Allenwood Prison Camp and the Lewisburg Penitentiary for the life of the Landfill Use Permit.

During the 1971 and 1972 landfill site search, it became evident that alternative technologies should be pursued. The Planning Commission Staff and County Commissioners investigated 45 systems throughout the United States ranging from composting, to epoxy fixation of waste materials into building blocks and other products, to pyrolysis, to incineration to generate steam and electricity, to methane gas recovery, to materials stabilization and conversion technologies. When the site investigation on the U.S. Prison land concluded that a high technology landfill would be required, and concurrently the investigation of various resource recovery technologies yielded the finding that a much larger volume of waste materials was needed to achieve economic viability, the 1970 Plan had to be adjusted.

THE SECOND GENERATION SOLID WASTE PLAN

The 1970 Plan was revised in April 1973 to incorporate compost stabilization , materials separation through shredding, magnetic separation, screening, and air separation for metals recycling, and pyrolysis for energy recovery with an elaborate processing system and high intensity magnetic separation to produce construction materials. White goods and packaged newsprint were to be pulled out at the transfer station to accomplish 9% volume reduction from the total waste stream. It was expected that 3% could be composted. Various metal separation processes were expected to yield 6% with 66% going into the pyrolysis incinerator and 10% diverted into the production of construction materials. The goal was 10.5% residuals to be landfilled.

The threshold of economic viability for a high technology landfill with liner, leachate collection, treatment, and leachate recirculation based upon 1973-74 prices was 450 tons per day. This represented a larger waste volume than was generated in Lycoming County and immediately shocked the County into a rethinking process for about 4 months. During this time, industries were starting to feel the problem acutely as advance indications of tighter regulations on the disposal of their waste raised considerable concern. Other counties were commencing to run out of space in their landfills as well. Consequently, a multi-county regional system was born in 1973-74.

PLAN IMPLEMENTATION-SYSTEM DEVELOPMENT

The preliminary design of the landfill site was undertaken during 1973-74 and a DER permit was issued in the fall of 1974. Concurrently, the design for a higher technology landfill aroused interest within U.S. EPA and a demonstration project with 1.3 million dollars grant funding was approved on the condition that the system be designed to accommodate industrial waste materials.

Immediately after issuance of the DER landfill permit, the White Deer Valley Citizens Committee transformed itself into the Organizations United for Ecology (OUE) and appealed the DER permit to the Pennsylvania Environmental Hearing Board. That permit challenge took about 7 months and resulted in a landmark ruling by the Hearing Board. This was the first case to affirm the validity of a lined landfill despite the fact that there were no design standards in the DER regulations for such facilities. The Hearing Board sustained the issuance for the DER Permit whereupon the OUE immediately appealed to the Commonwealth Court. In early 1976, the Commonwealth Court sustained the Hearing Board ruling in favor of Lycoming County and again the OUE appealed to the Pennsylvania Supreme Court. In mid-1976, the Supreme Court affirmed the lower court ruling in favor of the DER permit issuance.

With the legal challenges apparently out of the way, the Solid Waste Plan, as revised in 1973 to show the Allenwood Landfill site and various resource recovery options, was taken before the municipalities of the County. Despite the strong rulings by the Environmental Hearing Board, the Commonwealth Court and the Pennsylvania Supreme Court, the OUE continued to insist that this unproven landfill liner would not work. They successfully created such a visible storm of protest that municipal approval of the Solid Waste Management Plan became very difficult to secure. Some compromises and endless night meetings with each municipality became necessary to convince local officials that the County Solid Waste System was in their best interest. Nonetheless, a clear majority, some 30+ municipalities approved the Solid Waste Plan representing almost 70% of the waste generated within Lycoming County.

The major obstacles to municipal approval of the Plan were:

- Designation of only one facility (or system) owned by the County when state law allowed a municipality to designate any state permitted facility.
- Contracted commitment of all wastes generated within their municipality to the County System for an indefinite time period.
- Real obligations to enact a waste flow control ordinance AND enforce it.

Final design then commenced in 1976 and equipment procurement was undertaken. Construction commenced in April 1977 and was well underway when

OUE challenged the Use Permit issued by the Bureau of Prisons in Federal Court in July of 1977. By this time, the solid waste problems had grown accute and the County was at risk to suspend the construction contracts, so the County decided to continue with construction while the Federal Court case took its course. The Federal Court case concluded in March 1978 with a clear and strong ruling in favor of the Bureau of Prisons and the Lycoming County Solid Waste System.

While the Federal Court Case was in process, the site location search and negotiations were conducted by the Planning Commission for a transfer station in Williamsport. It should be noted that the decision to include a transfer station came as a result of pressure exerted on City Officials by the Refuse Haulers Association. Because approval of the plan by the municipalities was required, the City was able to exercise considerable leverage with the County Commissioners since the volumes controlled by the City were essential to economic viability of the system.

Landfill construction was completed and the facility opened in June of 1978 at $4.90 per ton. Construction was underway on the transfer station, but due to groundwater problems the completion was delayed until November 1978.

The landfill volume threshold for economical operation had been estimated at 450 tons per day. While the lack of a completed transfer station diminished landfill volumes to a minor degree the controversy organized by OUE directly discouraged needed waste volumes from other counties. Equally important factors included the policy reversal by DER to approve the landfill permit for a pre-existing dump which was located two miles from the Lycoming County Landfill gate. Also, the extremely cumbersome legal process and lack of legal staff for closure of old illegal dumps and unpermitted landfills allowed many such facilities to continue to receive waste volumes needed by the County. The extended court battle had also forced two other counties (which were to supply waste volumes) to proceed on their own to construct small unlined landfills. All these factors contributed to a devastating loss of needed volume so that initial tonnage received at the Lycoming County Landfill was barely half of the minimum threshold needed.

Early losses due to lack of volume were compounded in 1979 when DER instituted a new series of controls on industrial residual waste materials. The process required to secure approval for disposal (through the completion of a Module I disclosure form and the laboratory analyses) generally required at least six months for approval and elevated disposal costs for these waste materials dramatically. This further complicated the Lycoming County Solid Waste System cash flow problems, at least in the short run. To make matters worse a number of new regulatory requirements, inflation and economic forces in general escalated the minimum volume threshold of economic operation to 750 tons per day. The shock to the Lycoming County Commissioners of growing losses which reached the level of $750,000 plus cash flow loans from the County General

Fund totalling over $1.5 million, plus revenue sharing loans of $600,000 for the transfer station, and revenue sharing loans for further equipment purchases of over $250,000 was overwhelming.

The County Commissioners reacted by bringing in "stronger management" and inserted above the Solid Waste Department Facilities Manager a Director position to which they assigned the Director of the Redevelopment Authority.

When the new management attempted to increase rates $3 per ton in 1981, the haulers rebelled with a boycott which cut volumes in half. By the end of 1981, it was evident that major changes were necessary.

PLAN EVOLUTION

The Board of County Commissioners came back to the County Planning Commission and reminded all involved that "the planners got them into this mess, so it was now up to the planners to figure out a solution." Consequently, the Executive Director of the County Planning Commission was given the special assignment to manage the solid waste system for as long as necessary in order to solve the problems. During the next 13 months, the management structure was evaluated with management overhead cut in half, maintenance procedures were revamped and maintenance cost reduced 50%, and the Environmental Planner was assigned to negotiate with a wide variety of customers to secure commitments of waste, solve the industrial waste approval bottleneck, and to help negotiate an end to the hauler boycott. The customer base was not only recaptured but tripled.

A team of environmental compliance specialists and engineers was mobilized from member industries of the West Branch Manufacturers Association who served with no cost to the County as loaned executives and loaned technical specialists to advise on the management of the system. In addition, a paid engineering and hydrogeologic team was mobilized to reexamine literally every aspect of the facility design and operational concept. Concurrently, the total Solid Waste Management Plan was reevaluated as to the technology configuration, the waste flow assurances needed, the commitments of waste already agreed to but not delivered, as well as the future system direction. This led to a major Plan Update with DER funding during 1983-84. In fact, DER funding to mobilize the technical team to plan and design the System redirection was essential at a time when cash flow did not cover operations let alone extra engineering.

SYSTEM ADJUSTMENT

As a result of the redirection of the operations, marketing effort, and cost containment measures instituted during 1982, the losses were reduced to approximately $12,000.00. Those changes started to really pay off in 1983 and a modest "profit" of approximately $36,000 was achieved at year end. The 1984 "profit" was approximately $112,000.00 and at the end of 1985 over $380,000.00 "profit" had been realized. These "profits" are dedicated revenues restricted to the Solid Waste System enterprise fund and the repayment of subsidies and loans from the County General Fund. Likewise, all resource recovery revenues are restricted to the Solid Waste System enterprise fund.

RURAL WASTE COLLECTION & DUMP CLOSURE

Another important factor affecting volume is the degree of control on waste disposal, especially the number of alternative opportunities for uncontrolled disposal. The Lycoming County Planning Commission was shocked in 1979 when a staff survey of rural areas revealed almost 250 illegal roadside dumps or other open unattended disposal sites within Lycoming County. It was impossible to estimate the waste volume and revenues lost to these "free" options. Obviously, DER did not have the budget, staff, and legal counsel to effectively control and prosecute all these infractions.

Through planning funds from DER, a Rural Solid Waste Collection System and Open Dump Closure Program was designed and implemented. Appalachian Regional Commission funding was secured to pay 50-80% of the cost of open dump cleanup and closure and to purchase bulk containers for the rural collection system. All eligible rural townships within the County participated or took care of their own and 231 dumps were closed and cleaned up over 3 years. These cleanup efforts were very valuable for the tourism and economic development programs.

SPECIAL WASTES MANAGEMENT

Industrial residuals deserve special mention. The PA DER policy of strong waste-specific AND generator-specific controls is certainly an environmentally responsible approach. PA DER requires each waste generator to complete a Module I application for each waste from each company-owned facility listing the disposal facility for that waste. The range of waste quantities must be estimated on a daily, weekly, monthly and annual basis. All waste constituents must be listed and concentrations verified by laboratory testing of composite samples. Lycoming County goes further and requires disclosure of manufacturing process details sufficient to understand biochemical reactivity and byproducts. The Lycoming County Solid Waste Department retains a specialized engineering consultant to conduct a detailed review of liner and overall landfill compatibility of that waste. A computerized cumulative waste constituent loading analysis is performed to determine if a proposed residual would overload the waste mix and cause leachate treatability problems. Visits to the generator's factory and random grab samples are used to assure quality control and disposal contract compliance. All these procedures are invoiced to the applicant without guarantee of Module I approval by the County or DER.

Such advance testing and approval is not enough. Training and constant updated instructions to landfill personnel are crucial, along with a sincere policy commitment to enforce these controls. Lycoming uses a large chart for all approved residuals describing quantity approved, color, texture, odor, special handling precautions, etc. which is posted in the weighmaster's office and reviewed with all personnel involved at the landfill working face. Any Lycoming County equipment operator who knowingly buries a drum of waste without verifying that it was approved is subject to suspension or dismissal. To really be effective, there is a definite need for standardized Landfill and Transfer Station operator training and certification by a State regulatory agency.

The Module I process has worked. Although it initially caused loss of waste volumes and revenues it has recently worked in reverse fashion. As the DER industrial residuals policy and process became an established requirement generators began to realize that special handling with advance testing, disclosure of manufacturing process, and quality controls are actually measures which protect everyone in the long run. Since only lined landfills with tight controls on wastes accepted are approved by DER this has created a market advantage for such landfills. Industrial wastes are priced by the Lycoming County Solid Waste Department General Manager on a waste-specific case by case determination based upon:

1. Moisture content/impact upon leachate quantity
2. Waste constituents/impact upon leachate strength and treatability.
3. Long fiber dust/impact upon equipment cooling systems.
4. Wire or textile ribbon/impact upon equipment final drive seals.
5. Special handling precautions required to protect employees and users.
6. Special handling and scheduling cost impact.
7. Other operating or liability cost resulting from that waste.

Industrial residuals are priced from $10 to $50 per ton and account for 20-30% of total Lycoming County Solid Waste System revenues. Considering that residuals constitute only 15-20% of System volumes these residuals obviously represent a financial mainstay to the System and reveal a specialized role for modern high tech landfills.

Waste tires represent the most universally acclaimed pain to landfill operators of any type of waste. However, they also represent an economic opportunity of considerable magnitude. Consistent with the Lycoming County solid waste philosophy of resource management the approach to disposal of waste tires should be broadened to encompass the reprocessing of waste tires to recover the energy from them or a shredding and reuse of the tires as road material or for other applications. While the official solid waste plan does not define a specific solution to waste tires, the investigation of reuse in some other application and the recovery of energy continues as a part of the ongoing solid waste system planning being conducted by the Lycoming County Planning Commission.

Infectious, pathological, and other types of medical wastes represent an intense problem nation wide. They are equally difficult within the solid waste disposal facility. Employees are extremely cautious and fearful of working with these waste types, especially with the high profile recently emerging on the aids disease. Conventional incineration and autoclave technology is well proven in its ability to safely treat infectious pathological and most all types of waste. However, the difficulties inherent in handling these materials without proper quality control within the hospitals and without proper quality control in all of the sequence from generator to disposal facility have created extreme crisis-level problems for this waste category. Consequently, this problem is compounding the escalation of health care costs at a time when it is clearly in the public interest to do everything possible to contain the rise in health care costs. Therefore, solid waste systems which are governmentally owned have a special responsibility to address this category of waste problems. One proven management technique to help allay fears by the public and by the various persons involved through the chain of transactions in waste handling would be to incorporate a manifest system for control of medical waste from generation to ultimate treatment and disposal. This technique will add incremental additional costs to medical waste management but is one of the best methods to assure disposal privilege for what will otherwise become an extremely costly management problem. To take the opportunistic attitude, a solid waste system which can work with medical institutions in the development of proper controls and long term waste disposal solutions will not only serve the public interest but capture a market opportunity of tremendous financial proportions.

Construction and demolition debris represent that type of waste which is not per se a difficult waste to treat and or dispose of safely. However, the apprehensions of the public and isolated extreme cases of environmental hazard have combined to pressure many State regulatory agencies to establish vary stringent controls. Therefore, a somewhat arbitrary set of standards has been developed which will greatly escalate the cost of disposal for construction and demolition debris. For example, for most all categories of construction and demolition debris a lined landfill will be required. This reveals the real problem in most waste management systems, that is the lack of a proper control mechanism throughout the whole sequence of waste handling transactions. If construction and demolition debris had been properly controlled in the past the extreme cases which caused the stringent regulation would probably have not occurred.

RECYCLING AND REUSE

The Lycoming County Planning Commission and the Lycoming County Solid Waste Department have designed and developed a six-county regional recycling system. The program objectives are to save landfill space, conserve natural resources, save energy, fund local nonprofit service organizations, and promote economic development. The program is approached through meaningful involvement in the planning and execution of the program by private enterprise companies. Specifically, the beverage companies are involved in the recycling of aluminum containers. Traditional scrap dealers are involved in the recycling of ferrous metals and newspaper as well as other categories of paper.

The recycling system will embrace existing mandatory curbside collection in the town of Bloomsburg in Columbia County through the utilization of VRI (Volunteer Recycling Inc.). Other curbside systems have been identified for those communities mandated by Act 101 of 1988. In addition, any other municipality which chooses to mandate by ordinance curbside collection of recylcables may become involved in that part of the program. In addition, permanent drop off sites are being added to those already existing sites for daily volunteer drop off utilization, or for monthly drop off sites in the rural areas. Commercial generators, government agencies and institutions, and even some industry generators have been approached and are being solicited to institute source separation and recycling programs for specific categories of waste.

The principal value of the recycling program will be to instill a sense of citizen involvement and private company involvement in solving solid waste problems. This support by the average citizen is very valuable because they know then that other facilities are needed and other technologies must be employed beyond what can be accomplished with a sincere recycling program.

MARKET OPPORTUNITIES

The key to any user fee system is volume. The 1981 volumes of 200 to 300 tons per day have steadily grown through an improved image of the system, improved business practices, support from DER, and the vast array of changes taking place within the solid waste field in Pennsylvania and in the nation. Lycoming County landfill volumes now exceed an average of 1,300 tons per day. Approximately 20% of all wastes received are in the industrial residual and sludge category. Requests to accept waste from distant cities and industries are constantly received and rejected. Landfill long term life is judged more important by Lycoming County officials than short term financial gain.

STRATEGIC PLANNING THRUSTS

Opportunity capture and development represents the highest level of strategic planning and comprehensive solid waste system management which can be achieved among the many opportunities which arise. The most valuable is to be able to convert those opportunities for the solid waste system into a benefit for other dimensions of Lycoming County. The ability to convert these opportunities into benefits revolves around the organizational scope and capability of the solid waste management system. In a somewhat theoretical approach the Lycoming County Commissioners and Lycoming County Planning Commission decided in 1973 to create an operating department of the County government with the system planning role to be performed by the Planning Commission in a deliberate attempt to retain a mechanism or linkage for the identification and conversion of solid waste system opportunities into other community benefits.

Perhaps most exemplary of this opportunity capture is the fee strategy used in the last few years for the solid waste system. With the development of a price-elastic market condition throughout the nation for solid waste disposal the fee strategy for Lycoming County Solid Waste System has been revised. A three tier fee structure was designed to provide the most economical base rate fee to Lycoming County residents and businesses. The other 5 Counties within the Solid Waste System Service Area were provided the second tier fee. Beyond the service area all other municipal wastes and industrial wastes were surcharged in a deliberate third tier fee level in an attempt to secure a financial competitive advantage for our industries and businesses and in order to develop capital reserves for the solid waste system and other purposes. Basically stated, Lycoming County has no obligation, legal or moral, to accept the waste from distant cities and industries when the difficulty in siting and developing solid waste disposal facilities is so intense and expensive.

Consequently, whereas the solid waste system was initially designed to utilize depreciation for equipment and for field development in subsequent landfill phases, the creation of funded reserve accounts is now used. These reserve funds provide a cash asset which can be used to buy equipment in a more opportunistic manner, to undertake the improvements and developments necessary when new environmental regulations are promulgated, or when other opportunities arise.

TECHNOLOGY DIVERSIFICATION & INTERACTIVE SYNERGISM

As a result of many changes which have culminated in a sustained higher volume, the threshold of minimum volumes has been surpassed which has in turn opened up the possibility of pursuit of other solid waste technologies. The 1973 Lycoming County Solid Waste Management Plan called for the pursuit of various resource recovery technologies as volumes and economic considerations warranted. That concept was continued and refined in the 1984 Solid Waste Management Plan Update. Each waste technology is utilized to supplement the other technologies and enhance the total technology mix.

The Lycoming County Landfill has served the role of a base facility for the total solid waste system and has represented an essential vehicle to achieve regional solid waste volumes sufficient to allow pursuit of energy recovery and alternative waste management technologies. It is explicitly stated in the Lycoming County Solid Waste Management Plan that upon the implementation of the alternative waste management technologies, the role of the landfill will evolve toward a more narrow component of the total system. The landfill will continue to be the base facility for the regional system in that some materials will need to be delivered to the landfill which can not be separated or processed. Selected materials may require further processing but that will apply only to certain materials. For example, it is contemplated that plastics separation from the municipal waste stream may be accomplished at the landfill which may also become the site for a plastics reprocessing center, which will in actuality be a manufacturing operation. Exploratory discussions are underway now with feasibility studies contemplated for material recovery at the landfill.

The plan also proposes the continuation of the landfill role as the main facility for accommodation of hard-to-handle industrial residual wastes and other non-recurring waste receipts such as those which come from trucking accidents, fires, tornado damage, flood damage, etc.

The Lycoming County Solid Waste System will continue to embrace the county-owned transfer station located in Williamsport and other transfer stations (both municipally-owned and privately-owned) which are under contract as a network of feeder stations supplying waste volumes to the landfill and to other designated processing facilities.

METHANE RECOVERY - COGENERATION

In 1972-73 the concept of methane gas recovery for cogeneration of electricity or for other energy recovery purpose was identified by the Lycoming County Planning Commission and explored. Consequently the landfill design was modified to incorporate the ability to recover methane on a commercial scale. Methane quantities began to become commercially viable at the Lycoming County landfill in 1984-85. The methane recovery program has been structured into three phases:

Phase I - Centralized landfill gas collection and odor control
Phase II - Cogeneration of electricity
Phase III - Expanded cogeneration and/or alternative methane utilization

The methane Phase I program started in construction in the summer of 1988 with completion scheduled for December 1988. Phase II is scheduled to involve the construction during 1989 with sale of electricity to the Pennsylvania Power and Light Company under their Pioneer Rate Program. Additional feasibility studies are scheduled in 1988 and 1989 to capture and utilize the additional methane resources from the landfill. Enhancement of the methane generation is a deliberate strategy and has been pursued utilizing wet waste such as sludges.

LEACHATE RECIRCULATION

The Lycoming County landfill was designed on a wet system concept. The recirculation of leachate was deliberately intended to accelerate the decomposition of waste which therefore enhances the generation of methane. To the extent that regulatory safeguards can be incorporated it also makes sense to accept other wastes at less than 20% solids so long as they are able to pass the free liquids EPA paint filter test. Increased energy recovery from the landfill is not the only benefit which justifies the wet system concept. Of major significance is the reduction in long term perpetual care required after the majority of contaminants have been leached from the waste materials through the accelerated decomposition.

In addition, the Solid Waste Plan calls for leachate pretreatment at the landfill site prior to trucking of leachate to a municipally-owned sewage treatment plant (POTW). The pretreatment facilities as presently installed can be activated as needed to adjust pH or add flocculant materials. The Lycoming County Landfill is located approximately 2 miles from the Susquehanna River (West Branch). An NPDES Permit was secured as an alternative to allow direct discharge to the river in the event that the arrangements for treatment of leachate at POTW's in the area could not be continued.

Leachate trucking has worked well since 1982 for the Lycoming County Solid Waste System. With leachate recirculation, and during periods of extended dry weather, there is often no need for trucking of leachate to an off site POTW. Trucking of leachate is definitely preferable to the construction and operation of a dedicated leachate treatment plant on-site. From an economic standpoint alone, it is far less costly. From an environmental perspective, the trucking of leachate to a competently operated POTW assures a better level of leachate treatment than might be possible with a small scale leachate treatment plant on-site when the leachate constituents vary so frequently and substantially and when leachate volumes fluctuate dramatically.

Leachate recycle is not favorably regarded by most regulatory agencies throughout the United States. Much of this derives from the inability of most waste facilities to handle the additional leachate generated through a wet system design. In addition the enhanced generation of leachate and increased quantities of landfill gases pose a major odor problem. Nonetheless, if a facility is comprehensively designed to deliberately utilize a wet system and to control the odor problems and handle the additional quantities of leachate the concept makes a strong argument for the shortened period of environmental threat associated with the unmanaged liabilities in most waste facilities. In other words, accelerate the decomposition process and manage the total consequences and liabilities of the solid waste facility within the same time frame as the amortization provided for in the facility financing.

A major economic return from the leachate recirculation involves the accelerated settlement from the decomposition of waste. This creates the opportunity to recapture and reuse the vertical capacity of the landfill. With all the fixed capital investment generally in the at-grade or below-grade facility components, this represents a way to maximize return on the fixed capital investment. This acceleration can achieve, based on the Lycoming County Landfill experience, 40% to 50% acceleration in the decomposition rate for most waste materials. The benefits to this approach extend far beyond the individual landfill owners. To the extent the facilities have elongated life expectancy this reduces the crisis in which we have to find more disposal facility sites for good of society.

INDUSTRIAL DEVELOPMENT INFRA-STRUCTURE SUPPORT STRATEGY

Consistent with the approach of the Lycoming County solid waste system planning philosophy the opportunity to utilize the solid waste system resources to help solve other problems is evident in the following situation. From land use and growth forecast studies conducted by the Lycoming County Planning Commission in 1987-88 the need for sewage treatment in the Route 15 corridor has been documented. The possibility of new industry, a new federal prison, and the residential growth which is already occuring represent problems which can be addressed in part through a solid waste development initiative. Specifically, the Municipal Waste Regulations issued by DER in April 1988 require leachate treatment or the development of a treatment facility in a state of readiness which can implemented within a short time frame. This means that the Lycoming County Landfill, despite the success of its leachate trucking to POTW facilities, must be able to develop a leachate treatment plant within two years. Given this mandate, the Lycoming County Solid Waste System reserve funds which have been earmarked for leachate treatment can be utilized to help finance the development of a full range sewage treatment plant which can accommodate these other growth needs. In turn, this will help abate existing sewage malfunction problems and provide a stimulus to proper growth and development.

SELECTED WASTE INCINERATION - INDUSTRIAL COMPETITIVENESS

The Lycoming County Solid Waste Plan provides for selected waste incineration in conjunction with an industry or industries which need either steam or electricity and which are prepared to participate financially in developing and operating the incinerator with the County serving as the waste stream broker and overall waste stream manager. Active discussions have been pursued with several industries who have indicated strong interest in financial participation in this program. Waste volumes are just now approaching the stable volume necessary to start diverting segments of the volume into these other technologies. A direct benefit to industry participating in such a program is a competitive priced energy which enhances their competitiveness in national and world markets. In turn, this benefits the County through strong and stable employers.

A future opportunity related to selected waste incineration is the category of infectious, pathological, medical and veterinary wastes. As regulatory controls begin to tighten for this waste stream new opportunities are starting to emerge. Major medical institutions have been involved in the incineration of their waste for many years. However, the individual clinics and convenient care care centers, nursing homes, and doctor offices which generate substantial quantities of these wastes are not involved in a control program for these very difficult and potentially hazardous wastes. This suggests that the overall solid waste system planning must take into account these needs and develop opportunities to accommodate these waste streams in a manner which is environmentally sound. Obviously, this represents yet another future opportunity to utilize selected waste incineration to address a severe need in our society while generating economic and environmental benefits for the community as well.

COMPREHENSIVE WASTE RESOURCE MANAGEMENT

The Solid Waste Management Plan is founded upon the principle that the total waste stream can be more effectively managed by establishing a discreet management concept for each segment of the waste stream. This will entail the use of a mechanism by which to guarantee or assure that a particular segment of waste is directed to flow to the designated processing point or ultimate disposal/recovery facility. In order to achieve this level of waste flow control, contracts between the Lycoming County Commissioners and the generators of those Solid wastes (industries, institutions, businesses and municipalities) are used to accomplish separation of the materials at the point of generation so that they are handled on a discreet basis instead of being mixed in with the municipal waste stream.

This degree of waste flow management need not, and in the Lycoming County Plan does not, eliminate private haulers from waste collection or transfer haul. The Lycoming County Solid Waste Management Plan also utilizes waste hauler licensing on a county level to accomplish an additional degree of assurance in waste flow control. The hauler licensing is undertaken as a joint control function with cooperating municipalities. These townships, boroughs, and cities enact solid waste ordinances which specify that a hauler must obtain a County License in order to be allowed to collect waste within that municipality. The Lycoming County Solid Waste Department inspects the hauler's vehicles to insure that the waste collection components of the vehicle comply with the PA DER regulations and then issues an annual license.

CONCLUSIONS

The Lycoming County Solid Waste Management Plan approaches solid waste management as a resource management problem and opportunity. The System was deliberately focused on a land disposal technology as a mechanism for aggregating sufficient volumes of waste materials in order to achieve the quantity needed to pursue other technology approaches. The use of land disposal was also a deliberate strategy at the outset because of the need to achieve a level of acceptance by local municipalities and the public at large for a new County function using a cost-effective disposal technology. The planning approach used by Lycoming County must also be characterized as a selected waste management concept which applies the appropriate technology to a discrete segment of the waste stream in order to achieve a cost-effective system by which individual waste segments can be managed and to achieve an overall system which is likewise cost effective. Given the municipal-sovereignty governmental structure of Pennsylvania, the Lycoming County System is a living example of extreme effort to achieve cooperative solid waste management with literally hundreds of local officials involved in the Plan review and decision-making process. The importance of state planning funds and cooperative policy support to innovate and evolve a new high technology system cannot be emphasized enough.

Management of Hazardous Materials and Wastes: Treatment, Minimization and Environmental Impacts. Edited by S. K. Majumdar, E. W. Miller and R. F. Schmalz. © 1989, The Pennsylvania Academy of Science.

Chapter Eighteen

HAZARDOUS MATERIALS TRANSPORTATION

DEBORAH L. BOOTHE

United States Department of Transportation
Research and Special Programs Administration
Office of Hazardous Materials Transportation
400 7th Street, S.W.
Washington, D.C. 20590

OVERVIEW

Hazardous materials are an integral part of modern life. The United States economy requires the production of and distribution of increasing amounts of raw chemicals, basic energy sources, consumer goods, industrial by-products that are hazardous, i.e. flammable, corrosive, radioactive, etc. Even though the safety record of hazardous materials shipments is good, and the frequency of catastrophic accidents low, a single hazardous materials accident has the potential for loss of lives and millions of dollars in property damage.

Historically, state and local governments have had the primary responsibility for insuring the health and safety of their citizens. In the area of transportation, the federal government had been predominant with regard to interstate commerce. This was reinforced when Congress enacted the Hazardous Materials Transportation Act of 1974 (HMTA), which asserted the need "to protect the nation adequately against the risks to life and property which are inherent in the transportation of hazardous materials in commerce."[1]

In response to public concern, state and local officials have passed legislation which while intended to increase safety, sometimes has had a negative effect on commerce.

HISTORY

The Hazardous Materials Transportation Act (HMTA) of 1974 provides the primary legislative authority for the Department of Transportation (DOT) hazardous materials programs. A number of other related statutes guide the program as well. Among these are the Dangerous Cargo Act of 1940, Sections 601(c) and 902(a)(1) of the Federal Aviation Act of 1958, the Tank Vessel Act of 1936, the Ports and Tanker Safety Act of 1970, Section 311 of the Federal Water Pollution Control Act Amendments of 1972, and the Resource Conservation and Recovery Act of 1976. These statutes are specific to the various modes of transportation, and the administrations - United States Coast Guard (USCG), Federal Aviation Administration (FAA), Federal Highway Administration (FHWA), and Federal Railroad Administration (FRA) - retain jurisdictional responsibility for enforcement actions relating to transportation by water, air, highway, and rail, respectively.

The HMTA empowers the Secretary of Transportation to issue and enforce regulations necessary to ensure the safe domestic and international movement of hazardous materials in commerce. This authority has been delegated to the Administrator, Research and Special Programs Administration (RSPA).

In 1966, authority to regulate the transportation of hazardous materials was transferred from the Interstate Commerce Commission (ICC), the Department of Treasury, and the Civil Aeronautics Board (CAB) to a then new Federal Agency, The Department of Transportation (DOT). Within the DOT, separate modal administrations were retained to preserve organizational continuity. In order to carry out the changes authorized by Congress in the Hazardous Materials Transportation Act (HMTA), the Secretary of Transportation created the Materials Transportation Bureau (MTB), making it the lead agency in the department's hazardous materials transportation safety program.

In 1985, the Research and Special Programs Administration (RSPA) was reorganized. The Materials Transportation Bureau was abolished and its responsibilities were transferred to the Office of Pipeline Safety and the Office of Hazardous Materials Transportation within RSPA.[2]

The Research and Special Programs Administration (RSPA) is the primary element within the Department of Transportation responsible for developing a national regulatory program to protect against the risks to life and property inherent in the transportation of hazardous materials. Embodied in this regulatory program is the responsibility for the promulgation of all regulations of general applicability (entitled the Hazardous Materials Regulations (HMR) and codified at Subchapter C of Title 49, Code of Federal Regulations), and the issuance of related amendments and exemptions. Due to the multi-modal nature of hazardous materials transportation, RSPA relies extensively on the participation of the four modal administrations in the regulatory process.

The Hazardous Materials Regulations are a culmination of safety regulations implemented by several Federal agencies over a period of approximately eighty years. They address classification, packaging, handling, incident reporting and hazard communication requirements applicable to the transportation of hazardous materials. RSPA continually reviews and amends these regulations in order to address safety concerns, eliminate obsolete or unnecessary requirements, and convert the provisions of exemptions into regulations of general applicability to commerce today.

The rulemaking process, whether it leads to a new regulation or amends an existing requirement, evolves from one or more of four sources:

(1) Petitions for rulemaking received from the general public or other government agencies

(2) Rulemaking proposals from the four modal administrations;

(3) RSPA initiatives to improve existing regulations, convert exemptions into regulations of general applicability, or address unusual situations not presently covered by the rules; or

(4) The need to change existing regulations as a result of RSPA's enforcement experiences.

Whatever the impetus that drives the need for rulemaking, the process encourages the direct involvement of the public in formulating the regulations through the solicitation of comments at the time of issuance of Advance Notices of Proposed Rulemaking (ANPRM) and Notices of Proposed Rulemaking (NPRM). Publication of a final rule is the culmination of the rulemaking process and occurs only after analysis of all public comments, internal and external coordination, and examination of its potential economic and environmental impact.

INTERNATIONAL ACTIVITIES

The United States is a participant in the work of international standards setting organizations in an effort to promote a worldwide system of consistent modal and regional transportation safety requirements. The main U.S. objective is to ensure that shipments of hazardous materials move freely and safely by the various modes of transport throughout regions of the world. An active U.S. role in the development of international standards is essential to the economic interests of the domestic hazardous materials industry. Adoption of transportation safety standards by other nations and regional bodies throughout the world has a direct impact on U.S. shippers and carriers of exports, and inequitable or incompatible international requirements can have a profoundly negative impact on the chemical industry which has consistently earned a balance of trade surplus. Such impacts could affect the overall U.S. economy. The DOT participates in the work of the following international organizations:

(1) The Committee of Experts on the Transport of Dangerous Goods of the

United Nations Economic and Social Council (ECOSOC), including the Committee's two subsidiary bodies:
(a) The Group of Experts on Explosives, and
(b) The Group of Rapporteurs,

(2) The Subcommittee on the Carriage of Dangerous Goods of the International Maritime Organization (IMO);
(3) The Dangerous Goods Panel of the International Civil Aviation Organization (ICAO);
(4) The Group of Experts on the Transport of Dangerous Goods of the United Nations Economic Commission for Europe (ECE);
(5) The International Atomic Energy Agency (IAEA); and,
(6) The Highway Operations Committee of the Pan American Highway Congress.

The Committee of Experts on the Transport of Dangerous Goods of the United Nations Economic and Social Council (ECOSOC)

This Committee is the focal point of international activity regarding all transportation of packaged hazardous materials (except radioactive materials). It meets biennially to consider the work of its two subsidiary bodies, the Group of Experts on Explosives and the Group of Rapporteurs. RSPA is the agency charged by the Department of State to represent the U.S. on each of these committees.

International Maritime Organization (IMO)

The International Maritime Organization is a specialized agency of the United Nations concerned with the promotion of safety in shipping and the prevention of marine pollution from ships. IMO facilitates cooperation among governments on technical matters affecting marine safety and pollution prevention through the exchange of information under the auspices of its committees and the adoption of international agreements.

The U.S. Coast Guard participates in the work of IMO as a representative of the U.S. Department of State. The Coast Guard co-chairs the Department of State's Shipping Coordinating Committee (SHC), which coordinates all U.S. input to IMO. Technical matters concerning safety and prevention of pollution from hazardous materials are handled within four subcommittees:

Subcommittee on Bulk Chemicals (BCH)
Subcommittee on Carriage of Dangerous Goods (CDG)
Subcommittee on Containers and Cargoes (BC),and
Subcommittee on Fire Protection (FP).

International requirements developed by the technical subcommittees are then approved by the parent committees—The Maritime Safety Committee and the Marine Environment Protection Committee (MEPC). The Coast Guard provides the majority of technical expertise for U.S. delegations to IMO. In the case of the Carriage of Dangerous Goods Subcommittee, the Office of Hazardous Materials Transportation assists the Coast Guard in developing U.S. positions.

The CDG Subcommittee publishes and maintains the International Maritime Dangerous Goods (IMDG) Code. This code is recognized as the worldwide standard for the transportation of packaged hazardous materials by vessel. The U.S. regulations (49 CFR) governing the transportation of hazardous materials incorporate substantial portions of the IMDG Code by reference.

International Civil Aviation Organization (ICAO)

The Dangerous Goods Panel (DGP) of the International Civil Aviation Organization is responsible for periodic updating of Annex 18 to the Convention on International Civil Aviation. Annex 18 prescribes basic requirements for the safe transportation of dangerous goods by air and its supporting Technical Instructions provide the detailed requirements necessary to implement Annex 18. RSPA provides the panel member for the U.S.

International Atomic Energy Agency (IAEA)

International Atomic Energy Agency is an intergovernmental body chartered to foster the peaceful contribution of nuclear energy to mankind. A necessary element in the application of nuclear technology is the transportation of nuclear materials. The IAEA developed a regulatory system to help ensure the safe international transportation of such materials. RSPA represents the U.S. with respect to transportation and serves as the U.S. Competent Authority to both governmental and private parties involved in nuclear transportation, issues Certificates of Competent Authority relating to package and shipment approval required by the IAEA transportation regulations, and provides technical expertise to the IAEA.

RSPA participates as a member of the Standing Advisory Group for the Safe Transportation of Radioactive Materials (SAGSTRAM), the primary group advising the IAEA Director-General on transportation activities and initiatives involving radioactive materials. Knowledge of IAEA activities ensures appropriate and timely input by the US. in order to maintain compatibility of domestic and international regulations. RSPA's participation in SAGSTRAM ensures the U.S. is able to promote its goal of harmonization of international regulations affecting the transportation of radioactive materials.[3]

DOT's involvement in the development of international standards for the transportation of hazardous materials is of limited benefit unless steps are taken to recognize them within its domestic regulatory program. As more industrialized nations move toward the implementation of these international standards, U.S. industries involved in international sale and transportation of these materials may find they have to comply with differing domestic and international requirements. It becomes necessary to recognize these standards through our hazardous materials regulations in order to minimize the economic burden that compliance with such dual standards would impose on these industries. Title IV of the Trade Agreements Act of 1979 (19 U.S.C. 2532) the Department (DOT) has a statutory mandate to incorporate international standards, particularly those that are performance-oriented, in its hazardous materials regulations to the maximum extent consistent with safety. RSPA has taken action to implement some of these international standards through the rulemaking process such as:

— Amending certain areas of the HMR to align requirements for the carriage of hazardous materials aboard aircraft with those in the 1987-88 edition of the ICAO Technical Instructions on the Safe Transport of Dangerous Goods by Air, and

— Issuing a Notice of Proposed Rulemaking (NPRM) that proposes to align the hazardous materials regulations with the United Nations Recommendations on the Transport of Dangerous Goods. Under this proposal, most of DOT's non-bulk packaging specifications would be replaced with performance-oriented packaging standards.

While consistency with international standards is desirable, it may not be appropriate in all cases. Domestic transportation exceptions may be necessary due to unnecessary economic impacts or, more importantly, when an international standard does not provide for an adequate level of safety.

FEDERAL ADMINISTRATION

DOT's Research and Special Programs Administration has been encouraging increased interaction with the modal administrations, other Federal agencies and state agencies, and constituency groups involved in hazardous materials transportation safety.

RSPA's Office of Hazardous Materials Transportation (OHMT) provides formal and informal technical training and advice to, as well as coordination between RSPA and other elements of DOT, Federal agencies, State and local agencies and private sector groups relating to all aspects of hazardous materials transportation safety. OHMT furnishes technical assistance and support to state and local governments in their efforts to enhance public safety in their emergency response activities and to enforce nationally uniform requirements pertaining to the safe transportation of hazardous materials. OHMT also initiates and develops efforts aimed at avoiding potential conflicts involving hazardous materials issues at various levels of government.

This assistance takes many forms, one of which is the Cooperative Hazardous Materials Enforcement Development (COHMED) program. COHMED is a cooperative program which focuses on hazardous materials transportation. It is designed to enhance and complement the hazardous materials programs initiated under the Motor Carrier Safety Assistance Program (MCSAP), a grant-in-aid program for motor carrier assistance. COHMED doesn't fund states directly; MCSAP does. COHMED provides technical assistance, training, and information to states for development of their enforcement programs, and fosters state adoption and enforcement of the Federal Hazardous Materials Regulations. It serves to promote uniformity and consistency in regulation enforcement.

RSPA provides a forum for states to meet and discuss common interests/problems concerning hazardous materials transportation safety through a series of regional workshops. Representatives of other Federal agencies like the Nuclear Regulatory Commission, Department of Energy and Federal Emergency Management Agency, participate to provide information on policy and program activities.

Inspection and Enforcement

DOT conducts a comprehensive inspection and enforcement program to ensure industry compliance with hazardous materials transportation regulations and to reduce the potential for catastrophic accidents that may result from violations of the safety regulations. While responsibility for the development of hazardous materials transportation safety regulations lies with RSPA, enforcement is the responsibility of individual modal administrations.

Unlike RSPA, whose primary responsibility is hazardous materials transportation safety, modal administrations are primarily responsible for carrier safety. Monitoring the movement of hazardous materials is done in conjunction with those responsibilities. The primary focus of RSPA's inspection and enforcement program is packaging in all its forms, i.e. containers, cylinders, drums, etc., and related business such as retesting and reconditioning.

DOT promotes and relies heavily on States' enforcement programs for highway enforcement of the hazardous materials transportation rules. When state enforcement cases are encumbered by multiple state interactions with shippers, DOT's General Counsel will support or assist with the penalty actions. Cooperation between state and local law enforcement and the response community is encouraged to recognize violations during actual incidents so that evidence useful to penalty actions is identified and protected.

OUTREACH INITIATIVES

Hazardous Materials Information Exchange

DOT, in collaboration with the Federal Emergency Management Agency, developed the Hazardous Materials Information Exchange (HMIX). The HMIX is a hazardous materials information clearinghouse and exchange system designed to provide Federal, State, local and private sector organizations with a means of sharing timely and valuable information about the prevention of, preparation for, and mitigation of hazardous materials emergencies. The HMIX serves as a central source of reliable information to be used for planning for potentially dangerous occurrences. It is not intended to provide assistance during an actual emergency. HMIX is set up as an "electronic bulletin board" to provide information quickly and easily. Current information of immediate interest (i.e. current rulemaking actions) is posted on the main bulletin board. The rest of the bulletin board is divided into nine compartments, which HMIX calls conferences. Each conference contains information on a specific topic. The nine conferences cover the following topics:

(1) Federal training courses
(2) Public and Private Sector Hazardous Materials (HAZMAT) Information
(3) Calendar of Conferences
(4) Instructional Materials and Literature Listing
(5) Toll-free (800) numbers and On-line Data Bases
(6) Laws and Regulations
(7) Contacts
(8) Department of Transportation
(9) Federal Emergency Management Agency - Limited Access

To access HMIX, a personal computer, any communications software package, and a modem capable of transmitting at 2400, 1200, or 300 baud is needed.

A pilot test in 1987 of the HMIX has proven to be highly successful with the number of registered users rising steadily throughout 1987.

The HMIX is into its second year of operation and has proven to be a valuable source of information for emergency response training programs and activities. In a typical month, approximately 1200 users access the electronic system and an additional 600 calls are placed to the toll-free telephone numbers. Thus far, state and local agencies are the most frequent users of the system.

Emergency Response Guidebook (ERG)

RSPA undertook a major revision to the 1987 edition of the ERG since it was initially published in 1980. The ERG requires revision on a periodic basis (triennially) because changes in the UN identification numbers occur frequently, and an estimated several hundred new chemicals are developed each year, many of which meet the RSPA's definition of hazardous materials.

The Guidebook is used to identify hazardous materials by their respective four-digit UN identification numbers and to determine what immediate safety measures to take in the event of an accidental release. Intended for use by fire, police, and other emergency service personnel, the ERG is part of DOT's effort to provide state and local organizations with information critical to successful emergency response on an immediate or short term basis. The ERG is used by approximately 30,000 fire and police departments, and 10,000-15,000 other response organizations, i.e. private and public hospitals, public road maintenance departments, and debris removal firms. It is used in a multitude of situations ranging from assisting warehouse employees in identifying the contents of a package from the UN number displayed to potentially catastrophic accidents. Between 1980 and 1988, more than 2 million copies of the ERG have been distributed without charge to emergency services organizations.

Hazardous Materials Information System

An effective hazardous materials transportation management program depends greatly on its information base. Safety regulation development, enforcement, research, and training are guided in part by information on the safety performance of the hazardous materials transportation industry.

DOT's primary instrument for compiling, analyzing, and disseminating hazardous materials transportation safety data is the Hazardous Materials Information System (HMIS) which contains comprehensive system performance data on incidents, enforcement actions, exemptions and approvals and other elements of the regulatory program. In combination with other factors, data from the HMIS is used extensively within DOT and other Federal agencies, and by State and local governments, the press and the public. The data is useful for prioritizing program initiatives at the Federal and State level, and is a good basis for design of emergency preparedness exercises. The incident reporting system is undergoing a complete revision to improve the quality of incident information. The revised system should become effective during 1989.

Training

Training and education are an integral part of RSPA's responsibilities for emergency response to hazardous materials incidents in transportation. In order to promote compliance with the hazardous materials regulations, DOT sponsors a number of training and information dissemination activities designed to educate government inspectors and law enforcement personnel in enforcement procedures and familiarize industry personnel with the requirements of the regulations. The sources of training are DOT's (1) Transportation Safety Institute (TSI), located in Oklahoma City, Oklahoma, which receives financial and technical support from RSPA and the operating administrations to provide instruction on individual modal, as well as cross-modal, regulatory responsibilities; and (2) RSPA in Washington, D.C., which offers a wide array of training materials packages, videotapes and job aids to all segments of the hazardous materials enforcement and emergency response communities. The original concept for formal training in the transportation of hazardous materials to be conducted on-site at TSI was modified in 1983 when it became apparent that state and Federal financial constraints presented a barrier to state and local attendance at TSI.

RSPA launched an alternative approach to the original concept and made training available at state-selected sites around the country. RSPA's basic enforcement and compliance course has three phases: a self-study introduction that the student completes before attending class; a week of classroom instruction; and a field exercise to be completed independently once back on the job. Specific courses on cargo tank inspection and radioactive materials are offered as well. Modal-specific training courses are offered by the Federal Highway Administration's Office of Motor Carrier Safety (OMCS), Federal Railroad Administration, Federal Aviation Administration, and the U.S. Coast Guard. New emphasis is being placed on modular self study programs to integrate training into the workplace.

Historically, Federal emergency management assistance programs for States and communities, administered by FEMA, have been largely directed at improving civil defense and natural disaster preparedness. The Superfund and Reauthorization Act of 1986 (SARA) assigned responsibility for publishing unified Federal guidance for hazardous materials emergency planning and implementation to the National Response Team (NRT). The NRT is a national body composed of 14 Federal agencies having major responsibilities in environmental, transportation, emergency management, worker safety, and public health areas. The NRT "Hazardous Materials Emergency Planning Guide, NRT-1" is the primary guidance document, developed through the cooperative efforts of the NRT, that provides Federal guidance for the preparation and implementation of hazardous materials emergency plans. The guide provides a Federal consensus upon which future technical assistance, guidance, and training will be based. DOT/RSPA, as an active member of the NRT, is supporting training activities in cooperation with the other Federal agencies of the NRT.

New Training and Outreach Initiatives

RSPA, in cooperation with FEMA, will be participating in future teleconferences aimed at achieving unity in state compliance and enforcement programs, as well as emergency response involving hazardous materials transportation.

New enforcement training emphasis will be on (1) developing a state field program, based on a modular compliance and enforcement course, with orientation on management of the program for State COHMED coordinators, (2) delivery of an advanced hazardous materials inspection course, and (3) delivery of a course on management and administration of a hazardous materials enforcement program. In emergency response training DOT will (1) develop and deliver materials on response to rail yard incidents; (2) support distribution of guidelines for emergency medical personnel on handling chemically contaminated victims, and (3) develop materials for truck drivers on accident prevention, avoidance, and emergency response. New courses will use the National Fire Protection Association Standard for Professional Competencies for Responders as a guide for training objectives.

A new initiative anticipated is the development of workshops for regional, state and community level organizations involved in decisions about the routing of hazardous materials. The program will increase awareness of new federal routing regulations proposed to guard preemptive conflicts. The focus will be on (1) evaluating highways and arteries for the risk that hazardous materials transportation would pose to population and property, and (2) formulating alternative approaches. The workshop format will be suitable for roundtable discussions to encourage state officials to work on a regional basis to evaluate existing choices.

Joint curricula development initiatives will be undertaken with FEMA to advance the state-of-the-art in advanced tactical guidelines of the new National Fire Protection Association professional competency standard, and to upgrade hazardous materials plans to consider transportation issues. RSPA will work with FEMA on the development of alert and notification protocols to enhance public officials' capabilities to effectively warn the public of dangers from hazardous materials emergencies. Communities along interstate highways usually have less capability for public warning than communities planning for response near fixed manufacturing facilities. Outreach efforts are planned to stimulate the development and use of standard alert and notification protocols. RSPA product development to support these workshops will be regional profiles of chemicals likely to be transported through specific regions, including chemical behavioral characteristics and potential health hazards, and initial isolation and evaluation factors. These will be produced in a handy format as guideline for user by local officials.

RSPA has a variety of resources to be integrated in support of Federal, state and local hazardous materials exercises. Data from the Hazardous Materials Information System will be used to develop state profiles of incident activity. A series of scenarios for emergency exercises will be developed from these data, including information on types of vehicles, material involved, quantity, deaths and injuries, losses and property damage, cost of decontamination, and nature of packaging problems. Additional information emanating from enforcement activity, including accident investigations, will be provided to enhance the state data base on which decisions about exercise objectives are based. Additional guidance is planned on the involvement in exercises of participants such as law enforcement and transportation workers to provide opportunities for their interaction and experience in simulations. Additionally, OHMT will track and make available (in the HMIX) hazardous materials exercise plans and points of contact for people interested in participation such as neighboring jurisdictions, state agencies and private industry.

CURRENT ISSUES

Routing

The Hazardous Materials Regulations provide a comprehensive system of regulations covering the shipment and transportation of radioactive materials. Among the most significant regulations in that system is the rule governing the highway routing of radioactive materials. This rule prescribes general routing requirements for all placarded radioactive materials and specific requirements for highway route-controlled quantity shipments, such as spent fuel. Both routing rules have risk minimalization as their basic goal. A key feature to the rule is the requirement that highway route-controlled quantity radioactive materials be transported either over interstate highways selected to reduce time in transit (using interstate beltways or bypasses when available) or over a state designated alternative route.[4]

The subject of routing for non-radioactive hazardous materials was raised in an Advanced Notice of Proposed Rulemaking, HM-203, published in April 1988. This Notice invites public comment and announces a public hearing concerning the possible need to establish routing criteria, requirements, and methodologies for analyzing alternative routes for the highway transportation of non-radioactive hazardous materials. This new action is noteworthy because the current routing regulations concerning non-radioactive hazardous materials do not provide a framework to ensure consistency and clarity in their application.

For radioactive materials, state governments are given considerable latitude to carry out their highway routing functions. The routing criteria and requirements for the transportation of radioactive materials have been effectively applied by several state governments since 1981.

Presently, a large number of state and local routing requirements and control measures exist for the transportation of non-radioactive hazardous materials. In some cases these requirements and control measures (i.e. speed and time of day restrictions) can serve to impede the free flow of commerce, with little or no demonstrable positive effect on public safety. In some instances, they may result in the transference (export) of risk from one jurisdiction to other jurisdictions which may not be as prepared or as able to deal with such risks.

Toxic by Inhalation Materials

The tragedy in Bhopal, India, in which more than 2000 people died as a result of methyl isocyanate being released, greatly increased awareness of the risks presented by materials extremely toxic by inhalation. The major hazard of such toxic materials is their ability to disperse over a very large area in concentrations sufficient to cause death, thus placing a large number of people at risk. These toxic materials present substantial risks in manufacturing facilities, transportation and storage.

RSPA's approach to minimizing the risk associated with materials extremely toxic by inhalation began with the definition, identification, and classification of these materials. This resulted in the publication of: HM-196 "Materials Toxic by Inhalation" a quick response interim fix for classification and packaging deficiencies. More significant classification and packaging standards for liquids and gases that are extremely toxic by inhalation are proposed in notice of proposed rulemaking HM-181 "Performance Oriented Packaging Standards." This proposed classification standard expands a recommended standard in the Recommendations of the United Nations Committee of Experts on the Transport of Dangerous Goods.[5] Next, RSPA considered accident scenarios that could result in a release of substantial quantities of materials extremely toxic by inhalation and identified the factors that are likely to cause or contribute to these accidents. Factors include equipment, operator training and performance, and operational procedures. Control of these factors can reduce the probability of a transportation accident occurring. Routing can reduce risk in transportation. However, due to the fact that these toxic materials often have to be moved through heavily populated areas, routing can't be considered the primary means in reducing this risk.

RSPA investigated the relationship between the quantity of materials transported and the area at risk from a release of materials extremely toxic by inhalation. Present regulations for some of these materials allow them to be transported in large volumes. This investigation found the quantity limit necessary to reduce risk in transportation was much smaller than quantities commonly used in commerce. Because there is a vital need to transport large volumes of toxic materials, the investigation concluded that the ability to package these materials so they can survive an accident with little or no release of material is vital to their safe transportation. Thus, an accident survivability performance standard for packaging is the most effective way to control the risks of materials toxic by inhalation. Additional investigation is required concerning the risk and cost involved before accident survivability performance criteria or engineering performance standards can be developed. DOT is continuing this work in the research and development of accident survivability criteria for materials extremely toxic by inhalation.[6]

Inconsistency Rulings

To promote national uniformity and increased enforcement of the regulations affecting hazardous materials transportation, DOT encourages states to adopt and enforce the Federal regulations. Apart from improvements in public safety that result from uniform regulation, there are several distinct economic benefits that result from state adoption of Federal regulations.

State governments are spared the expense of maintaining costly and duplicative regulatory agencies, and industry is spared the expense of complying with multiple, diverse, and potentially inconsistent requirements.

Section 112(a) of the HMTA provides that any state or local requirement that is "inconsistent" with the HMTA or the Hazardous Materials Regulations (HMR) issued under it is preempted, and therefore, invalid as a matter of law. To assist in the interpretation and application of this provision, RSPA has issued regulations establishing procedures for the issuance of advisory "inconsistency rulings" in response to formal requests from affected shippers, carriers, or state and local governments. To date, RSPA has issued 24 such rulings, setting forth the Department's views regarding the validity of various state and local requirements under the preemptive scheme of the HMTA.

In issuing inconsistency rulings, RSPA applies two tests that have been borrowed from judicial precedents: first, whether or not it is possible to comply with both the Federal regulations and the nonfederal requirement (the "dual compliance" test), second, whether the nonfederal requirement presents an obstacle to the purpose of the HMTA and the HMR's ("obstacle test"). For example, Washington State shipping paper requirements, IR-4. State law requiring intrastate shipments of hazardous materials carried by motor vehicles to be accompanied by red or red bordered shipping papers was *inconsistent* with the HMTA. In IR-7, the Governor of New York issued an order advising a company to suspend spent nuclear fuel shipments on two non-Interstate highway routes which was found *consistent* with the HMTA because it required compliance with Federal regulations requiring use of the Interstate Highway System.[7] In applying the obstacle test, which is usually the critical test, RSPA looks at two previous objectives of the HMTA, as expressed in the statute and its legislative history.

First, Congress' overriding purpose in adopting the HMTA was to enhance public safety (49 USC 1801). If the effect of a nonfederal requirement is to reduce overall safety, it is inconsistent with the HMTA. Second, the legislative history of the preemptive provisions indicates that Congress' purpose in adopting it was "to preclude a multiplicity of state and local regulations and the potential for varying as well as conflicting regulations in the area of hazardous materials transportation." (S. Report No. 1192, 93rd Cong., 2nd session, 37 (1974)). Therefore, if the effect of a nonfederal requirement is to interfere substantially with the uniform regulatory system, then it is inconsistent with the HMTA.[1]

Section 112(b) of the HMTA provides that an inconsistent state or local requirement will not be preempted if DOT determines that the requirement affords an equal or greater level of protection from the Federal requirements and does not unreasonably burden commerce.

Training of Hazardous Materials Transportation Personnel

DOT will be publishing a notice of proposed rulemaking proposing to require training for all personnel working in the proximity of hazardous materials in transportation (HM-126F). Based on information provided to RSPA through its hazardous materials incident reporting system, human error is the probable cause of most transportation incidents, and associated consequences, involving the release of hazardous materials. By requiring adequate training of persons involved in the transportation of hazardous materials (i.e. warehousemen, drivers, shipping personnel, etc.), awareness of safety considerations and regulatory requirements involved in transporting hazardous materials would be increased and the number and severity of hazardous materials incidents reduced.

RSPA Policy (HM/EM Response)

It is the policy of the Research and Special Programs Administration (RSPA) to improve the Nation's preparedness for and response to transportation emergencies involving hazardous materials through a Federal, State, local and private sector partnership in safety. In furtherance of this policy, RSPA's rulemaking, enforcement, research, analysis, and education programs all contribute to the pursuit of the following objectives:

- Development of hazardous materials transportation safety regulations which are uniform nationwide;
- Development of a nationally uniform system of hazard communication for hazardous materials transportation safety regulations;
- Development of technical guidance for State and local governments on hazard evaluation and risk assessment associated with hazardous materials transportation;
- Development of training and assistance programs to support Federal, state and local capabilities to respond to large state hazardous materials transportation emergencies;
- Fostering cooperative planning by Federal, State, local and private sector organizations in the prevention and mitigation of emergencies involving hazardous materials transportation;
- Development of information which supports public understanding of the Department's system of HAZMAT transportation, emergency prevention and mitigation techniques, and response capabilities in place for life and property protection.

SUMMARY

The Federal, State and private sector partnership is the key to a successful national hazardous materials transportation safety program. RSPA, the Federal agency primarily responsible for hazardous materials transportation, will continue to work with appropriate state agencies and localities, by providing technical assistance on routing, regulatory uniformity, enforcement, emergency planning and response, accident prevention, and training.

It is essential that the various levels of government work closely together to assure that the public is provided the highest level of protection. In effort to support a successful Federal/State partnership, RSPA is encouraging this type of cooperation with other Federal agencies having hazardous materials responsibilities, such as the Nuclear Regulatory Commission, Federal Emergency Management Agency, the Environmental Protection Agency, and Department of Energy.

It is hoped that through this cooperation, a sound partnership among the various levels of government will be developed and strengthened. The result will be a national hazardous materials transportation safety program which provides the public with the highest level of protection possible.

RESOURCES

1. National Conference of State Legislatures, *Hazardous Materials Transportation, a Legislator's Guide*, February 1984
2. United States Congress, Office of Technology Assessment, *Transportation of Hazardous Materials,* OTA - SET 304, Washington, D.C.: United States Government Printing Office, July 1986
3. Office of Hazardous Materials Transportation, Research and Special Programs Administration, "Annual Report of Hazardous Materials Transportation," United States Department of Transportation, Washington, D.C., Calendar Year 1986
4. Federal Register, Thursday, April 7, 1988, 49 CFR Part 177, Highway Routing Standards for Hazardous Materials, ANPRM
5. Federal Register, Friday, November 6, 1987, 49 CFR Parts 171 through 179, Performance Oriented Packaging Standards
6. J.K. O'Steen, "Transportation Accident Survivability Performance Standards for Materials Extremely Toxic by Inhalation" Article, United States Department of Transportation, May 1988
7. E. Bonekemper, "Inconsistency Rulings and Applications Under Section 112(a) of the Hazardous Materials Transportation Act", January 8, 1988, U.S. Department of Transportation, Office of the Chief Counsel, Washington, D.C.

Management of Hazardous Materials and Wastes: Treatment, Minimization and Environmental Impacts. Edited by S. K. Majumdar, E. W. Miller and R. F. Schmalz. © 1989, The Pennsylvania Academy of Science.

Chapter Nineteen

TRANSPORTATION OF HAZARDOUS MATERIALS

MICHAEL S. BRONZINI
Professor and Head
Department of Civil Engineering
The Pennsylvania State University
University Park, PA 16802

INTRODUCTION

There are an estimated 180 million annual shipments in the U.S. of some 30,000 different materials which are regulated by the Department of Transportation as hazardous materials.[1] These materials are produced and used in every part of the nation and are vital components of nearly every sector of a modern industrial society. The transport of hazardous materials (hazmat) occurs without incident most of the time, however, the potentially catastrophic consequences of a major hazmat transportation accident have caused the public to be very concerned about regulating and controlling hazmat transportation.

The public debate over hazmat transportation could proceed more rationally if the circumstances of such transport were more widely known. Thus the next section presents data on the kinds and quantities of hazardous materials shipped in the U.S. This is followed by discussions of hazmat transportation accidents and incidents, regulation of hazmat transportation, and emergency preparedness. Recommendations for improving the hazmat transportation system are then offered.

HAZMAT TRANSPORTATION FLOWS

One of the major problems hindering the development of effective hazmat transportation policy is the lack of a comprehensive data base on the quantities and patterns of movement of these materials. The most authoritative in-

TABLE 1

Estimated Transportation of Hazardous Materials by Mode in 1982.

		Quantity of Hazmat Transported			
Mode	Vehicles/Vessels Used for Hazmat Transport	Tons (millions)	%	Ton-Miles (billions)	%
Truck	370,000 dry freight or flatbed 130,000 cargo tanks	927	59.83	93.6	11.95
Rail	115,600 tank cars	73	4.71	53.0	6.76
Water	4,909 tank barges	549	35.44	636.5	81.23
Air	3,772 aircraft	0.29	0.02	0.46	0.06
Total		1,649.29	100.00	783.56	100.00

Source: Office of Technology Assessment[1]

formation available to date is a 1986 report prepared by the Office of Technology Assessment (OTA) of the U.S. Congress.[1]

Table 1 shows the OTA estimates of hazmat transportation by each mode, excluding pipeline, in 1982. As shown, hazmat movements totaled over 1.5 billion tons and 780 billion ton-miles that year. (A "ton-mile" is a unit of transportation equal to one ton moved a distance of one mile.) The inclusion of pipeline transport would more than double these totals.[1] Trucks moved nearly 60 percent of the tonnage but only 12 percent of the ton-miles, reflecting the short haul intraregional nature of most truck movements. Conversely, water carriers moved 35 percent of the tons but over 80 percent of the ton-miles. The average haul by water exceeded 1,100 miles, which is more than 10 times longer than the average truck haul of 101 miles. The rail and air average haul distances were 726 miles and 1,586 miles, respectively.

The average hauls of the transportation modes reflect their respective roles in hazmat transportation. Barges are used to move large quantities of petroleum and chemical products over long distance. Rail transport is also used for long distance moves, but in smaller quantities. Trucks are used primarily for regional and local distribution, as in moving gasoline from a jobber's storage tanks to a service station. Trucks also make long distance hauls of quantities that are too small to be moved efficiently in rail tank cars, and handle moves for shippers not served by rail. Air hazmat shipments typically consist of very small quantities (55 gallons or less) moving long distances. Typical modal shipment sizes are as follows:

Truck	6,000 gallons
Rail	15,000 gallons
Water	450,000 gallons
Air	55 gallons

Most people would find it difficult to name more than a dozen of the hazardous materials being moved in commerce, and would probably overlook the single commodity involved in the largest number of hazmat accidents, gasoline. In fact, thousands of hazardous materials are transported in the U.S. every day. Table 2 lists the hazard classes that are defined by the U.S. Department of Transportation (DOT) for regulatory purposes, and gives examples for each class. Poison A materials, class A explosives, and radioactive materials are generally regarded to be the most hazardous, and are subject to the utmost care in handling and transport. Flammables move in the largest quantities. Highlights from the OTA analysis[1] of the materials moving by each mode of transportation in 1982 are as follows:

- The two largest hazard classes transported by truck were flammable liquids and Poison B.
- Petroleum and gasoline shipments accounted for almost half of all truck hazmat transport.

TABLE 2

Department of Transportation Hazard Classes

Hazard Class	Examples
Flammable liquid	Ethyl alcohol, gasoline
Combustible liquid	Fuel oil, ink
Flammable solid	Nitrocellulose (film), phosphorus
Oxidizer	Hydrogen peroxide, chromic acid
Organic peroxide	Urea peroxide
Corrosive	Bromine, hydrochloric acid
Flammable gas	Hydrogen, liquefied petroleum gas
Nonflammable gas	Chlorine, anhydrous ammonia
Irritating material	Tear gas, monochloroacetone
Poison A	Hydrocyanic acid, phosgene
Poison B	Cyanide, disinfectants
Etiologic agents	Polio virus, salmonella
Radioactive material	Uranium hexafluoride
Explosives	
Class A	Jet thrust unit
Class B	Torpedo
Class C	Signal flare, fireworks
Blasting agent	Blasting cap
ORM (Other Regulated Materials)	
ORM-A	Trichloroethylene, chloroform
ORM-B	Calcium oxide, potassium fluoride
ORM-C	Cotton, inflatable life rafts
ORM-D	Small arms ammunition
ORM-E	Kepone, polychlorinated biphenyls

Source: 49 CFR 172.101 and 173, as reported by OTA.

- The largest hazard classes transported by rail were flammable liquids (26 percent of the tonnage), corrosive materials (25 percent), and flammable gases (12 percent).
- Rail hazmat flows were geographically concentrated within and between the southwestern and southeastern states.
- Hazardous commodity shipments constituted 55 percent of all domestic waterborne tonnage, and 85 percent of the hazmat tonnage was crude or processed petroleum.
- The geographic pattern of water hazmat flows was similar to that of the rail flows.
- Chemicals accounted for 80 percent of the hazmat shipped by air.

TRANSPORTATION RISK

With what degree of safety is the transport of hazardous materials accomplished? To answer this question, it would be desirable to review the recent record of accidents and incidents. Unfortunately, as with the transportation flow data, there is no complete data source on transport releases.

The DOT regulations define a transportation-related incident as any unintentional hazmat release during transportation, loading or unloading, or temporary storage incidental to transportation. All such releases, except those from bulk water transport, intrastate trucking, and specified minor spills, must be reported to DOT by the carrier within 15 days of discovery of the release. These release reports are entered into the Hazardous Materials Information System (HMIS) database. Although releases must be reported under the hazmat regulations, OTA[1] and others[2] have found that the actual number of incidents may be twice the number reported. Nonetheless, the HMIS database is the best source available for tracking the frequency, severity, and causes of hazmat transportation accidents and incidents.

Table 3 summarizes the hazmat transportation incidents recorded in the HMIS database for the period 1976-1984. A total of 78,082 incidents were reported during that period, for an average of 8,676 per year. The number of reported incidents has decreased in recent years to around 5,000 per year.[1] Highway transport accounted for over 86 percent of the incidents, which would be expected given the large number of small shipments handled. Rail transport was next, with 11.8 percent of the reported incidents.

To assess the true relative safety records of the transportation modes, it is necessary to compare the incidents with an appropriate measure of exposure, such as ton-miles or number of shipments. Unfortunately, no reliable source of consistent data is available for this purpose. However, taking the number of releases that were reported for 1982 and doubling that number to account for underreporting, and using the exposure estimates given in Table 1, the following approximate release rates can be calculated:

	Approximate Number of Incidents per	
	Million Tons	*Billion Ton-Miles*
Truck	12.2	121
Rail	23.0	32

The water and air data are too sparse to permit even this level of analysis. Although the gross nature of the assumptions made renders these estimated truck and rail release rates far from authoritative, they are somewhat indicative of the relative safety records of the two modes. Trucks are involved in a very large number of small hazmat shipments. Even though they have seven or eight times as many incidents as rail, the truck incident rate is lower than that of rail on a tonnage basis. The much longer average haul for rail transport, however, gives it the safety edge when exposure along the length of the route is considered. Glickman[3] uses similar data sources but factors in average shipment size estimates, and concludes that truck hazmat shipments experience lower releases per vehicle-mile than rail shipments. Given the large uncertainties and disparaties in the various data source, it is best to conclude that track and rail have comparable hazmat safety records, and that situation-specific analyses are far more revealing than systemwide averages.

TABLE 3

Reported Hazardous Materials Transportation Incidents, 1976-1984

	Mode					
Location	Air	Highway	Rail	Water	Other	Total
U.S. Total	905	67,567	9,185	126	299	78,082
Five Highest States						
Pennsylvania	28	6,718	322	3	38	7,109
Ohio	16	4,947	328	—	22	5,313
Illinois	49	3,465	828	—	11	4,353
Texas	48	2,854	1,265	6	18	4,191
California	76	2,900	817	15	25	3,833

Source: Office of Technology Assessment[1].

TABLE 4

Severity of Reported Truck and Rail Incidents, 1982

Mode	Total Incidents	Significant Spills (a)	Casualty Related (b)
Truck	5,672	1,654 (29%)	76 (1.3%)
Rail	838	326 (39%)	27 (3.2%)

Notes: (a) Reported release exceeded 5 gallons or 40 pounds.
(b) Release resulted in a fatality or reportable injury.

Source: Glickman[3].

Table 3 also shows the five states that had the highest number of hazmat releases reported to the HMIS for the period 1976-1984. Collectively, these states accounted for 32 percent of the reported incidents. Although state-level exposure data are lacking, it is probably true that these five states also recorded the highest levels of hazmat traffic, since they are all major industrial states and serve as traffic corridors for materials destined to other states. Nevertheless, Pennsylvania's number one ranking on this list, with 34 percent more reported incidents than neighboring second place Ohio, has prompted Pennsylvania officials to initiate a statewide hazmat transportation assessment.[4]

While the number of reported hazmat releases is high enough to cause concern, the severity of most releases is rather low. Table 4 shows the number of reported truck and rail incidents in 1982 that resulted in significant spills (those exceeding 5 gallons or 40 pounds) or casualties.[3] Only 1 to 3 percent of the incidents produced hazmat related injuries or fatalities. Allowing for underreporting, this means that the odds that a hazmat shipment will be involved in a personal injury accident are lower than 1 in 500,000, and the chances that a death will occur are much less than 1 in a million. The odds that a significant spill will occur are on the order of once in every 100,000 shipments.

According to OTA,[1] human error is the principal cause of 62 percent of the hazmat transport releases, followed by package failure (27 percent) and vehicular accident (6 percent). The top five proximate causes of the incidents reported to HMIS were:[1]

external puncture
loose fittings
damaged by other freight
load/unload spill
defective fittings

Collectively, these causes produced 60 percent of the hazmat releases.

The analysis and abatement of transportation risk is an important element of the shipper's hazmat transport operations plan, and the community's or state's emergency preparedness planning. Risk in this context is defined as follows:[5]

$$\text{Hazmat Transport Risk:} = \begin{pmatrix}\text{Probability of} \\ \text{a Release}\end{pmatrix} \times \begin{pmatrix}\text{Consequences of} \\ \text{a Release}\end{pmatrix}$$

Thus risk can be reduced by adopting safer packaging and operating practices, and by minimizing the exposure of people and property to a hazardous substance after it has been released.

A number of techniques are presently being used to estimate risk.[5] Some approaches estimate absolute risk, which is usually expressed as the expected numbers of fatalities and injuries and the expected value of the property damage resulting from a particular hazmat transportation operation. The difficulty in estimating risk lies in identifying all of the possible ways that the system can fail, thereby creating a potential release situation, and determining the probability and severity of a release in each case. Other approaches[6] estimate the relative risk, which is usually expressed as a risk index that can be used to compare transport options and identify problems. Both absolute and relative risk assessment methods are best used to analyze specific hazmat transportation alternatives, for the purpose of guiding decision-making and public policy formulation.

REGULATION

The Hazardous Materials Transportation Act of 1975 focuses much of the Federal regulation of hazmat transportation through the DOT. The Act empowers the Secretary of Transportation to designate those materials deemed to be hazardous and to develop regulations prescribing how such materials must be moved. Most of the Federal regulations that govern hazmat transport appear in 49 CFR 171-179. The regulations define the DOT hazard classes (see Table 2), list the materials subject to regulation, specify the packaging, handling, labeling, marking, placarding, loading, and routing procedures to be used, and provide additional instructions pertaining to the various modes of transport. The Federal regulations are rather complex, and it takes considerable training and experience to apply them correctly.

Proper identification of the material being shipped is of particular concern, since this information is vitally important to the emergency response personnel who arrive at the scene of an incident. All packages and shipping containers and, in some cases, vehicles must be marked with the material's proper shipping name, hazard class, and United Nations/North American hazard identification number. The shipping papers must also carry this information, and must be carried in a standard location so that emergency responders will know where to find them.

Packaging is another major concern of the hazmat regulations, since proper packaging is the only real defense against the release of hazardous materials. The general requirement is that the packaging must prevent release during normal transport, including minor accidents. Beyond this, the regulations provide detailed specifications for the design, construction, and testing of the approved packaging for each hazard class. These specifications are of the prescriptive variety, which has tended to stifle innovation in hazmat packaging.[1] In view of this, there is presently a move afoot to replace these specifications with performance standards for packaging.

Packaging standards for highly radioactive nuclear materials are an exception to the general rule. In this case, the packaging must prevent the release of its contents even during a very severe accident. Packages for highly fissible materials such as spent fuel assemblies from nuclear reactors, referred to as Type B packaging, must successfully survive the following sequence of tests:

1. A free drop of 9 meters onto a flat, unyielding, horizontal surface;
2. A free drop of 1 meter onto the top of a vertical, cylindrical, mild steel bar, 15 centimeters in diameter;
3. Exposure to a fire of 800°C for 30 minutes, with no artificial cooling of the package;
4. Immersion under 15 feet of water for not less than 8 hours.

Shipping casks which survive this test sequence are so durable that, when involved in an accident, they inflict considerable damage on other vehicles and structures, while sustaining little or no damage themselves. In fact, there were 50 transportation accidents involving Type B containers during the period 1971-1985, none of which resulted in failure of the package or release of the material.[1] After reviewing all of the technical data available on Type B packaging, OTA concluded that these containers "provide an extremely high level of public protection, much greater than that afforded in any other current hazardous materials shipping activity." (1, p. 109)

It is incongruous that, despite the high degree of protection afforded by spent nuclear fuel shipping containers, the public continues to believe that the risks associated with radiation exposure are great. In fact, risk assessments of nuclear waste transport[7] have shown repeatedly that the nonradiological risk—the traffic hazards inherent in the vehicle movements, without regard for the cargo carried—is higher than the radiological risk by many orders of magnitude.

The public's concept of how best to regulate hazmat transport so as to minimize risk to population is to exercise control over the routes traveled. Route regulation is most often expressed in terms of prohibiting vehicles carrying hazardous materials from moving through population centers. This risk mitigation strategy attacks the second factor in the risk equation, the consequences of release, by minimizing the number of people who would be in the immediate vicinity of a release. However, this strategy ignores the first factor, the probability of a release. To avoid population centers, vehicles must often be routed over secondary highways or railroad tracks, which are of poorer quality than the primary routes connecting population centers. These secondary routes will also often be more circuitous, requiring a longer travel time. The combination of poorer quality routes that are longer increases the probability that a hazmat shipment will be involved in an accident. It is imperative that routing regulations be enacted only with full knowledge of the risk tradeoffs that are involved.[8,9]

No discussion of hazmat regulation would be complete without mentioning the many agencies that have responsibilities in this area. Federal agencies and their activities are listed in Table 5. As can be seen, there is a considerable amount

TABLE 5

Federal Activities in Hazardous Materials Transportation

	Regulation of:									
	Hazardous Materials*	Containers	Vehicles and Vessels	Operators	Planning	Record-keeping	Inspection	Enforce-ment	Training	Emergency Response
DOT:										
RSPA	X	X	—	—	X	X	X	X	X	—
FHWA	—	X(a)	X(b)	X	X	X	X	X	X	—
FRA	—	X(a)	X	—	X	X	X	X	X	—
FAA	—	X(a)	X	—	X	X	X	X	X	—
USCG	X	X(a)	X	X	X	X	X	X	X	X
FEMA	—	—	—	—	X	—	—	—	X	X
EPA	X	—	—	—	X	X	X	X	X	X(c)
NRC	X	X	—	—	X	X	X	X	X	X(d)
DOE	X(e)	X	—	X	X	X	X	X	X	X(d)
DOD	X(e)	—	—	X	X	X	X	X	X	X

KEY: DOT—Department of Transportation; RSPA—Research and Special Programs Administration; FHWA—Federal Highway Administration; FRA—Federal Railroad Administration; FAA — Federal Aviation Administration; USCG—United States Coast Guard; FEMA—Federal Emergency Management Agency; EPA—Environmental Protection Agency; NRC—Nuclear Regulatory Commission; DOE—Department of Energy; DOD—Department of Defense.

*This category includes hazardous substances, hazardous wastes and radioactive materials, and the tools for communication of those hazards such as shipping papers, placarding, and marking.

(a) Package/container design.
(b) In addition, National Highway Transportation Safety Administration issues requirements for new vehicles.
(c) EPA responds to accidents involving the release of products regulated under the Comprehensive Environmental Response, Compensation, and Liability Act (CERCLA) and oil spills in coastal and ocean water.
(d) Depends on the type of radioactive material, severity of the accident, and the adequacy of State and local response programs.
(e) In cases of national security, DOD and DOE are not required to comply with DOT regulations provided they follow standards affording equal protection.

Source: Office of Technology Assessment.[1]

of overlap between the activities of these agencies. Furthermore, many states and localities have enacted their own laws and regulations, and have an array of agencies and activities that overlap the Federal roles. The OTA study observes that: "Hazardous materials enforcement officers and transportation industries—manufacturers, shippers, and carriers—find this welter of regulations inefficient, confusing, and difficult to comply with and enforce." (1, p. 6)

EMERGENCY PREPAREDNESS

Hazardous materials shipments reach their destinations without mishap a very high percentage of the time. However, an estimated 10,000 incidents each year result in a hazmat release, which is an average of 27 incidents per day. Release incidents are concentrated at the terminals where shipments originate and terminate, but can and do happen anywhere along the route. Consequently, the readiness of emergency response personnel to arrive at the scene quickly and take effective action to mitigate the hazard is another focal point of public policy.

As is the case in the regulatory area, there is an almost bewildering crazy quilt of Federal, state, local, and industry groups involved in emergency response. These groups are engaged in planning for emergencies, providing emergency response services, and training the responders. The OTA study[1] provides a good summary of the roles and responsibilities of the various parties.

Local fire or police department personnel are usually the first to arrive at the scene of a hazmat transportation incident. Other first responders include ambulance drivers, paramedics and hospital emergency room staff. OTA estimates[1] that there are approximately 2 million people in the emergency response network, but that only 25 percent of them have received adequate training. Although an abundance of training is available, it is not being delivered to those who need it, due to a variety of institutional and funding problems. There is currently no consensus plan for organizing the needed training effort.

Many communities have prepared emergency response plans to deal with hazmat incidents.[5,10] Such planning was made a national priority in the Superfund Amendments and Reauthorization Act of 1986 (SARA). Title III of SARA requires each state to establish emergency response commissions to oversee the development of local emergency response plans. The *Hazardous Materials Emergency Planning Guide*,[11] prepared under the auspices of Title III, presents an excellent summary of the Act's planning requirements and gives detailed guidance for implementing these provisions.

CONCLUSIONS

The public debate over transportation of hazardous materials seems to have intensified in recent years. This is probably because the assumed risk is increasingly viewed as involuntary and unacceptably high. A general lack of the technical knowledge needed to understand the true risks inhibits informed debate. For example, what is almost certainly the safest hazmat transport activity, the movement of spent nuclear fuel in approved Type B containers, excites widespread opposition, while the activity that produces the most fatalities, moving gasoline, is accepted with hardly a second thought. Finding a way to accurately portray the relative risks involved in hazardous materials transportation would do much to improve public policy.

A widespread lack of basic information concerning the kinds and quantities of hazardous material moving and the number of releases that occur also blocks effective policy formulation. As noted by OTA, "data and information about shipments are so poor and difficult to acquire that state and local regulations are often developed with little or no understanding of the magnitude or nature of the problems to be controlled." (1, p. 6) The basic mechanisms to collect better data are already in place, and coverage of water and rail movements is excellent. The data on truck and air hazmat shipments need to be improved. Also, a vigorous program to correct the underreporting of incidents to the HMIS and to expand its coverage to intrastate shipments should be instituted.

Increased training for personnel who may be involved in responding to a hazmat incident is clearly needed. It will take a comprehensive and coordinated national training program to achieve this end.

The overlapping activities and programs of the many agencies that have hazmat transportation responsibilities creates unnecessary, and potentially dangerous, confusion about who should be consulted about what. Streamlining, coordinating, and simplifying the regulation of hazmat transport would be of considerable benefit.

REFERENCES

1. U.S. Congress, Office of Technology Assessment, 1986. *Transportation of Hazardous Materials,* OTA-SET-304, U.S. Government Printing Office, Washington, D.C., 265 pp.
2. Harwood, D.W. and E.R. Russell, 1987. Present Practices of Highway Transportation of Hazardous Materials, Task B Interim Report, Literature Review. Federal Highway Administration, U.S. Department of Transportation, Washington, D.C., 84 pp.
3. Glickman, T.S., 1988. Benchmark Estimates of Release Accident Rates in Hazardous Materials Transportation by Rail and Truck. Transportation Research Board, 67th Annual Meeting, Washington, D.C., 25 pp.
4. Pennsylvania Department of Transportation, 1988. Hazardous Materials Movement. Request for Proposals, Research Project 87-26, Harrisburg, PA.

5. Rowe, W.D., 1983. Risk Assessment Processes for Hazardous Materials Transportation. *Synthesis of Highway Practice 103,* National Cooperative Highway Research Program, Transportation Research Board, Washington, D.C., 27 pp.
6. Russell, E.R., et al., 1986. *Risk Assessment/Vulnerability Users Manual for Small Communities and Rural Areas.* DOT/OSP/P-34/86/043, U.S. Department of Transportation, Washington, D.C., 135 pp.
7. Bronzini, M.S., et al., 1986. *Monitored Retrievable Storage of Spent Nuclear Fuel: Transportation Studies.* Transportation Center, The University of Tennessee, Knoxville, TN, 350 pp.
8. ICF Technology, Inc., 1988. *Risk Assessment of Nitrogen Tetroxide Transportation Routes.* San Antonio-Air Logistics Center, Directorate of Energy Management, Transportation Office, Kelly Air Force Base, San Antonio, TX, 90 pp.
9. Abkowitz, M. and P.D. Cheng, 1988. Developing a Risk/Cost Framework for Routing Truck Movements of Hazardous Materials. *Accidents Analysis and Prevention* 20(1):39-51.
10. U.S. Congress, Office of Technology Assessment, 1986. *Transportation of Hazardous Materials: State and Local Activities.* OTA-SET-301, U.S. Government Printing Office, Washington, D.C., 92 pp.
11. National Response Team, 1987. *Hazardous Materials Emergency Planning Guide.* NRT-1, U.S. Government Printing Office, Washington, D.C., 123 pp.

Management of Hazardous Materials and Wastes: Treatment, Minimization and Environmental Impacts. Edited by S. K. Majumdar, E. W. Miller and R. F. Schmalz.

Chapter Twenty

"HOW CLEAN IS CLEAN?" CONTAMINANT REMEDIATION LEVELS IN SOIL

DONALD KILLIAN

Scientific Services Section
Division of Special Investigations
Bureau of Waste Management
Pennsylvania Department of Environmental Resources
P.O. Box 2063
Harrisburg, PA 17120

INTRODUCTION

In the late 1970's, the American public was made explicitly aware of the widespread practices used for decades to dispose of industrial chemical wastes in the United States. Such practices included illicit dumping and burial of chemical wastes as well as abandoning such wastes in trucks and "permanent" storage in tanks and containers (or at least until the tanks and/or containers corroded and leaked their contents into the environment). Recollection of sites such as Love Canal, Knickerbocker Landfill, the Valley of the Drums, the Wade Site and many others recreate disturbing visions of leaking drums; puddles of variously colored liquids; smoldering, fuming liquids oozing down the face of a dump; dead vegetation; and chemicals oozing into basements. As one experiences such recollections, one must also begin to realize the enormous magnitude of remedial problems at these sites caused by the simple act of dumping liquid chemicals into the environment where they can disperse immediately both vertically and laterally. In an attempt to control the management of industrial chemical wastes and realizing the extreme problems at many sites where chemical wastes have been disposed of improperly, two federal laws were enacted in 1976 and 1980. These were, respectively, the Resource Conservation and Recovery Act (RCRA) and the Comprehensive Environmental Response Compensation and Liability Act (CERCLA or also commonly referred to as Superfund). Both Acts have been reauthorized since their initial enactment.

Since the late 1970s, remediation or cleanup of contamination in environmental media has become an increasing concern of the Pennsylvania Department of Environmental Resources. Obviously, cleanup is predicated on the assumption that there are cleanup goals or contaminant concentrations which, if not exceeded in the applicable environmental medium, are considered acceptable from some perspective. Commonly, these cleanup goals or acceptable contaminant concentrations apply to soils and have been euphemistically labeled "How Clean Is Clean?" levels. Exactly what they are (numerically), what they mean, and how they are developed is largely based on the perspective of the person or persons applying them. They are often perceived differently by different groups of society (regulatory agencies, legislators, other government officials, environmental groups, industry, environmental consultants, academia, and the general public) and by different individuals in each of these groups. Because of this, they are often sources of disagreement. This disagreement makes the application of "How Clean Is Clean?" levels extremely difficult for regulatory agencies. Some of the reasons for these perceptual differences and disagreement will become more evident.

"HOW CLEAN IS CLEAN?": THE LOGIC

One of the major tasks of the Department is to protect human health and the environment, and this is usually the ultimate goal in determining a "How Clean Is Clean?" level for contaminants in soil. With this goal in mind, the logic for determining "How Clean Is Clean?" levels in soil is very straightforward and is contained in the following simple statements:

1. Soil is contaminated at some concentration.
2. The contaminants may move from the soil to one or more other environmental media (surface water, groundwater, air, biota) by various transport mechanisms.
3. Organisms (man, animals or plants) may be exposed to the contaminants in the soil or to contaminants which have been transported from the soil into other environmental media.
4. The exposure of organisms to one or more of these contaminants at certain concentrations may produce one or more toxic or otherwise undesirable effects.

This logic, when applied to specific scenarios using a risk-based approach, should ultimately result in "protection of human health and the environment"; however, there are several obstacles to be overcome before achieving this result. First, while the logic for determining "How Clean Is Clean?" levels is quite simple and understandable, each of its four parts, as previously stated, requires the use of diverse technical concepts and information which are based on interpretation of scientific data and other information from several different disciplines. Unfortunately, not all scientists apply these concepts identically nor do they interpret information in the same way.

Second, "protection of human health and the environment" is directly related to the level of risk to human health and the environment which one is willing to allow. Therefore, if the allowable level of risk is not held constant, "How Clean Is Clean?" levels become "moving targets" and the probability that they will be applied inconsistently increases significantly.

Third, even if the first two obstacles are overcome, the final "How Clean Is Clean?" level which is adopted may be highly dependent on other factors considered by the individual or individuals applying the level. Such factors are not based on risk and may include regulatory, analytical, technological, and/or financial constraints. Consideration of one or more of these factors could result in "How Clean Is Clean?" levels which are either more stringent or less stringent than levels which are based solely on risk to human health and the environment.

Fourth, Article I, Section 27 of the Pennsylvania Constitution states:

> "The people have a right to clean air, pure water, and to the preservation of the natural, scenic, historic and esthetic values of the environment. Pennsylvania's public natural resources are the common property of all the people, including generations yet to come. As trustee of these resources, the Commonwealth shall conserve and maintain them for the benefit of all the people."

This essentially charges the Department "to preserve the natural values of the environment". This tends to complicate the "How Clean Is Clean?" issue even further. Taken to its most idealistic conclusion, this is interpreted by some to mean that all "How Clean Is Clean?" levels should be "no contamination" above that which is "natural". In reality and in most cases, cleanup of soils to such levels is virtually impossible to achieve. Similarly, with respect to hazardous waste, some Pennsylvania regulations require remediation of soil contaminated with hazardous waste constituents to "background" levels, i.e. the removal of all hazardous waste constituents.

Obviously, while the logic for determining "How Clean Is Clean?" levels is fairly simple from the perspective of protection of human health and the environment, its application often becomes very complicated. Currently, the risk-based approach to the "How Clean Is Clean?" issue is the most widely accepted and most common approach used by regulatory agencies. The concepts of this approach warrant further discussion.

"HOW CLEAN IS CLEAN?": THE RISK-BASED APPROACH

The determination of purely risk-based "How Clean Is Clean?" levels in soil is based on three general considerations. These include the following:

1. Environmental fate and transport of contaminants.
2. Exposure assessment.
3. Risk assessment.

Environmental fate and transport of contaminants

In order to determine the risk that a contaminant or contaminants in soil poses or will pose to human health and the environment (and ultimately formulate a "How Clean Is Clean?" level), one must first determine where the contaminant(s) will be located and at what concentrations as functions of time. To do this, one must apply the concepts of environmental fate and transport. These concepts are generally used to "predict" the movement of contaminants among the various environmental media, (i.e., soil, surface water, sediment, groundwater, air and biota) and are based on site-specific characteristics and physical/chemical characteristics of the contaminant(s).

Site-specific characteristics pertain to the characteristics of the actual environment in which the contaminants exist and the environment to which they could migrate due to various transport mechanisms. Such characteristics include meteorological conditions (temperature, wind velocity, humidity and precipitation rate), unsaturated zone soil properties (soil type(s), porosity, bulk density, soil water content, clay content, organic matter content, soil pH, and cation exchange capacity), geological/hydrogeological conditions (depth to bedrock and/or groundwater, type of bedrock and saturated zone soil properties, presence of shallow and/or deep aquifers, hydraulic conductivity, hydraulic gradient, effective porosity, aquifer thickness, and recharge rate), site topography/topology (surface contours and gradients, proximity of surface water bodies, and types and amounts of surface vegetation), surface water characteristics (type of surface water bodies, body dimensions and flow rate of surface water), sediment characteristics (organic matter content, particle sizes, and volume or amount of sediment), and biota characteristics (proximity to site, bioconcentration factors and position in food chain).

Physical/chemical characteristics of contaminants pertain to the characteristics of the specific contaminant(s) in the soil. These characteristics include concentration in soil, aqueous solubility, vapor pressure, saturated vapor concentration, latent heat of vaporization, diffusion coefficient, molecular weight, Henry's Law constant, dissociation constants, boiling point, melting point, valence state, biodegradation rate, various partitioning coefficients, bioconcentration factor, and solubility in non-aqueous phases.

These site-specific and contaminant-specific characteristics are numerous and represent most of the characteristics which could be needed to predict environmental fate and transport for a specific situation. Rarely would they all be used for a specific case. Generally, as the number of transport mechanisms being considered increases and/or as the complexity of the model used to predict environmental fate and transport increases, the number of characteristics considered also increases. Conversely, as the number of transport mechanisms being considered becomes smaller and as the transport model becomes simpler, the number of characteristics considered also becomes smaller.

A number of transport mechanisms are relevant in determining "How Clean Is Clean?" levels for soil contaminants. These mechanisms include migration of contaminants through the soil unsaturated zone into the saturated zone (groundwater); migration through the saturated zone; diffusion and volatilization into the atmosphere; overland migration of contaminants via precipitation runoff and erosion into surface waters; and uptake of contaminants by plants and animals. While any one or all of these mechanisms can be very significant on a site-specific basis, with regard to soil contamination, the Department has focused most often on the migration of contaminants through the soil unsaturated zone to groundwater. This is usually (but not always) the most significant route of migration with respect to "How Clean Is Clean?" levels in soil.

Several factors dictate the speed of contaminant migration through the soil unsaturated zone to groundwater as well as the contaminant concentration at any point (both spatial and temporal) during migration. These factors include concentration of contaminant in the soil, pore water velocity, dispersivity, depth to groundwater, contaminant degradation rate, soil bulk density and water content, soil organic matter content, cation exchange capacity, contaminant aqueous solubility, organic carbon partition coefficient, contaminant dissociation constant, and contaminant cosolubility in non-aqueous phases which may co-exist with the contaminant in the soil. A variety of models currently exists to predict the migration of contaminants through the soil unsaturated zone to the groundwater. Some very complicated computer models exist which predict this migration. Unfortunately, these models often require input values which are unavailable, and they may assign input values that are not applicable to the geographical location being considered. In cases where site-specific values are unavailable, "default" values are often used as input values. Generally, default values are extremely conservative and may not represent reality very closely. As the use of default values increases, uncertainty also increases thus raising the probability of error. Most of the complex, computer models are not particularly user friendly" so that they are usable by a very limited number of individuals. This severely limits the accessibility of such predictive models. Other relatively "user friendly" less complex computer models exist to predict contaminant migration through the soil unsaturated zone. These models require fewer input values (thus necessitating the use of fewer default values) and are easily usable by a variety of individuals. There also exist extremely simple predictive models which require no computer program and very few input values. Such models are environmentally protective because they are extremely conservative (in some cases by as much as two to three orders of magnitude compared to more complex models) although they may not give a reasonable representation of reality.

Similarly, once the migration of contaminants through the soil unsaturated zone has been predicted, models exist to predict migration of contaminants through the groundwater using similar input factors. These models also range from the very complex to the very simple.

In addition, models do exist that can predict the migration of contaminants from soil into the air, surface waters and biota. In the context of soil contaminant remediation, such models have historically not been used routinely by Department technical staff to develop risk-based "How Clean Is Clean?" levels.

Exposure assessment

Assuming that one has successfully modeled environmental fate and transport of the contaminant(s) in the soil (using an appropriate model) and has determined how much of the contaminant will be where and at what point in time, an exposure assessment can be performed. An exposure assessment is essentially a prediction of how much of a contaminant or contaminants various individuals or receptors (including humans and/or other living organisms) could potentially absorb from exposure to the contaminant via contact with various contaminated environmental media. The contact mechanisms (routes of exposure) that are generally considered are inhalation, ingestion and dermal contact. The applicable environmental media include air, soil, surface water, sediment, groundwater, fish and, less typically, plants and food products from animals other than fish (i.e., milk, beef and poultry). The routes of exposure and environmental media that are considered are largely dependent upon the results of the environmental fate and transport modeling as well as the proximity and location of receptors relative to the applicable environmental media.

After determining the applicable routes of exposure and environmental media, one must then make some very basic assumptions concerning the behavior and physiology of the potential receptors. These include (among other things) the amount of air inhaled over a given period of time; the amount of contaminant absorbed via inhalation; the amount of water ingested daily; the amount of contaminant absorbed via ingestion; the amount of soil ingested incidentally over a given period of time; the amount of fish ingested over a given period of time; the amount of soil dermally contacted over a given period of time; the amount of contaminant absorbed via dermal contact with soil; the lifespan and body weight of receptors; and the number of exposures over a given period of time. Standard assumptions are commonly used for many of these giving some consistency to exposure assessment. For example, an adult receptor is generally assumed to weigh 70 kilograms, drink two liters of water daily, inhale 20 cubic meters of air daily, and live 70 years. For several of the others, there is much less agreement among the experts concerning what assumptions are appropriate. This can significantly affect consistency as well as the validity of specific exposure assessments.

Another area which requires the use of assumptions concerns the future activities on and around the site which is being assessed. For example, will land use at the site change in the future which would change the exposure of receptors currently living in the vicinity of the site? Will access to the site be changed at some point in the future? Will additional receptors move onto the site in the future? Will future land use be restricted? Certainly, anything is possible, but the reasonableness in attempting to predict such future events becomes extremely crucial in an exposure assessment. The consideration of future site activities thus is a major source of uncertainty and may become a source of disagreement and inconsistency in exposure assessment.

Yet another factor which can affect the consistency of exposure assessment concerns the prediction of behavior of receptors who know that various environmental media around them are contaminated. For instance, receptors who believe that they are potentially being exposed to some contaminant may take voluntary action to reduce such exposure by moving, buying bottled drinking water, or restricting fishing and swimming in streams near the site.

Obviously, there are many assumptions inherent in exposure assessment. Extremely conservative exposure scenarios can be created which would have a conservative effect on "How Clean Is Clean?" levels (i.e., lower them). Similarly, extremely liberal exposure scenarios would raise "How Clean Is Clean?" levels in soil. The ideal scenario is one which most reasonably represents reality and therefore could be used to determine "How Clean Is Clean?" levels which are acceptably protective of human health and the environment. However, before such a determination can be made, risk assessment must also be performed.

Risk assessment

After the amount of contaminant which may potentially be absorbed by a receptor is predicted during the exposure assessment, one more piece of the puzzle is needed to determine the risk to the receptor from such exposure. This is the risk assessment. The risk assessment basically applies the toxicology of the contaminant or contaminants to the dose or amount of contaminant absorbed by the receptor (the result of the exposure assessment). This is most often done by comparing the dose of contaminant to various applicable standards, criteria, health advisories, etc. which are generally developed by the "scientific community" and are deemed to be protective of human health. While this concept is very simple, its actual application can be extremely complicated.

One of the major complicating factors is that thousands of chemical contaminants are found in the environment for which no such standards exist and for which little, if any, toxicological data are available to develop such standards. This severely limits risk assessment to a few hundred chemical contaminants which represent a relatively small fraction of the universe of contaminants detected in environmental media.

Another major complicating factor in risk assessment is that chemical contaminants can have a variety of toxic or adverse health effects including systemic, carcinogenic, teratogenic, and mutagenic effects. This, in turn, can result in multiple standards or criteria for a specific contaminant. A determination of which effects are most significant in the context of risk assessment must then be made. Similarly, many toxic health effects apply to only one specific route of exposure to the contaminant. No data may be available for the contaminant with regard to other routes of exposure so that extrapolations (which include assumptions) for other routes of exposure must be made. In many cases, several contaminants must be considered simultaneously and, therefore, the manner in which they affect each other toxicologically must be considered. There is currently a paucity of toxicological data in this area.

Additional problems concern those contaminants which are considered carcinogens. Currently, there are approximately a dozen chemicals which are classified as "known human carcinogens" based on human epidemiological studies. There are many other chemicals which are classified as carcinogens based primarily on laboratory animal studies. The data from such studies are extrapolated (using several assumptions) from animals to humans in order to determine carcinogenic risk levels for humans. Extrapolation of animal data to humans has become extremely controversial within the scientific community often casting considerable doubt on the validity of human carcinogenic risk derived from them. In addition, the most currently accepted thought on carcinogenesis states that there is no "absolutely safe" dose of a carcinogen, i.e., any dose of a carcinogen, no matter how small, could cause cancer in an individual. Unfortunately (or perhaps fortunately), this is not provable using existing data because doses of carcinogens administered to laboratory animals are extremely high when compared to the extremely low (i.e., approaching zero) doses which would be required to prove such an assertion. As a result, dose versus cancer incidence curves must be extrapolated at very low doses. Generally, it is assumed that the curve is linear at very low doses and that the curve passes through the origin resulting in the assumption that there is no safe dose for carcinogens. Both of these assumptions are currently being questioned by certain members of the scientific community.

Another complicating factor concerns the dynamic nature of the field of toxicology. Currently, toxicological data are being generated very rapidly and by several agencies to such an extent that it is very difficult to maintain familiarity with the most current available data. This often results in risk assessments which are based on outdated toxicological information which renders them technically incorrect.

As with exposure assessment, there are clearly many difficulties and sources of potentially large error in risk assessment. These problems are created by the extrapolation of data, lack of health standards/criteria, lack of toxicological data, and difficulty in accessing the most current toxicological data.

Risk management or "this is how clean clean is"

Before a "How Clean Is Clean?" level can be specified for contaminants in soil using the tools of environmental fate and transport, exposure assessment, and risk assessment, one seemingly simple decision must be made. This is purely a risk management decision which addresses the simple question "What risk to human health and the environment is deemed to be acceptably protective of human health and the environment?" Assuming that this question is answered in the form of some numerical level, "How Clean Is Clean?" levels for contaminants in soil can then be developed "scientifically" on a site-specific basis. The development of such "How Clean Is Clean?" levels begins with applying the acceptable risk (from the risk management decision) to all of the environmental media for which there is a concern that exposure to receptors could reasonably occur. The allowable risk for each of these media is then either stated as an already existing standard/criterion (if one exists) or is converted into an allowable dose using the concepts of risk assessment. The allowable dose is then converted into an allowable contaminant concentration in the applicable environmental medium using the appropriate exposure assumptions. (This conversion is not necessary for standards/criteria since such assumptions have generally been used already to develop those standards/criteria.) Environmental fate and transport modeling can then be done "in reverse" (using site-specific and contaminant-specific characteristics) to determine what contaminant concentrations in soil will result in the allowable contaminant concentrations in the applicable environmental media. These "allowable contaminant concentrations" in soil are then the "How Clean Is Clean?" levels for soil.

This procedure, if accomplished correctly, should allow the assessor to input the "How Clean Is Clean?" levels into the environmental fate and transport models, perform exposure assessment and risk assessment with the result being a risk value identical to that deemed to be acceptable in the risk management decision. While this procedure has been described here in very simplified terms, its application can be very complicated especially if more than one environmental medium is considered.

"HOW CLEAN IS CLEAN?": THE REALITY

With all of the problems (extrapolations, assumptions, lack of site-specific data, lack of toxicological data, failure to decide on an acceptable level of risk, etc.) currently inherent in environmental fate and transport modeling, exposure assessment, risk assessment, and risk management, risk-based "How Clean Is Clean?" levels can vary by several orders of magnitude depending on the various assumptions and data used by the assessor. This is most clearly apparent in a current environmental consulting advertisement which offers "scientifically defensible risk assessments that favor our client". Apparently, it does not matter what the "science" is in the mind of such a consultant because he or she can make it "say" whatever the client wants.

Admittedly, the discussion to this point does not create a picture that the risk-based determination of "How Clean Is Clean?" levels is very precise, and this may be somewhat misleading. The science of risk assessment obviously spans several scientific disciplines (geology, hydrogeology, toxicology, physiology, chemistry, biology, soil science, human behavior and, in some cases, chemical engineering, computer science and meteorology) and requires values for many input parameters. The key to accomplishing a proper risk assessment is to assign values which most closely represent reality. Clearly, it would be a very difficult task to choose the "correct" values for every parameter; however, there are mechanisms to determine which parameters are the most critical in the accomplishment of a risk assessment. Perhaps the most widely used mechanism is the sensitivity analysis. This is essentially the practice of doing the assessment while keeping all input values constant and changing one (or limited combinations of more than one input value) at a time. In this way the assessor can determine which input values will change the final result most significantly (i.e., which input parameters the assessment is most sensitive to). The assessor can then concentrate available resources on the task of determining the most realistic values to assign to the parameters to which the assessment is most sensitive.

To put the risk-based approach to the "How Clean Is Clean?" issue into perspective, it is important to realize that risk assessment is a relatively new science which requires an understanding of several disciplines as well as an understanding of the overall concepts of risk assessment. If used properly by individuals who possess an understanding of its capabilities as well as its limitations, risk assessment can be (and likely will be) a very powerful tool in determining "How Clean Is Clean?" levels in soil. It is extremely important that those using risk assessment have a proper understanding of it. It is very simple to assign values to input parameters in environmental fate and transport models, and exposure and risk assessments, but, if these values do not represent reality, the resulting answer will likewise not truly represent what the assessor is seeking (i.e., a contaminant concentration which is "protective of human health and the environment"). Unfortunately, the assessor will not be aware that the resulting value or values are not realistic. This is where the extreme danger in using risk assessment lies — unqualified individuals attempting to perform valid risk assessments.

CONCLUSION

Currently, the policy of the Department concerning the "How Clean Is Clean?" issue for contaminants in soil is to approach it on a site-specific basis. Remediation of contaminants in soil has been occurring in Pennsylvania throughout the 1980s using this approach. These remedial activities have largely incorporated the risk-based approach, to the extent possible, and have sometimes relied on a "cleanup to background" approach when feasible.

Ideally, "How Clean Is Clean?" levels should be risk-based (unless remediation to background levels is realistically feasible) in order to "protect human health and the environment". While there are currently a number of "experts" in this field, it must be understood that environmental fate and transport modeling, exposure assessment, risk assessment, and risk management are relatively new concepts and scientific tools which will require time to refine and improve to a point at which they can be used consistently by the scientific and regulatory communities at large and with relative confidence in the results.

Recently, risk assessment has gained increased use and popularity particularly within regulatory agencies and the field of environmental consulting. As time goes on, more and more individuals are using risk assessment. Unfortunately, the *use* of risk assessment if often *misuse.* Risk assessment must be used properly and its results viewed with extreme caution, otherwise those who rely on its results to make crucial management decisions may make decisions which are truly not protective of human health and the environment or which may actually be overly protective.

Management of Hazardous Materials and Wastes: Treatment, Minimization and Environmental Impacts. Edited by S. K. Majumdar, E. W. Miller and R. F. Schmalz. © 1989, The Pennsylvania Academy of Science.

Chapter Twenty-One

RISK ASSESSMENT AND MANAGEMENT[1]

BRUCE MOLHOLT

U.S. Environmental Protection Agency
Region III
841 Chestnut Street
Philadelphia, PA 19107

INTRODUCTION

Risk assessment has become a major tool in the analysis and management of hazardous materials and wastes. Central in the risk assessment process is the quantification of potential harm inherent in the constituents of hazardous wastes, which potential harm must be coupled with the ability of these same constituents to reach receptor populations, including humans. Although the processes employed by risk assessment and risk management may be applied to both human health and environmental endangerments, for the purposes of this discussion, I will address only human health concerns.

In general, two types of analyses are performed during quantitative risk assessment for endangerment to human health from hazardous materials and wastes. Most toxicants which adversely affect human health have thresholds below which no toxicity is apparent, such as the central nervous system toxicity from exposure to hydrogen cyanide gas. For hydrogen cyanide, the threshold for adverse human health effects is established by the affinity of the enzyme cytochrome oxidase in neuronal cells of the cerebral cortex and corpus callosum. If hydrogen cyanide emanates as a hazardous waste in groundwater at less than 770 ppb, there should be no evidence of toxicity to any individual in a receptor population. However, unlike hydrogen cyanide and other acute or chronic toxicants with thresholds, there is at present no evidence to show that carcinogenic initiators have thresholds below which they fail to induce cancer. Therefore, unlike noncarcinogenic toxic constituents of hazardous wastes, carcinogenic initiators are treated as if they

---

[1]The opinions expressed in this Chapter are those of the author and should not be construed as the official position of EPA.

continue to induce cancer until infinitely diluted and in proportion with the remaining concentration of carcinogen upon dilution.*

In this Chapter I will discuss some features of quantitative risk assessment as practiced by the Environmental Protection Agency (EPA) and other regulatory agencies. The emphasis of this Chapter will be upon carcinogenic risk and the utility of quantitative carcinogenic risk assessment as a tool in decision-making by risk managers. I will also discuss the limitations of carcinogenic risk assessment and proffer suggestions for reducing uncertainties in the risk assessment process by differentiating carcinogenic initiators from carcinogenic promotors.

For additional aspects of the health effects of hazardous wastes, the reader is directed to two articles which appeared in a previous addition of this series.[1,2]

CARCINOGENESIS

Recent advances in molecular genetics have afforded our contemporary understanding of the mechanisms whereby cancer is induced.[3] Carcinogenesis is a multistage process which may take as long as 40 years between initial exposure to a carcinogenic substance and frank development of a malignant tumor. The steps of carcinogenesis may be divided into three stages:[4]

1. *Initiation*—A minimum of two genetic events which result in the overexpression of oncogenes.
2. *Promotion*—A nongenetic event which results in cellular proliferation or differentiation.
3. *Progression*—Further genetic events which result in the overexpression of additional oncogenes and other genes which aid in the invasion and metastasis of evolving cancer cells. (See *Box*)

Genotoxic carcinogens are active in initiation and progression stages of carcinogenesis whereas nongenotoxic carcinogens may act as either promotors or immunosuppressors or both. Carcinogens with both genotoxic and nongenotoxic modes of action have been called "complete carcinogens".[5]

Differentiation of genotoxic from nongenotoxic carcinogens is important for the risk assessment process. It now appears that genotoxic carcinogens continue to cause DNA damage in proportion to their concentration down to infinite dilution, whereas nongenotoxic carcinogens have thresholds below which they fail to induce cellular proliferation or differentiation.[6] Short-term test systems, such as the Ames test, which measure mutagenesis as an end point

*The linear no threshold treatment of carcinogens and risk assumes a constant risk product of carcinogen concentration times exposed population. Hence, the following scenarios represent equivalent risks in terms of total numbers of cancers: 1) 100 persons exposed daily to 100 ppm carcinogen, 2) 10,000 persons exposed daily to 1,000 ppb carcinogen, and 3) 1 million persons exposed to 10 ppb carcinogen.

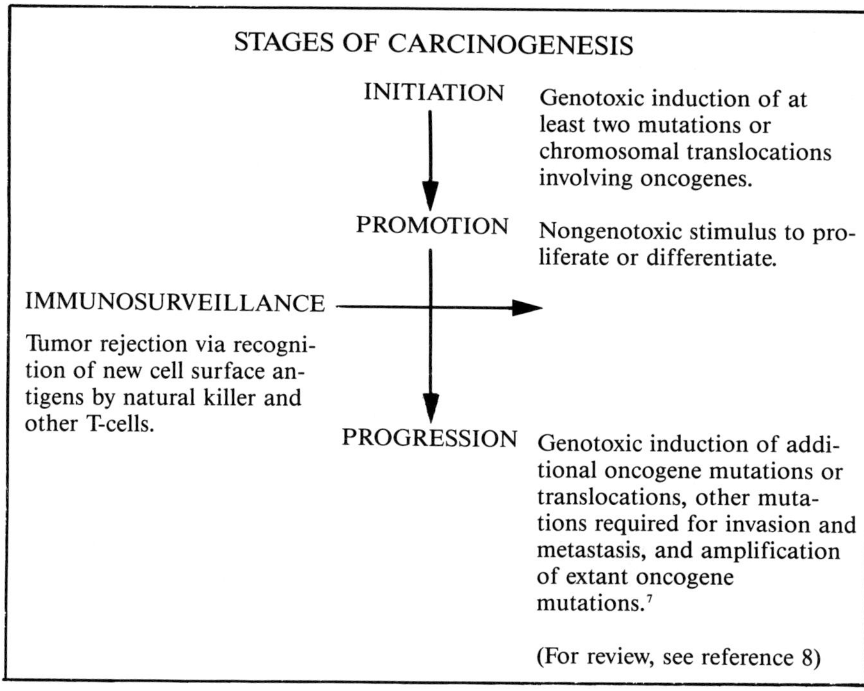

are useful in distinguishing genotoxic from nongenotoxic carcinogens. At present, in the quantitative carcinogenic risk assessment process, both genotoxic and nongenotoxic carcinogens are treated equivalently. However, it is anticipated that in the near future nongenotoxic carcinogens will be treated like noncarcinogenic toxicants for which safe thresholds have been established (e.g., reference doses or RfDs).

Quantification of Carcinogenesis

Although at the individual level carcinogenesis is frequent, afflicting 25 percent of our population in their lifetime and being a lethal affliction at approximately half that rate, carcinogenesis is a cellular disease and at the level of the cell is extremely rare. It has been estimated that an adult human contains 100 trillion (10^{14}) cells. Virtually all known cancers are clonal, i.e., result from the malignant transformation of one original cell which after initiation, promotion and progression goes on to form a tumor comprised of n generation descendants of the originally transformed cell, where the final tumor population is 2^{n-1}. Hence, at the cellular level, the frequency of carcinogenesis is $0.25/10^{14} = 2.5 \times 10^{-15}$, or less than three tumor-initiating events per quadrillion human cells per lifetime.

Chances are that carcinogenic initiation events occur much more frequently than 2.5×10^{-15} per cell. The genetic events (mutations or chromosomal translocations) which mitigate carcinogenic initiation are known to occur much more frequently in individuals with DNA repair defects, such as Xeroderma pigmentosum, Ataxia telangiectasis, Bloom's syndrome and Fanconi's anemia. Secondly, many individuals who have died of other diseases are found to have occult cancers at autopsy. It would appear that the immune system, specifically natural killer T-lymphocytes, keeps many microscopic cancers in check which are only able to progress during immunosuppression. Our experience with increased cancer incidence in organ transplant recipients would indicate that the immune system is responsible for rejecting about 99 percent of all induced tumors.* Carcinogenic initiation is also much more frequent in individuals who inherit one of the two genetic events required for initiation at birth. In the case of retinoblastoma, inheritance of one defective allele *rb-1* at the 14th band of the long arm of chromosome 13 (13q14) results in a carcinogenic frequency of about 10^{-6} per retinoblast.[11] Since each eye contains millions of retinoblasts, it is 95 percent certain that a *rb-1* carrier will develop at least one and most likely four retinoblastomas. Hence, this cancer normally strikes bilaterally during childhood.

Classification of Carcinogens

From the discussion of genotoxic initiators and nongenotoxic promotors above, it is obvious that all carcinogens cannot be treated equivalently in the risk assessment process. At the present time EPA and other regulatory agencies have categorized carcinogens not as to genotoxic v. nongenotoxic modes of action, but according to the adequacy of data in proof of carcinogenicity. The present EPA scheme for classifying carcinogens is as follows:

TABLE 1

EPA Carcinogen Classification[12]

	Strength of carcinogenic evidence		
Class	Human epidemiology	Animal experimentation	Carcinogen category
A	Strong	Strong	Definite human
B1	Suggestive	Strong	Probable human
B2	Insufficient	Good	Probable human
C	Insufficient	Weak	Possible human
D	Insuff. or Neg.	Negative	Noncarcinogenic
E	Insufficient	Insufficient	Unknown

*Organ transplant recipients require profound immunosuppression with agents such as the cyclosporins in order to prevent heterologous graft rejection. The subsequent cancer frequency in these patients is 10-100 times that expected.[9,10]

Additional data are considered in the above classification, such as the number of species affected, number of organ sites affected, whether both sexes are affected and the results of mutagenicity and other short-term tests.

Based upon valid human epidemiologic studies and tests in experimental animals carcinogenic potencies have been established for several Class A, B1, B2 and C carcinogens by various groups within EPA*. Obviously, the validity of extrapolating carcinogenic potencies from studies in order to assess potential risks for human populations through exposure to hazardous waste constituents depends heavily on the quality of the data, with epidemiologic results for Class A carcinogens being preferable to experimental animal data in that *Homo sapiens* is the species studied. On the other hand, it would be foolish to perform risk assessments only on those compounds for which there exists valid human data (a tactic not in line with the best tenets of human health protection). Hence, carcinogenic potencies from rodent studies are frequently extrapolated to humans, with less and less assurance as to uncertainty as one descends the classification scheme through Classes B1, B2 and C. In that only 13 percent of carcinogens are classified by EPA as Class A or B1 at present, the majority of carcinogenic risk assessment depends upon the translation of animal data to humans.[14] About ¾ths of all carcinogenic potencies are derived from Class B2 carcinogens as tested in rodent species, often with liver tumors as the only positive indication of carcinogenicity.

Carcinogen Risk Assessment

Risk assessment for carcinogens as presently employed by EPA[12] utilizes a linear multistage model of carcinogenesis which allows direct extrapolation of lifetime cancer risk from daily exposure times carcinogenic potency in compatible units. Usually exposure rates are quantified as mg/kg/day and carcinogenic potencies expressed in $(mg/kg/day)^{-1}$ such that the product of exposure and potency is a pure number representing the lifetime risk of carcinogenesis to an individual. Since all evidence presents a range of possible carcinogenic potencies and dose extrapolation is frequently over orders of magnitude, the potencies derived represent the upper-bound 95 percent confidence level determination, i.e., there is a 19/20 chance that calculated carcinogenic potencies are too high (they could in fact be orders of magnitude too high for Classes B2 and C) and only a 1/20 chance that they have been calculated too low.

An example of risk assessment for a common carcinogen in urban water supplies is didactic. Due to the necessity for disinfection all the way from the treatment plant to the end of the line, most urban water companies employ chlorine

* Carcinogen Assessment Group (CAG), Health Effects Assessment (HEA) and Integrated Risk Information System (IRIS).

gas as a disinfectant. Unfortunately the action of chlorine on humic acids from decaying organic debris in surface waters produces trihalomethanes, mainly chloroform. Chloroform is a Class B2 carcinogen with a carcinogenic potency of 0.081 $(mg/kg/day)^{-1}$.[14] Not long ago chloroform levels in many city water supplies generally exceeded 100 ppb, especially in the summer. EPA established a maximum contamination level (MCL) for chloroform at 100 ppb and has enforced this MCL by fining municipalities which exceed the limit.

How safe is the chloroform MCL of 100 ppb? If one assumes consumption of 2 liters of water per day, the daily ingestion of 100 ppb chloroform (100 μg/l) becomes 200 μg or 0.2 mg. For a 70 kg adult, this intake is thus $0.2/70 = 0.00286$ mg/kg/day. Lifetime cancer risk from this source for an individual may be then computed as 0.00286 mg/kg/day x 0.081 $(mg/kg/day)^{-1} = 0.00023$ or 2.3×10^{-4}. If a population of 100,000 were to drink water containing 100 ppb chloroform for a lifetime, this risk assessment would predict that at most twenty-three persons in this population could develop cancer from chloroform contamination at the MCL limit.*

Other carcinogens have much higher carcinogenic potencies than chloroform. Some of these more potent carcinogens, and their lifetime cancer risk at 100 ppb in water are shown in the following table.

TABLE 2

Lifetime Cancer Risks from 100 ppb in Water

Vinyl Chloride	6.6×10^{-3}
Benzidine	6.6×10^{-1}
Methylnitrosourea	8.6×10^{-1}

RISK ASSESSMENT FOR INITIATORS

Carcinogenic initiators act by inducing mutations or chromosomal translocations within or near the 35-50 oncogenes which have been found to be widely scattered over the 22 autosomes of human cells. Initiators may be easily screened in short-term test systems, such as the Ames test and other microbial or cell culture tests which use mutagenesis or evidence of chromosomal damage as endpoints. In short-term test systems, carcinogenic initiators induce mutations or chromosomal damage in direct proportion to their concentration (linear dose-response curve). Similarly, in animal test systems for carcinogenesis, dose-response curves are linear over the response which is measurable above background. In order to attain statistical significance (at the 95 percent confidence level), experimental animal population sizes necessarily increase as less and less animals contract cancer through exposure to diminishing concentra-

*This calculation does not include risk from any volatilization of chloroform from cooking, showering, washing clothes, etc., which risks may exceed those from drinking water.

tions of carcinogens. Hence, the largest study of dose-response relationships for a carcinogenic initiator, the ED_{01} or "megamouse" study, employed 24,000 mice and 8,000 of them in that part of the study which showed 1 percent induction of cancer above the control (untreated) group.

All evidence for carcinogenic initiation indicates that the dose-response curve is linear for as far as one can measure with statistical confidence. However, for test animals this is a cancer incidence of 10^{-2} as a lower limit and considerably higher for human epidemiologic studies. On the other hand, regulatory agencies such as EPA frequently attempt to protect public health from carcinogenic exposures in the 10^{-7}-10^{-4} range, which is two to five orders of magnitude below available dose-response data. While the assumption of linearity is prudent and has not been disproved, nevertheless linear extrapolation of animal or human data to several orders of magnitude less than can ever be measured remains an enterprise which engenders considerable skepticism. A careful examination of the literature and understanding of initiation mechanisms, however, has convinced this reviewer that the linear dose-response curve for carcinogenic initiators is essentially correct. Furthermore, in that further mutagenic responses are required for oncogenic progression of initiated and promoted tumors through invasion and metastasis, carcinogenic initiators are active in both ends of the multistep sequence of events necessary for induction of frank (potentially lethal) malignancy. The situation with promotors, however, is far from clear.

RISK ASSESSMENT FOR PROMOTORS

In some of Ames', original studies with carcinogens and back mutation of histidine-requiring *Salmonella typhimurium* strains, he and his coworkers found that roughly 90 percent of those carcinogens tested gave a mutagenic response. Many of these carcinogens required activation with rat liver microsomal oxidases in order to become mutagenic and, similarly, it is known that many carcinogens must be metabolized by these same enzymes in vivo in order to become ultimate carcinogens (e.g., the metabolism of benzo(a)pyrene to benzo(a)pyrene diol epoxide). Since the advent of rodent liver tumor assay systems, viz., hepatocarcinogenesis in B6C3F1 mice and F344 rats, many more carcinogens have been added to the list which fail to induce mutations in short-term test systems. It would appear that the 10 percent of Ames originally tested carcinogens and nearer 50 percent of those carcinogens added to B2 and C Classes over the past decade do not act to mutate DNA or to directly mitigate chromosomal translocation. These nongenotoxic carcinogens are promotors.*

*There is no consistent definition of carcinogenic promotion. Here I am defining promotion as nongenotoxic exacerbation of carcinogenesis by mechanisms other than immunosuppression. Whereas most promotors enhance carcinogenesis only after initiation, this broad definition of promotion includes activities which precede and predispose initiation.

Promotors are bizarre in the range of biological activities they display and in the ways they exacerbate the carcinogenic response in contaminated tissue.[20] Here I briefly describe four promotor mechanisms with some insight as to why they increase carcinogenesis:

1. Increased DNA replication and/or reduction in DNA repair,
2. Increased numbers of target cells for carcinogenesis,
3. Increased opportunity for DNA translocation, and
4. Decreased intercellular communication through closure of gap junctions.*

1. *Increased DNA replication and/or reduction in DNA repair*

DNA repair enzymes travel up and down DNA helices scavenging base mispairs, damaged bases and deoxyribose-phosphodiester backbone cleavage. In normal human cells there is ample time between DNA replications (ca. 20 hours) to repair most DNA damage via strand scission, stepwise nucleotide excision and strand replacement through faithful polymerization of new DNA using the opposite, intact strand as template. Upon exposure to genotoxic agents (carcinogenic initiators), there essentially is a "race" between replication and repair DNA polymerases as to which will reach the induced DNA lesions first. If replication polymerases arrive at the lesion prior to repair polymerases, the defect is replicated into daughter DNA such that both strands now contain the induced mutation. The newly replicated region, even though mutant, looks essentially normal to repair enzymes, and, hence, the genotoxin-induced mutation is entrapped in one daughter DNA strand and, eventually, in one daughter cell. If this mutation is in or near an oncogene, the unrepaired lesion may be one of the two genetic events required to initiate cancer.

Many promotors enhance cellular replication directly by mimicking mitogenic factors at the cell surface, e.g., the strong affinity of 2,3,7,8-TCDD for epidermal growth factor receptor.[22] Other promotors may inhibit DNA repair through binding metallic cofactors. Any substances which increase the rate of DNA replication relative to the rate of DNA repair will be active in this type of promotor activity.

2. *Increased numbers of target cells for carcinogenesis*

It appears that oncogenes are tissue-specific, most probably reflecting which growth factors and receptors were important in embryologic and fetal development. Hence, the opportunity to form a malignant clone which survives immunorejection is dependent upon the target cell population size which is sensitive to carcinogenesis.

*These mechanisms are by no means exclusive. Others include induction of P450 enzymes, peroxysomes and protein kinase C. Even iodine-deficiency promotes thyroid gland carcinogenesis.[21]

Many mechanisms exist for differential expansion of one subset of human cells with respect to other subsets. Some of these mechanisms are in common with those enhancers of DNA replication mentioned above, such as mitogenic agents. Additional stimulators of cellular replication may not grossly enhance DNA replication rates, but allow cells to replicate for a longer period of time, substantially increasing the target cell population for carcinogenesis. Hormones definitely fall into this category. For example, it is known that mothers who took diethylstibestrol (DES) for control of adverse side-effects during pregnancy gave birth to daughters with increased risk for genital cancer. Premature expansion of genital epithelium predisposed carcinogenesis in these daughters who had been affected by DES in utero.

Solid state carcinogenesis also proceeds via increased populations of carcinogen-sensitive cells. Brand's experiments with carcinogenesis induced by inert plastic sheets inserted under mouse skin showed the resultant tumor response was a direct function of the plastic sheet surface area, i.e., the number of host cells which attempted to wall off this foreign insert.[23] Chemically inert human carcinogens such as asbestos may work in this manner with millions of fibers in the asbestosis lung predisposing the induction of bronchiogenic carcinoma or mesothelioma by other carcinogenic initiators. Asbestosis is characterized by thickenings of pleural mesothelium (plaques) and virtually all mesotheliomas have been linked to asbestos exposure.

3. *Increased opportunity for DNA translocation*

Chromosomal translocations are required for the generation of immunoglobulin diversity during B-lymphocyte development in bone marrow. For each of the light and heavy chains which in doublet form will comprise final immunoglobulin molecules there exists a multitude of genes for variable and constant portions. Chromosomal breakage and reunion fuses various variable and constant regions together in the major mechanism for generation of diversity in immunoglobulin production, allowing for roughly one million different specificities. These chromosomal translocations which are required for B-lymphocyte diversity occur in the pre-B-cell stage of differentiation in bone marrow.

Development of B-lymphocytes in bone marrow is orchestrated by intracellular signals from other marrow cells. In particular, stromal cells elicit the signal which moves the pre-B-cell along towards a mature B-lymphocyte which will leave bone marrow and enter the bloodstream as an immunocompetent cell. Wierda and his colleagues have found that stromal elicitation of pre-B-cell maturation factor is inhibited by hydroquinone, which in turn is a metabolic product of benzene catabolism.[24] Hence, daily exposure to benzene "freezes" B-lymphocyte development at the pre-B-cell stage and allows for abnormally long time for chromosomal translocation in and near immunoglobulin genes.

Benzene is known to induce leukemias in exposed worker populations. Among the leukemias induced; acute myelogenous leukemia (AML), chronic myelogenous leukemia (CML) and multiple myeloma (MM); one, CML, has an associated chromosomal translocation which is practically diagnostic for the disorder. Patients with CML almost always bear numerous B-lymphocytes with a 9q x 22q translocation which results in a smaller chromosome 22 which carries at its long arm (q) terminus a piece of the q terminus derived from chromosome 9. The smaller translocated chromosome 22 is easily spotted in metaphase spreads of lymphocyte cultures from peripheral blood and is called the Philadelphia chromosome, or Ph^1, due to its discovery at the University of Pennsylvania School of Medicine.[26] This specific translocation which is seen in virtually every CML case juxtaposes a very strong immunoglobulin gene, C_λ, which encodes formation of the constant region of the light chain lambda, with a normally quiescent oncogene, *abl*. This rearrangement of chromosomes activates the *abl* gene which in turn causes clonal expansion of that cell with the original 9 x 22 translocation, eventually flooding peripheral circulation with progeny which all contain Ph^1, leading to the CML phenotype (disease).

4. *Decreased intercellular communication through closure of gap junctions*

Cytoplasmic substances of normal cells are able to suppress the carcinogenic phenotype of fully initiated cancer cells. This dominance of normality can be seen in monolayers of cell cultures in which the cancer phenotype leads to a focus of non-contact-inhibited cells. The reason that a cancer cell surrounded by noncancer cells is suppressed in overzealous replication is that juxtaposed cells communicate cytoplasms via holes in lamenated outer membranes called gap junctions. These gap junctions allow suppressors of the cancer phenotype to flow freely from normal cells to cancer cells, causing the cancer cell, even though genetically initiated (two mutations in oncogenes), to behave normally. Once these gap junctions are closed, however, which closure has been demonstrated with many carcinogenic promotors (e.g., dieldrin), the cancer cell in culture no longer receives suppressors from neighboring cell cytoplasms and, hence, is relieved of normal controls.[27,28] Now the cancer cell can multiply even while in contact with its normal neighbors and clonally expand to two daughter cells, then four, then eight, etc. Soon a knot of clonally expanding cancer cell nucleus is surrounded not by normal tissue, but by other cancer cells of the same clone such that once gap junctions are reopened (through removal of promotor substance), suppression of continued expansion by elements of normal cytoplasm is no longer possible.

These four different mechanisms for carcinogenic promotion are by no means exhaustive and have not considered some of the classic promotion models, such as two-stage carcinogenesis in mouse skin. The purpose of this somewhat lengthy discussion has been to demonstrate the variety of metabolic activities which work in concert with carcinogenic initiators to induce human cancers. In addition to the promotor-type activities listed here, other nongenotoxic agents, such as immunosuppressors certainly have an exacerbating effect on the frequency of carcinogenesis. Unlike genotoxic agents which initiate the carcinogenic process through mutagenesis or DNA damage in or near oncogenes, nongenotoxic promoting agents have thresholds below which they fail to mitigate requisite metabolic functions (increased DNA synthesis, increased cellular replication or inhibition of cellular communication). In the next section I consider the implication for these two different dose-response models for risk management of carcinogenic initiators and promotors.

RISK MANAGEMENT OF CARCINOGENIC INITIATORS AND PROMOTORS

At present, regulatory decisions for carcinogenic substances have been based upon the linear multistage model of risk assessment regardless of whether the carcinogens in question are initiators or promotors. As discussed above, in that carcinogenic initiators and promotors have different modes of action in target cells, I advocate that these two classes of carcinogens be managed differently.

To summarize, carcinogenic initiators act via DNA mutation or chromosomal damage. Available data for initiators indicate that they continue to act in genotoxic fashion in direct proportion to their concentration until diluted to the point where their activity becomes indistinguishable from background effects. By contrast, carcinogenic promotors, although acting in a wide variety of fashions, all have in common the lack of genotoxic effect. Available evidence indicates that carcinogenic promotors have thresholds below which they fail to induce metabolic derangement and exacerbate carcinogenesis by initiators.

The present EPA classification for carcinogens is based upon available quantification of carcinogenic proof in human epidemiologic studies and in experimental animal tests. It does not differentiate between carcinogenic initiators and promotors, and guidelines for risk assessment with carcinogens as proposed by EPA utilize the linear multistage model for determining risk for all carcinogens. A close look at carcinogens show promotor-type substances in every EPA Class (Table 3).

It may be seen that many carcinogens which are definite in humans (Class A) are promotors, such as asbestos, benzene and DES, whose promotor-type mechanisms have been discussed on previous pages. Based upon its lack of genotoxic activity, it would also appear that beryllium (Class B1) and perhaps other carcinogenic metals as well act more as promotors than initiators (and much stronger when inhaled than when ingested). By far the largest group of

TABLE 3

Some Examples of Carcinogenic Initiators and Promotors

Class	Initiators	Promotors
A	Benzidine Vinyl chloride Radon-222	Asbestos Benzene Diethylstibestrol
B1	Acrylonitrile	Beryllium
B2	Benzo(a)pyrene Bis(chloroethyl)ether Nitrosamines	Dieldrin PCBs TCDD
C	?	Saccharin

carcinogens falls into Class B2 (73 percent, at present), for which many, such as dieldrin, PCBs and TCDD have little or no genotoxic activity. Finally, all listed Class C carcinogens lack demonstrable genotoxic activity, and, like some of the nongenotoxic Class B2 entries, are deemed carcinogenic solely upon their responsiveness in rodent liver tumor assays.

From a risk management perspective, regulation of carcinogenic initiators versus carcinogenic promotors should logically employ two different tactics regarding low concentrations such as those entering groundwater from a hazardous waste site. Whereas the present linear multistage nonthreshold model applies to carcinogenic initiators, carcinogenic promotors are most amenable to statistical treatment as if they had thresholds below which there is no adverse effect. Hence, by virtue of their mechanisms of action and observed dose-response curves, carcinogenic promotors should be treated in risk assessment equivalently to non-carcinogens, which have reference doses (RfDs) assigned rather than quantitative risk based numbers such as 10^{-5}, 10^{-6}, and 10^{-7}.

PROPOSAL

I propose that the present classification of toxic agents into carcinogenic and noncarcinogenic agents be revised into three categories:

1. Non-carcinogens
2. Carcinogenic initiators
3. Carcinogenic promotors

For purposes of risk assessment, non-carcinogens and carcinogenic promotors will both be assigned reference doses (RfDs) below which they will have no adverse health effects in exposed human populations. Carcinogenic initiators will continue to be quantitatively assessed according to extant risk assessment guidelines. In complex sites where both carcinogenic initiators and promotors coexist, cancer risk assessment will be the product of initiator risk and multiples of RfDs for promotors. (The use of RfD multiples is similar to Hazard Indices for non-carcinogens as employed by EPA).

This proposal will require two innovative policy revisions by risk managers at EPA:

1. The establishment of RfDs for all known carcinogenic promotors.
2. A revision of carcinogen classification to clearly differentiate promotors from initiators.

In most cases, identification of carcinogenic promotors is facilitated by their lack of genotoxic activity in short-term test systems and their sole carcinogenic response in experimental animals limited to liver tumor induction. Perhaps a committee (such as CAG, HEA, IRIS or Risk Assessment Forum Project) can establish RfDs for carcinogenic promotors from extant rodent liver data, or request the proper tests when data are insufficient for determination of threshold concentrations. Any carcinogen with insufficient data to categorize it as initiator or promotor should be treated as if it is an initiator for public health protection. Some carcinogens possess both initiation and promotion activities, although these complete carcinogens are among the most potent, and therefore already known and firmly established as initiators.

REFERENCES

1. Heath, C.W., Jr. 1984. The effects of hazardous waste on public health. *In* Majumdar, S.K. and E.W. Miller (eds.). *Hazardous and Toxic Wastes: Technology, Management and Health Effects,* Pennsylvania Academy of Science, 376-387 pp.
2. Lumb, G. 1984. Health effects of hazardous wastes, *In ibid,* 388-405 pp.
3. Bishop, J.M. 1987. The molecular genetics of cancer. Science *235:* 305-311.
4. Farber, E. 1981. Chemical carcinogenesis. New Engl. J. Med. *305:* 1379-1389.
5. Weinstein, I.B., M. Wigler and C. Pietropaolo. 1977. The action of tumor-promoting agents in cell culture, 751-772 pp., *In* Ref. 8.
6. Stott, W.T. and P.G. Watanable. 1982. Differentiation of genetic versus epigenetic mechanisms of toxicity and its application to risk assessment. Drug Metabol. Rev. *13*: 853-873.
7. Nowell, P.C. 1986. Mechanisms of tumor progression. Cancer Res. *46:* 2203-2207.

8. Hiatt, H.H., J.D. Watson, and J.A. Winsten (eds.). *Origins of Human Cancer.* Cold Spring Harbor Press, Cold Spring Harbor, NY 11724.
9. Penn, I. 1975. The incidence of malignancies in transplant recipients. *Transplant Proc.* 7: 323-335.
10. Stutman, O. 1977. Immunological surveillance, 729-750 pp., *In* Ref. 8.
11. Knudson, A.G., Jr. 1977. Genetic predisposition to cancer, 45-52 pp., *In* Ref. 8.
12. U.S. Environmental Protection Agency. 1986. Guidelines for carcinogenic risk assessment. Fed. Reg. *51,* 33992-34002.
13. Allen, B.C., A.M. Shipp, K.S. Crump, B. Kilian, M.L. Hogg, J. Tudor and B. Keller. 1987. Investigation of cancer risk assessment methods. Clement Assoc., Inc., 1201 Gaines St., Ruston, LA 71270.
14. U.S. Environmental Protection Agency. 1986. Superfund public health evaluation manual. Off. Emerg. Remed. Resp. Washington, D.C. 20460.
15. Littlefield, N.A., J.H. Farmer and D.W. Gaylor. 1979. Effects of dose and time in a long-term, low-dose carcinogenic study. J. Environ. Path. Toxicol. *3,* 17-34.
16. Waldren, C., L. Correll, M.A. Sognier and T.T. Puck. 1986. Measurement of low levels of x-ray mutagenesis in relation to human disease. Proc. Natl. Acad. Sci. US *83,* 4839-4843.
17. Birnboim, H.C. 1982. DNA strand breakage in human leukocytes exposed to a tumor promoter, phorbol myristate acetate. Science *215,* 1247-1249.
18. Trosko, J.E. and C.C. Chang. 1985. Role of tumor promotion in affecting the multihit nature of carcinogenesis. *In* Woodhead, A.D., C.J. Shellabarger, and V. Pond (eds) *Assessment of risk from low-level exposure to radiation and chemicals,* Plenum Press, N.Y. 10013.
19. Goldsworthy, T., H.A. Campbell and H.C. Pitot. 1984. The natural history and dose-response characteristics of enzyme-altered foci in rat liver following phenobarbital and diethylnitrosamine administration. Carcinogenesis *5,* 67-76.
20. Boutwell, R.K. 1974. Function and mechanisms of promoters of carcinogenesis. CRC Crit. Rev. Toxicol. *2,* 419-459.
21. Ohshima, M. and J.M. Ward. 1986. Dietary iodine deficiency as a tumor promoter and carcinogen in male F344/NCr rats. Cancer Res. *46,* 877-883.
22. Madhukar, B.V., D.W. Brewster and F. Matsumura. 1984. Effects of in vitro-administered 2,3,7,8-tetrachlorodibenzo-p-dioxin on receptor binding of epidermal growth factor in the hepatic plasma membrane of rat, guinea pig, mouse and hamster. Proc. Natl. Acad. Sci., US *81,* 7407-7411.
23. Brand, K.G., L.C. Buoen, K.H. Johnson and I. Brand. 1975. Etiological factors, stages and the role of the foreign body in foreign body tumorigenesis: a review. Cancer Res. *35.* 279-286.
24. Wierda, D. 1987. Seminar at Toxicology Update '87, Johns Hopkins Univ. Baltimore, MD.

25. Nowell, P.C. and C.M. Croce. 1986. Chromosomes, genes and cancer. Am. J. Pathol. *125*, 8-15.
26. Nowell, P.C., J. Jensen and F. Gardner. 1975. Two complex translocations in chronic granulocytic leukemia involving chromosomes 22, 9 and a third chromosome. Humangenetik *30,* 13-21.
27. Wade, M.N., J.E. Trosko and M. Schindler. 1986. A fluorescence photobleaching assay of gap junction-mediated communication between human cells. Science *232,* 525-528.
28. Oh, S.Y., B.V. Madhukar and J.E. Trosko. 1988. Inhibition of gap junctional blockage by palmitoyl carnitine and TMB-8 in a rat liver epithelial cell line. Carcinogenesis *9,* 135-139.

Management of Hazardous Materials and Wastes: Treatment, Minimization and Environmental Impacts. Edited by S. K. Majumdar, E. W. Miller and R. F. Schmalz. © 1989, The Pennsylvania Academy of Science.

Chapter Twenty-Two

OCCUPATIONAL SAFETY AND HEALTH GUIDANCE: RULES, REGULATIONS, AND PROTECTIVE EQUIPMENT

THADDEUS E. TOMCZAK
Safety Training Programs Manager
Eastek Corp.
3543 Baldwin Drive
Easton, PA 18042

AN OLD PROGRAM

Protecting workers from exposure to hazardous materials is a relatively simple matter if the decision maker has available and uses the necessary information. We will not consider radioactive materials in this section, although the basic methods for protecting personnel are the same; it is certainly easier to know both if a person has been contaminated and when they have been thoroughly decontaminated.

Protection from the hazards of exposure to only a very limited number of substances was required by law. Guidance for selecting proper forms of protection was vague. OSHA relied on a hierarchy of methods: engineering control was the method of choice; second was administrative control; personnel protective equipment (PPE) was a poor third choice, to be used when all else failed or could not be used. Hazardous waste sites, unless they can be operated or remediated by discardable robots, are an area where engineering controls have relatively limited applicability and PPE becomes the first line of protection.

FEDERAL GUIDANCE

A number of Federal Agencies have an interest in hazardous chemicals, either as freshly produced materials or as waste materials. Four of these Agencies,

NIOSH, OSHA, Coast Guard, and EPA, recognized several facets of the problem. These Agencies pooled some of their resources to produce an "Occupational Safety and Health Guidance Manual for Hazardous Waste Site Activities." The quality of the guidance given improves, and information it contains increases with each new issue. It is available from the Superintendent of Documents, U.S. Government printing Office, Washington, DC 20402. This material is for guidance; the advice contained is not mandatory.

FEDERAL REGULATION

President Ronald Reagan signed the Superfund Amendments and Reauthorization Act of 1986 (SARA) into law on October 17, 1986. SARA directed the Occupational Safety and Health Administration to promulgate regulations (safety standards) to protect workers at Superfund sites, most RCRA sites, and certain responders at hazardous waste emergencies. OSHA published its "Interim Final Rule" in the Federal Register on December 1, 1986. The "Interim Final Rule" was to have been followed and superseded by a "Final Rule" within one year of publication of the "Interim Final Rule", even though the "Interim" will remain in effect for one year after publication of the "Final" rule. The "Interim" will be the rule for some time yet. As of December 1988, OSHA has not published the "Final Rule".

SUBSTANCE OF THE OSHA REGULATION

"Hazardous Waste Operations and Emergency Response" is the title of the OSHA regulation published as 29 CFR 1910.120 in the Code of Federal Regulations. In essence, this regulation requires hazardous waste site employers to gather all the information possible about any hazards their employees may be exposed to, including the identity of the materials at the site and the severity of the hazards these materials may present; the topography of the site as it may affect dispersion of hazardous gas or vapor clouds and escape routes from the site; atmospheric conditions as these affect employee exposure to released materials. The employer is required to devise a program for operating the site in such a way that employees are not exposed to harmful levels of these materials, and to document that they are not overexposed by monitoring and recording all atmospheric exposures, and by providing medical surveillance for workers.

GENERAL REQUIREMENTS

Providing effective personal protective equipment for each employee is an important requirement of the standard. (Due to the large number of factors influencing the selection of equipment, and the numerous combinations of equipment that can be used as an effective ensemble for providing this protection, PPE selection will be discussed after the summary of the requirements.) The standard describes the factors that determine the acceptable choices of equipment in appendices, not in the mandatory text.

SITE SAFETY AND HEALTH PROGRAM

A written Site Safety and Health Program is the first and most important requirement of OSHA's regulation. This program has twelve sub-programs which are mandated for each hazardous waste site in subparagraphs (c) through (n) of 29 CFR 1910.120. These sub-programs are:

- (c) Site characterization and analysis
- (d) Site control
- (e) Training
- (f) Medical surveillance
- (g) Engineering controls, work practices, and personal protective equipment for employee protection
- (h) Monitoring
- (i) Informational programs
- (j) Handling drums and containers
- (k) Decontamination
- (l) Emergency response
- (m) Illumination
- (n) Sanitation of temporary workplaces

Essentially, the employer must have a site safety and health program if any employee of that employer will enter a hazardous waste site in any capacity, and the employee must be covered by that program.

SITE CHARACTERIZATION AND ANALYSIS

The employer must maintain current information, including topographical maps and inventories of all materials known or suspected to be on the site indicating the quantities, identity, and condition of the materials as they exist on the site. In addition to the map of the site, its location and approximate size must be stated, the accessibility of the site by road and by air must be described along with possible pathways for dispersion of hazardous substances. The status and capabilities of emergency response teams that would provide assistance to on-site employees in event of emergency must also be stated and kept current.

Employees performing a site survey (initial characterization) must also be protected even during the first entry to the site so that their exposures do not exceed permissible exposure limits for substances for which PELs have been established, or if PELs have not been established they are not exposed to any level otherwise known to be harmful. (PELs are defined by OSHA in Subpart Z of the General Industry Standards, Toxic and Hazardous Substances. The "otherwise known to be harmful" addresses more materials than the PELs. This information is most easily available from material safety data sheets (MSDSs) which are required by OSHA's Hazards Communication Standard, 29 CFR 1910.1200. However, unless exposure limits on the MSDSs are not obtained from the OSHA PELs, the ACGIH's published TWAs or NIOSH "RELs", their reliability may be questionable.) Employees must be informed of all the hazards found, and the risk these materials may pose to them.

SITE CONTROL

This program must be developed during the planning stages of a hazardous waste site operation clean-up. It will include a site map indicating work zones, standard operating procedures, communications to be used on site and to summon emergency assistance when needed, and practices such as a "buddy system" for assuring employee safety. Arrangements must be made for providing emergency medical assistance, and the procedure for getting this assistance must also be included in the site control program.

TRAINING

This is one of the best features of this Standard. Every person employed at a hazardous waste site must be trained to know the hazards that may confront him or her, including their nature, their severity, and the consequences of exposure to the hazards, what means are provided for their protection (including engineering controls, work practices, and PPE), and how they must use these "means of protection". Employees must know the symptoms and signs of overexposure to hazards known or suspected at that site. Employees must also know to whom hazards will be reported, including symptoms or signs which could indicate over-exposure to any of these hazards. All employees must receive at least 40 hours of initial training before they enter the site. Then they must have at least three days of actual field experience under the direct supervision of a trained, experienced supervisor. Managers and supervisors must receive before they begin the supervisory or managerial assignment, at least eight additional hours of specialized training on managing such operations in addition to having the training other employees receive. Employees must also be given any special training needed for special assignments. All training must be documented and presented by a trainer who has received a higher level of instruction which includes the subject matter being presented. Eight hours of refresher training must be provided for each employee every year. This refresher training does not need to be presented in a single session. The documented training a worker receives at one site may be at least partially applicable to other sites. Specific site training is also required.

MEDICAL SURVEILLANCE

This program requires employers to make available to specific employees, at no charge to the employee and at times and locations convenient to the employee, medical examinations and consultations. Frequency and content of examinations and consultations are determined by the physician, based on information supplied by the employer. All examinations must include questions concerning the employee's exposure. Information the employer must provide the examining physician includes:

A copy of 29 CFR 1910.120 and its appendices;
A description of the employees duties as they relate to potential exposures;
A statement of known/anticipated exposure levels;
A description of all PPE used by that employee;
Information from previous medical examinations of the employee not readily available to the examining physician.

Employees who must be provided medical examinations include:

- Employees actually/potentially exposed to recognized health hazards or hazardous substances above established PELs for 30 days, or more, per year regardless of whether or not they use respirators;
- Employees who wear a respirator 30 days or more per year;
- Employees who may respond/be involved in a hazardous material emergency or spill clean-up operation.

Schedule of examinations:

- Prior to assignment to hazardous waste operations;
- At least once each year thereafter, but at more frequent intervals if the examining physician deems it necessary;
- As soon as possible after an employee is exposed;
- As soon as possible after an employee develops/reports signs or symptoms indicating possible exposure;

Upon termination of employment or reassignment that would remove the employee from inclusion in the medical surveillance program unless the employee has had an examination.

The physicians written opinion furnished to the employer must not reveal specific findings or diagnoses unrelated to occupational exposure. Employee medical records must be preserved and maintained for at least the duration of employment plus thirty (30) years. (These requirements for medical record keeping already existed in 29 CFR 1910.20, Access to employee exposure and medical records.

ENGINEERING CONTROLS, WORK PRACTICES, AND PERSONAL PROTECTIVE EQUIPMENT FOR EMPLOYEE PROTECTION

Wherever feasible, engineering controls are the preferred method for reducing employee exposure to acceptable limits. This applies to all hazardous materials, regardless if PELs are established or if the hazards are otherwise known. Examples of feasible engineering controls are the use of pressurized cabs or control booths or the use of remotely operated material handling equipment. Examples of feasible work practices are removing all nonessential employees from potential exposure areas and drum opening operations; wetting down dusty operations; and locating employees upwind of possible hazards. Specifically forbidden is the practice of rotating employees as a means of compliance with permissible exposure limits. Personal protective equipment which protects employees from hazards and potential hazards they are likely to encounter on the site must be provided based on an evaluation of the performance characteristics of the PPE relative to the requirements and limitations of the site, the task-specific conditions and duration of use, and the hazards and potential hazards identified at the site. Only positive pressure self-contained breathing apparatus (SCBA), or positive pressure air-line respirators with an escape air supply may be used in IDLH conditions.

The level of protection offered by the PPE is determined on the basis of hazards known/suspected present. The level of protection used may be adjusted according to monitoring data. Level A protection (totally encapsulating, chemical protective suit, SCBA worn inside the suit) must be used whenever the presence of a substance which can cause an IDLH condition by skin contact is known/suspected.

Properties required of totally-encapsulating chemical protective suits:

Protect wearer from the particular hazards identified during site characterization and analysis;

Can maintain positive air pressure;
Are capable of preventing inward test gas leakage of more than 0.5 percent.

The PPE program must be followed. The program must include these elements:

Hazards present/suspected at the site;
Criteria for PPE selection;
Proper use of PPE, especially maximum duration of work mission;
Procedures for maintenance and storage of PPE:
Decontamination procedures and specifications for PPE;
Proper fitting of PPE;
Procedures for donning and doffing PPE;
Inspection procedures for PPE;
Monitoring of PPE in use;
Limitations during temperature extremes;
Evaluation of effectiveness of the PPE program.

Training in the use of PPE must include all the elements of the PPE program.

MONITORING

A satisfactory air monitoring program must identify and quantify airborne levels of hazardous substances and health hazards to determine the selection of the appropriate level of protection for employees at the site. The first objective of monitoring is to identify any IDLH or other especially dangerous situations such as flammable atmospheres, oxygen deficient atmospheres, radioactive materials, or toxic levels of airborne contaminants. Air monitoring is necessary whenever:

Work begins on a different portion of a site (or a new site);
There is a change in the substances being handled;
There is a change in the type of operation;
Leaking drums are being handled or there is obvious liquid contamination in the area.

Personnel monitoring are required for employees having the highest potential for exposure to hazardous substances and health hazards as soon as clean-up operations begin.

INFORMATIONAL PROGRAMS

Requires posting the current site safety and health plan on the site so it is available for inspection by employees, their designated representatives, and OSHA personnel. This plan addresses the safety and health hazards of each phase of site operation and includes the corresponding requirements for employee protection.

Stated in the posted plan are:

> Names of personnel (and alternates) responsible for site safety and health;
> A safety and health risk analysis for each site task and operation;
> Employee training assignments;
> PPE to be used by employees for each site task and operation;
> Medical surveillance requirements;
> Specific and detailed statement of air, personnel, and environmental monitoring;
> Site control measures and decontamination procedures;
> Decontamination procedures;
> Contingency plans for the site and any special procedures or programs needed for site operation.

Pre-entry briefings held prior to initiation of new activities at the site are also considered informational programs.

HANDLING DRUMS AND CONTAINERS

Requires appropriate procedures to be developed and used to determine their condition and minimize handling. Special care is required for excavating buried materials. All personnel not actively involved in the drum handling or opening operation should be removed from the area. Remote control procedures are preferred. Drums and containers containing radioactive wastes may not be handled until their hazards are properly assessed.

DECONTAMINATION

Decontamination must be done only in areas which minimize exposure of uncontaminated employees and equipment to potential contamination. Employees must be thoroughly decontaminated before leaving a contaminated area. All clothing and equipment leaving a contaminated area must first be decontaminated or packaged for appropriate safe disposal. Decontamination procedures must be monitored, and any deficiencies corrected immediately. Equipment and solvents used for decontamination must be decontaminated or disposed of properly. Impermeable protective clothing must be decontaminated before being removed from the employee. Non-impermeable clothing wetted with hazardous material must be removed immediately. The wearer must decontaminate (shower) immediately after removing the clothing. Protective clothing or equipment may be removed from change rooms only by authorized employees. Finally, commercial laundries or cleaning establishments that decontaminate protective clothing or equipment must be warned of the potential effects of the contaminants.

EMERGENCY RESPONSE

This requires development and implementation of an emergency response plan to handle foreseeable emergencies before hazardous waste operations begin. Elements of an acceptable plan include at least:

Pre-emergency planning;
Designation of individual duties, lines of authority, training, and communication;
Emergency recognition and prevention;
Safe distances and places of refuge;
Site security and control;
Evacuation routes and procedures;
Decontamination;
Emergency medical treatment and first aid;
Emergency alerting and response procedures;
Critique of the response and follow up of critique;
PPE and emergency equipment.

ON-SITE RESPONSE PROGRAM

Training: previously described in the training section.
Additional elements include: Site topographical map, including current meteorological data;
Procedures for reporting incidents to local, state, and federal governmental agencies: This is a separate section of the Site Safety and Health Plan, and it must be compatible and integrated with the disaster, fire/emergency response plans of local, state, and federal agencies. The on-site response program must be rehearsed regularly as part of the overall training program.

OFF-SITE RESPONSES (NOT MANDATORY)

Illumination: Minimum illumination intensities are specified for the following areas whenever any work is in progress:

Foot candles	*Area or Operation*
3	Excavation and waste areas, accessways, active storage areas, loading platforms, refueling, and field maintenance areas.
5	General site areas. indoors: warehouses, corridors, hallways, and exitways. tunnels, shafts, and general underground work areas; (Exception: minimum of 10 foot candles is required at tunnel and shaft heading during drilling, mucking, and scaling. Bureau of Mines approved cap lights shall be acceptable for use in the tunnel heading.
10	General shops (e.g., mechanical and electrical equipment rooms, active storerooms, barracks or living quarters, locker or dressing rooms, dining areas, and indoor toilets and workrooms.
30	First aid stations, infirmaries, and offices.

SANITATION OF TEMPORARY WORKPLACES

This describes how potable water must be provided, how this water differs from non-potable water, at what ratio toilets must be provided (employees per toilet), washing facilities, and temporary sleeping quarters. Food handling is also discussed.

CRITERIA FOR SELECTION OF LEVELS OF PROTECTION AND PROTECTIVE GEAR

LEVEL A: To be selected when the greatest level of skin, respiratory, and eye protection is required. Level A equipment:

1. Pressure-demand, full face-piece self-contained breathing apparatus (SCBA), or pressure-demand supplied air respirator with escape SCBA, approved by NIOSH.
2. Totally-encapsulating chemical-protective suit.
3. Coveralls.*
4. Long underwear.*
5. Gloves, outer, chemical-resistant.
6. Gloves, inner, chemical-resistant.
7. Boots, chemical-resistant, steel toe and shank.
8. Hard hat (under suit).*

9. Disposable protective suit, gloves and boots (depending on suit construction, may be worn over totally encapsulating suit).
10. Two-way radios (worn inside encapsulating suit).

*Optional, as applicable.

LEVEL B: The highest level of respiratory protection is necessary but a lesser level of skin protection is needed.

1. Pressure-demand, full-facepiece self-contained breathing apparatus (SCBA), or pressure-demand supplied air respirator with escape SCBA (NIOSH approved).
2. Hooded chemical-resistant clothing (overalls and long sleeved jacket coveralls; one or two-piece chemical splash suit, disposable chemical-resistant overalls).
3. Coveralls*
4. Gloves, outer, chemical-resistant.
5. Gloves, inner, chemical-resistant.
6. Boots, outer, chemical resistant steel toe and shank.
7. Boot covers, outer, chemical resistant (disposable).*
8. Hard hat.
9. Two-way radios (worn inside encapsulating suit).
10. Face shield.*

*Optional, as applicable.

LEVEL C: The concentration(s) and type(s) of airborne substance(s) is known and the criteria for using air purifying respirators is met. Level C equipment:

1. Full-face or half-mask, air purifying, cannister equipped respirators (NIOSH approved).
2. Hooded chemical-resistant clothing (overalls; two-piece chemical-splash suit; disposable chemical-resistant overalls).
3. Coveralls*.
4. Gloves, outer, chemical-resistant.
5. Gloves, inner, chemical-resistant.
6. Boots (outer), chemical-resistant steel toe and shank*.
7. Boot-covers, outer, chemical-resistant (disposable)*.
8. Hard hat.
9. Escape mask*.
10. Two-way radios (worn under outside protective clothing).
11. Face shield.

*Optional, as applicable.

LEVEL D: A work uniform affording minimal protection; used for nuisance contamination only. Level D equipment:

1. Coveralls.
2. Gloves*.
3. Boots/shoes, chemical-resistant steel toe and shank.
4. Boots, outer, chemical-resistant (disposable)*
5. Safety glasses or chemical splash goggles*
6. Hard hat.
7. Escape mask*.
8. Face shield*.

*Optional, as applicable.

Management of Hazardous Materials and Wastes: Treatment, Minimization and Environmental Impacts. Edited by S. K. Majumdar, E. W. Miller and R. F. Schmalz. © 1989, The Pennsylvania Academy of Science.

Chapter Twenty-Three

OCCUPATIONAL SAFETY: REGULATIONS IN PERSONAL PROTECTIVE EQUIPMENT[1]

JAMES E. JOHNSTON

Regional Industrial Hygienist
OSHA, U.S. Department of Labor
Room 2100
3535 Market Street
Philadelphia, PA 19104

The Occupational Safety and Health Act of 1970 (hereafter referred to as the "Act") was passed on December 29, 1970 by Congress to "assure safe and healthful working conditions for working men and women; by authorizing enforcement of the standards developed under the Act; by assisting and encouraging the States in their efforts to assure safe and healthful working conditions; by providing for research, information, education, and training in the field of occupational safety and health; and for other purposes." The Act became effective on April 18, 1971. The Act further states under section 5(a) that "each employer:

(1) shall furnish to each of his employees employment and a place of employment which are free from recognized hazards that are causing or are likely to cause death or serious physical harm to his employees;
(2) shall comply with occupational safety and health standards promulgated under this Act."

The Act also places the following responsibilities on employees under section 5(b):

---

[1]The views expressed in it are solely those of the author and not necessarily the opinion of the Department of Labor.

> "Each employee shall comply with occupational safety and health standards and all rules, regulations, and orders issued pursuant to this Act which are applicable to his own actions and conduct."

Section 6(b) (5) of the Act further states that the Secretary of Labor "in promulgating standards dealing with toxic materials or harmful physical agents under this subsection, shall set the standard which most adequately assures, to the extent feasible, on the basis of the best available evidence, that no employee will suffer material impairment of health or functional capacity even if such employee has regular exposure to the hazard dealt with by such standard for the period of his working life."

It is the responsibility of the Occupational Safety and Health Administration (OSHA), a division of the Department of Labor, to carry out and administer the responsibilities of the Labor Secretary as established by the Occupational Safety and Health Act of 1970. The rules adopted by OSHA are concerned with the elimination or avoidance of hazards either proven by research or experience to be harmful to the safety or health of the employee. Rules adopted by OSHA apply either to general categories of hazards or to specific hazards applicable to a specific industry or process.

OSHA had adopted standards applicable to the construction industry, the agricultural industry, the maritime industry, and general industrial processes. The purpose of the standards is to provide methods to reduce workplace hazards to the extent feasible. In addition, for hazards not specifically addressed in an OSHA standard, section 5(a) (1) of the Act places the general responsibility to reduce hazards to the extent that such hazards are known and can be minimized.

OSHA places primary emphasis on the reduction of hazards by the elimination of the hazard, with secondary emphasis placed on the usage of personal protective equipment. It is obvious that elimination of the hazard at the source through engineering controls or substitution would thereby eliminate the need for personal protective equipment. Such equipment has inherent problems due to the need of proper usage as well as the possibility of creating physiological stresses on the employee. Personal protective equipment can be defined simply as "all clothing and accessories designed to create a barrier against workplace hazards."

The OSHA General Industry Standards discuss personal protection equipment in Subpart I of the Department of Labor Code of Federal Regulations (CFR). Section 29CFR 1910.132 states:

> "Protective equipment, including personal protective equipment for eyes, face, head, and extremities, protective clothing, respiratory devices, and protective shields and barriers, shall be provided, used, and maintained in a sanitary and reliable condition whenever it is necessary by reason of hazards of processes or environment, chemical hazards, radiological hazards, or mechanical irritants encountered in a manner capable of caus-

ing injury or impairment in the function of any part of the body through absorption, inhalation or physical contact...All personal protective equipment shall be of safe design and construction for the work to be performed."

The OSHA standards proceed to discuss protective equipment for the eyes, face, head, feet, respiratory protection and other parts of the body which may be exposed to hazardous conditions in the occupational setting. For the equipment to be effective, it must be readily available and suitable for hazards to be encountered. It must have an employee acceptance and understanding over its proper use, it and it should be properly fitted for the employee. But most importantly it should be used with care along with awareness of its limitations. Simply using personal protective equipment without an understanding of its purpose or limitations may sufficiently sacrifice the effectiveness of the equipment rendering the usage of the equipment meaningless in protection of the employee. For example, for head protection, there are two types of hard hats, one with a full brim and the other brimless. There are also three classes of hard hats, one for general service, one for protection against high voltage hazards, and one for lightweight comfort and light impact protection. Respiratory protection varies depending on the contaminant for which the respirator was designed to provide protection as well as the portion of the face or head protected by the respirator. For example, a respirator designed exclusively for protection against inhaled dust will not provide protection against solvent vapors. A half-mask respirator that covers only the nose and mouth will not typically provide as much protection as a full-face respirator, which as its name implies, covers the full facial area. Therefore, it should be apparent that in choosing protective equipment, one should evaluate the hazard to which the employee needs protection. One should choose protection designed to protect against the hazard in question and which properly fits the employee. One must also be aware that protective equipment must be properly maintained in order to be effective. In several instances, the OSHA standard itself details the selection process to be followed in choosing the appropriate protection. For example, the OSHA standard for acrylonitrile, 29CFR 1910.1045, specifies that for concentrations of airborne acrylonitrile less than or equal to 20 ppm, a chemical cartridge respirator with organic vapor cartridges and half-mask respirator is permissible. The standard specifies the permissible respiratory protection at various exposure levels. It is incumbent upon the employer to determine that the potential for physiological stress on the employee is ascertained prior to the employee using any protective equipment. This is particularly important where the protective equipment may compromise the employee's respiratory or circulatory systems. If full body suits are required due to the hazardous environment, one must consider if heat stress will be a potential problem. Also, if the employee must use respiratory protection, then an evaluation must be done to determine if the respiratory protection will place sufficient physiological stress on the employee to dangerously impair the employee's respiratory system. Therefore, one must assess all personal protective equipment in terms of its stress on the wearer.

SUMMARY

In any personal protective equipment program, the following points should be considered:

(1) The program should be in writing. A written program is necessary to ensure consistency in its purpose and application as well as detailing responsibility for the various phases of the program's implementation.

(2) Limitations of personal protective equipment should be considered. As previously stated, all protective equipment is designed to protect against a particular hazard, Wire-mesh gloves may be needed for protection against nip-point hazards while a different type glove may be needed for protection while handling high temperature objects.

(3) The reason for the use of the protective equipment must be understood in order to assure its proper usage; therefore, employees should be trained in the protective features of the equipment.

(4) The equipment should be properly suited to the employee. Most protective equipment is available in various sizes to assure proper fitting for employees. An uncomfortable piece of protective equipment will not be worn.

(5) The program should include an assessment of the physiological stress placed on the employee by using the equipment. This must be assessed in view of each employee's medical status. Respiratory protection which would not compromise a healthy employee, may place undue stress upon an employee with a respiratory ailment or cardiac dysfunction.

(6) The program should be reviewed periodically to determine the adequacy of implementation as well as potential changes required in the equipment due to changes in the hazards or more protective equipment available.

Employers who rely upon a personal protective equipment program are required by OSHA to assure that the equipment is appropriate for the protection required. An evaluation must be made concerning potential physiological stress, and that the program is properly implemented.

REFERENCES

Occupational Safety and Health Act, Public Law 91-596, U.S. Congress, December 29, 1970.

Code of Federal Regulations, Department of Labor, 29CFR, Parts 1900 to 1910.

Management of Hazardous Materials and Wastes: Treatment, Minimization and Environmental Impacts. Edited by S. K. Majumdar, E. W. Miller and R. F. Schmalz. © 1989, The Pennsylvania Academy of Science.

Chapter Twenty-Four

GENETIC TOXICOLOGY TESTING OF HAZARDOUS MATERIALS AND WASTES

ARCHANA SHARMA and ARUN K. SHARMA

Centre for Advanced Study in Cell & Chromosome Research
Department of Botany, University of Calcutta
35 Ballygunj Circular Road
Calcutta 700019, India

INTRODUCTION

The exposure of the human system to a wide variety of agents—physical, chemical and biological—following increasing release of hazardous wastes in the environment, has considerably enhanced the importance of the study of genetic toxicology. A large number of protocols has been developed for short term assays, both *in vitro* and *in vivo,* of the toxicity of different agents acting either on the genetic architecture, that is, on DNA skeleton, or as promoters or inhibitors of genotoxic effects. Over the past decade, these researches have become oriented towards precise risk assessment and risk benefit analysis as applied to human populations.[1,2,3] The test systems designed to detect point mutations and chromosome anomalies cover a wide range, from *Salmonella* and *E. coli* to plant systems (including yeast), *Drosophila,* submammalian and mammalian cells *in vivo* and *in vitro.* The number of short-term assays exceeds 74 and almost every major laboratory group has its modified schedules. As a part of E.P.A. Genetox Assessment Panel program, 2622 screened were listed.[4] For correct assessment, both single test systems with multiple endpoints and mixed batteries of tests have been employed.

A large number of compounds have been tested to detect their potential for inducing aneuploidy or alterations in chromosome number, initially developed as an ancillary to clastogenesis. This aspect is gaining importance following the observation that aneuploidy has been associated with stillbirth, abortions and certain human syndromes.

Genotoxic effect as reflected in carcinogenicity is the other facet of the problem. The relationship between mutagenic and carcinogenic effects is, however, not very simple; particularly since several compounds have been shown to behave as a carcinogen not necessarily through a direct action on DNA, but through their crucial role on growth and development of transformed cells.

Specific chromosome translocations have been located in human lymphomas and leukemias. Such translocations, which can be induced by different agents, involve breakage points where cellular oncogenes responsible for carcinoma have been located. The translocation may lead to alterations or activation of oncogenes. The discovery of induced segmental rearrangement of chromosome which otherwise leads to position effect, contributing to malignancy, has opened up a novel approach to genotoxicity. This rearrangement can be accelerated through mobile sequences capable of insertion at different loci, which may be modified through exposure to genotoxic compounds. All these advances emphasize the need for understanding the relationship between functional segments of chromosomes and genotoxicity.[5]

Several human recessive disorders have been associated both with increased spontaneous frequencies of chromosomal aberrations and/or sister chromatid exchanges and with one or more classes of mutagens. Such sensitivity to specific mutagens may be employed in the cytological detection of such syndromes. For example, Fanconi's anemia shows both increased frequency of spontaneous chromosomal aberrations and enhanced susceptibility to X-rays and chemicals like mitomycin C and acetaldehydes.

The availability of a large number of protocols,[6] a variety of test systems and tremendous advances in the last decade on the regulatory sequences in chromosomes have ushered in a new era on researches in genetic toxicology.

Various techniques have evolved for assessment of the environmental status, the hazards to which an organism is exposed and the extent to which such studies can lead to remedial measures. Such techniques are based on the principle of the relationship between internal levels of exposure and the effects caused thereby and may be utilized for both indication and monitoring of environmental pollutants. The tests principally relate to an analysis of toxicity, mutagenicity and carcinogenicity of the different types of pollutants found in the environment.[2,3,7,8,9,10,11,12,13,14]

The principal endpoints for monitoring are:

(i) mutational events, involving DNA, and

(ii) chromosomal aberrations, involving breaks and interchanges at different levels. These changes may occur in the somatic cells or in the germ cells. The methods include both *in vivo* and *in vitro* systems.

TEST SYSTEMS USED

Bacteria:

Bacterial *in vitro* mutagenicity tests include *Salmonella typhimurium*[15] and *Escherichia coli*[16] reverse mutations.

Of these, the series of *Salmonella typhimurium* strains developed by Ames for use as testers for back mutation at the histidine locus, constitutes the most widely used and most useful systems for identifying genotoxic agents. The substance to be tested is added, in suitable concentrations and under suitable conditions, to several strains of the bacterium carrying different molecularly defined mutations in the histidine locus. Back mutations are identified. From the type of the strain exhibiting the presence of back mutations and from the conditions required to induce this response, it is possible to characterize the type of mechanism underlying the genotoxic action. One limitation is that the system can not easily be used for the quantitative analyses of mutagenic potency. However, it forms a sensitive qualitative probe for genotoxic activities of single chemicals or mixtures of different toxicants.

The bacterial strains are characterized by their specific and different mutations—base substitution or frameshift—in the histidine locus. In addition, all strains have genetic constitutions which make them more susceptible to penetration and attack by genotoxic agents.

Mutation *rfa* increases the susceptibility of the cell wall to penetration by substances added to the medium. Mutation *uvrB* blocks certain normal repair functions, which would ordinarily reduce the yield of back mutations. The plasmid *PKM101* enhances an error-prone repair mechanism. The amplification system thus introduced again increases the sensitivity of the parent strain to genotoxic damage. Strains have also been engineered for specific purposes. For example, TA97 is similar to TA1537 but more sensitive to certain chemicals while TA102 is susceptible to some agents poorly detected by the other strains like ultraviolet rays, hydrogen peroxide and formaldehyde. Any response to the test treatment is dose dependent. In the context of single identifiable substances, exposure is needed to a series of concentrations ranging from zero to slightly toxic levels.

In higher organisms, many substances non-toxic in the original form, acquire genotoxic properties during or after metabolic degradation. Classical examples of biodegradants with genotoxic properties are the degradation products of DDT and malathion. These complexities may be simulated *in vitro* by adding a cell extract to the medium, containing the relevant enzyme complexes (p-450) together with the necessary energy sources, constituting the so-called S-9 mix. The cell extracts usually contain the microsomal fraction of mammalian (rat) liver from animals that have been pretreated with a substance (e.g. Aroclor 1254) which induces/stimulates the production of such degradation enzymes. This method has been further modified by TLC/Salmonella technique involving the separation of complex samples on thin layer chromatography and registering of mutagenic effects directly on the plates by the bacterium.[17]

Yeast and fungi

Strains of yeast have been utilized *in vitro* as indicators of reverse mutations, gene conversion, mitotic recombination, aneuploidy and other forms of chromosome damage.[18,19] Growing cells of different species of *Saccharomyces* produce the enzymes necessary to activate many promutagens which would not be able to induce mutations *in vitro* without S9 mix.[20,21] Mutation studies have also been carried out with species of *Neurospora* and *Aspergillus.* These systems have a relatively limited scope.

Plants

Higher plants are used extensively for monitoring genotoxicity of chemicals due to the complexity of their genomes, their capacity in some cases to activate promutagens to mutagens and the ease of their use. The more commonly used ones are different species of *Vicia* and *Allium, Tradescantia, Hordeum* and maize. For the localization of chromosomal alterations, karyotypes with marker chromosomes are required. Pretreatment of root tips with other chemicals prior to exposure alters the susceptibility in some cases.[22,23]

Gene mutation studies have been carried out on *Arabidopsis, Glycine, Hordeum, Tradescantia* and *Zea.* Of these, *Zea mays* has been found to be a very good bioindicator because of the convenience of detecting mutation in the specific gene locus in identified chromosomes.

1. *Maize*

The two tests which are widely employed involve chromosome 9.

1.1. *Waxy locus pollen test:* The gene is located on the short arm of chromosome 9 and is responsible for the synthesis of carbohydrate amylose. The starch in individuals carrying the dominant allele *Wx* is a mixture of amylose and amylopectin, whereas amylose is absent in the double recessive individuals.[24,25,26] Iodine test colors amylose blue-black. Its absence gives a red or tan color. Since the pollen grain is haploid, containing either *Wx* or *wx,* the color can be used as monitor for this test.

1.2. *Yg-2 locus* on short arm of chromosome 9 in maize induces pale yellow green and slowly maturing leaves in homozygous recessive plants (Yg-2/yg-2). The dominant Yg-2 causes dark green coloration. In the heterozygous state

(Yg-2/yg-2) leaves are a normal green in color. In a heterozygous individual, mutation of the dominant allele leads to deficiency in the production of chlorophyl in the leaf sector. If a leaf primordial cell is involved, yellow green sectors develop in the green leaf.[26,27]

1.3. *Micronucleus test:* In Early-Early synthetic type of maize, the presence of micronuclei in root tips has been taken to indicate chromosome aberrations produced as a result of environmental toxicant.[28] Micronucleus test has also been utilized in other plants, e.g. in root-tips of *Vicia faba* following exposure to radiation. Micronuclei are taken to indicate non-disjunction, resulting in whole chromosomes or acentric fragments being left out of the spindle.[29]

2. *Tradescantia*[30]

1. *Staminal hair method:* A change in color in high mutagen sensitive staminal hairs has been utilized to monitor air pollutants as well as physical and chemical mutagens in *Tradescantia.*[31,32] The mutational events change the color from blue to pink, which is observed in the dissected stamen under the microscope.

2. The microspore has been used to measure the extent of radiation as absorbed by the plant when exposed to radioactive cloud.[33] In general, the microspores of Tradescantia are excellent materials for scoring the effects of pollutants as shown by the formation of chromosomal aberrations during mitosis of pollen grains.

3. *Others*

Barley, *Vicia faba*[34] and species of *Arabidopsis* offer good systems for environmental monitoring. In the latter, due to the short life cycle, the effect of the pollutant can be measured in all phases of growth. Barley serves as an efficient environment monitor as well.[35,36]

4. *Allium cepa, Allium sativum*

The *Allium* test, initially worked out by Levan[37] is one of the most convenient methods of monitoring genotoxicity.[38] The chromosomes of *Allium*, being quite large, allow a detailed analysis. The bulbs can be easily handled and roots with meristems grow out profusely at regular intervals at a rapid rate. Such regular and successive growth enables an analysis of both long and short term effects. The method involves the culturing of healthy bulbs on the top of small jars containing the solution to be tested for genotoxicity with the growing roots immersed in the solution. The roots are cut at regular intervals to study the effects of different periods of treatment at different concentrations. The analysis of genotoxicity at the chromosome level is carried out following acidulated alcohol fixation and orcein staining of the root-tips. Orcein staining has to be applied carefully as overheating may lead to chromosome breakage.[7] In order to study the recovery from the toxic effects, if any, the bulbs with roots after treatment are transferred to Knop's nutrient medium for observation at successive periods.

The effects normally are classified as subnarcotic, narcotic and lethal. The subnarcotic effects include chromosomal alterations and consequently gene alteration as well. The narcotic effect is principally on the spindle, leading to spindle arrest and polyploidy, without any consequent lethality. Certain vital processes are switched off but enough of the basic processes are maintained so that the tissue can recover when the external influence is removed. On the other hand, certain compounds are highly toxic, leading to pycnosity, liquefaction of chromatin and subsequent lethality.

Animals

A wide range of animals, from *Drosophila* to fish and rodents like mouse, rabbit and Chinese hamster, has been utilized for indication and monitoring of agents present in the environment. Both *in vivo* and *in vitro* systems have been employed. A major addition is peripheral blood lymphocyte culture from human systems. The latter is extensively used, partly due to the lower cost and partly to the relative ease of extrapolation of the results obtained from human leukocytes *in vitro* to human subjects *in vivo.*

The usual endpoints observed in mammalian test systems *in culture* are: (i) chromosomal aberrations in metaphase and anaphase; (ii) micronuclei in interphase; (iii) sister chromatid exchanges in metaphase and (iv) unscheduled DNA synthesis as measured by microauto-radiography. These endpoints, particularly the first three, are common to *in vivo* test systems as well.

1. *The fruitfly Drosophila melanogaster* is a very good indicator for *in vivo* testing of genetic damages from the chromosome to gene level.[39] It has a great potential for use in monitoring biologically active substances in the environment. By the use of the wild type as the exposed species, the response becomes more realistic than the use of genetically manipulated species. The latter may be highly efficient registrars of gene changes, but the quantification of implications for other organisms or the ecosystem is difficult.

A. *Dominant lethal test:* This effect does not accumulate and is thus more suitable for the tracking of immediate response to a genotoxic substance. A dominant lethal may act at any stage prior to reproduction in the F_1 and includes F_1 sterility. The test is performed by mating an exposed male or female to an unexposed individual of the opposite sex. Eggs are collected over a suitable period and counted and unhatched eggs are counted after 24 hours. To avoid unfertilized eggs, exposed virgin females are mated to 2-3 unexposed males and conversely, exposed males are given at least 24 hours to mate with one unexposed virgin female.

B. Testing for *recessive sex-linked lethals* (SLRL) implicates only those offspring which survive and remain fertile to produce an F_2 generation. The principle of the test is to isolate single X-chromosomes and register if this chromosome allows survival at the hemizygous state. The test is excellently suited for quantification of genotoxicity of toxicants and may be used to reveal damage which has accumulated over several generations in chromosomes inherited through the female line and thus not subject to selection in the hemizygous condition.

In Muller 5 test—individual X-chromosomes are sampled either in exposed virgin females or in exposed males. By mating to Basc (or M5) or Binsey flies of the opposite sex, F_1 females with one M5 and one exposed X are produced. Each such heterozygous female represents one exposed X to be tested. When she is mated, usually to M5 females, she will produce M5 sons but, if a lethal has been induced in the other X-chromosome, no wild type sons. The measurement of the effect thus emerges from the number of F_1 females that do not produce wild type sons, relative to the total number of fertile F_1 females tested. This test is simple and rapid, since wild type males have large round red eyes, while M5 males have narrow pale apricot colored ones and thus can be easily distinguished.

In an other sex-linked recessive lethal test CLB, the male fly is generally exposed, in order to avoid the possibility of pre-existing lethal mutation in the X-chromosome of female. As indicators of point mutation, the exposed male fly is then bred with a specific tester strain.

C. A somatic mutation/mitotic recombination assay (SMART) measures the frequencies of single and twin spots in the eye and is found to be sensitive.[40]

D. Non-disjunction or maldistribution of chromosomes may be utilized to identify changes at chromosomal level. Survival of the aneuploid offspring is the endpoint recorded. The test therefore is limited to the X-chromosome or the small fourth chromosome. In order to distinguish the offspring resulting from non-disjunction, one set of chromosomes must carry markers.

E. Tests for specific locus mutations consist simply in crossing the exposed individual to a tester which is homozygous for a maximum number of recessive markers. If an exposed chromosome has a dominant mutation, or a recessive one at one of the marker loci, or a non-lethal deficiency uncovering one of the marker loci, it can be recognized in the individual carrying that exposed chromosome.

2. *Fishes* have been utilized for monitoring of the aqueous environment and chromosome damages caused. The damage is principally measured through chromosome breaks, bridges and other aberrations. The organs utilized are the gill cells as well as kidney and intestine though less frequently. Laboratory fish such as *Umbra pygmaea* (mudminnow) can be exposed for different periods in various aqueous environments.[41,42] Short-term bone marrow culture technique has also been employed.[43]

3. *Mammals:* Almost all rodents have been used as test systems. Of these, the mouse has been most extensively employed in monitoring the mutagenic potentials of hazardous chemicals and wastes. Toxicity can be measured in terms of the production of lethals or of chromosome aberrations *in vivo.* Embryonic lethality can be judged by scoring of dead implants in rodents (see 39, 44, 45). Chromosome aberrations can also be studied following culture techniques, either in the bone marrow or in leukocytes.

3.1. *Dominant lethal* in mouse is a quick *in vivo* test representing a mutation either in the sperm or the egg, which is lethal to the zygote produced. Most dominant lethals can be related to chromosomal aberrations and they reduce the litter size in different degrees. Nearly 150 chemicals have been tested for dominant lethals.[46]

3.2. *Heritable translocation test in mouse* has been used to identify partial or full sterility amongst the first generation offspring of mutagen treated animals. Heterozygosity for a reciprocal translocation leads to partial sterility. This test also gives an estimate of the transmissibility of induced chromosomal exchanges.

3.3. *Specific locus and spot tests in mouse:* The mouse specific locus test detects mutations transmitted from treated germ cells and recovered in the offspring of the next generation. Usually coat color markers are screened simultaneously at 7 different loci. In the spot test, heterozygotes are treated *in utero* and new spots are identified on the skin of the newborn mice.

3.4. *In vivo chromosomal alterations* induced by external agents can be scanned in all dividing cells.

A routine test involves the *bone marrow,* following a short term colchicine pretreatment—hypotonic— air drying schedule. The aberrations scored include breaks of different types—chromatid and chromosome — depending on the phase (G_1 or G_2) of the cell treated and the nature of the clastogen — whether dependent on DNA synthetic (S) phase or not. Translocations may also be seen. Breaks not included in the daughter nuclei later form micronuclei. Such micronuclei persist in the polychromatic erythroblasts formed from bone marrow and can be scored easily to give the frequency of breaks induced. In liver micronucleus assay, treated animals are partially hepatectomized and the hepatocytes from the regenerating liver are analyzed for the presence of micronuclei. Metaphase and micronucleus analyses may also be combined in the same animal, e.g. in rat.

3.5. *In vitro cytogenetic tests:* The popularity of the *in vitro* systems stems principally from the ease and low cost of performing these tests. The most frequently used systems for *in vitro* screening for mutagens are established Chinese hamster cells or stimulated human peripheral lymphocytes. Agents which induce DNA double strand breaks, lead to chromosome breaks in G_2, chromatid breaks in G_1 and both types in S phase of the cell cycle. Other mutagens induce only chromatid breaks irrespective of the cell cycle stage treated.

3.6. *Point mutations in mammalian cells in culture* are frequently detected as resistance to selective agents like 8-azaguanine and ouabain, which occur at different frequencies. Mutations of hypoxanthine-guanine-phosphoribosyl transferase (HGPRT) locus, induced by X-rays or EMS, are shown to be also associated with rearrangements involving the X-chromosome.

3.7. *DNA repair tests in mammalian in vitro systems* involve detection of the incorporation of tritiated thymidine in the nuclei of the cells exposed to the environmental agent, which are not at the S phase of the cell cycle. This incorporation indicates the extent of repair of damaged DNA and is known as unscheduled DNA synthesis (UDS). It may be measured by microautoradiography or scintillation counting and can be best estimated in non-proliferating cells like rodent hepatocytes or human lymphocytes.

SISTER CHROMATID EXCHANGE (SCE) IN ANIMALS AND PLANTS AS INDICATORS OF TOXICITY

Sister chromatid exchanges are cytogenetic processes related to DNA damage and repair. It is an effective cytogenetic parameter for monitoring the damage.[48] These are due to unrepaired lesions present at the S phase followed by chromosome replication. It can be induced by different compounds capable of alkylating DNA. The SCE for visualization requires at least two replications in the presence of Bromodeoxyuridine (Brdu). This leads to DNA of one chromatid of metaphase chromosome to become bifilarly substituted by Brdu, the other remaining singly substituted. Following the use of fluorochromes such as Hoechst 33258 and observation under fluorescent microscopy or Giemsa staining, the contrasting color of the two chromatids can be clearly noted. This test has been applied to most mammalian systems and to higher plants.

SCE in Human System

Sister chromatid exchange can be studied in human leukocyte culture. The test is very sensitive since SCEs can be induced by even 1/100th of the exposure required to cause aberrations.[50] It has been utilized to monitor individuals exposed to potentially dangerous compounds of the environment. In the usual method for leukocyte culture of exposed individuals, Brdu should be added 24 hours after initiation and reincubated for additional 48 hours. The culture should be grown in dark. Two hours of colcemid treatment (0.05 ug/ml) before harvesting for usual air dried preparation is needed.[1,51,52]

In higher plants, SCE's have been used mainly for identifying genotoxic events in long chromosomes. Species with low number and with long chromosomes, such as species of *Allium* and *Vicia,* have been used extensively (vide 53).

RELEVANCE TO HUMAN EXPOSURE:

The identification of risk factors is an essential requisite for the primary prevention of damages through exposure to hazardous wastes. "Risk" may be qualitative or quantitative, which is an estimate of the probability of a hazard at a given exposure level and often includes statistical modeling for extrapolations from high to low doses.[8] The term *hazard* implies the existence of a threat, whereas *risk* indicates both the existence of a threat and its potential for affecting the living system.

During the 1970s, various scientists attributed 80-90% of all human cancers to environmental factors. However, later studies have shown that "human environment" should be taken to include not only occupational, iatrogenic and environmental exposures to chemicals, but also human behavior patterns that are influenced by cultural habits (e.g. tobacco smoke, alcohol), diet, sexual habits and other lifestyle factors (e.g. exposure to sunlight). For example, proportions of cancer deaths attributed to tobacco and diet exceed those due to occupation, drugs and environmental pollution.[54]

Short-term *in vitro* testing has contributed immensely to the development of the field of genetic toxicology. Guidelines for Mutagenicity Risk Assessment as proposed by EPA (1984), however, places greater weight on tests conducted in germ cells rather than on somatic cells, on tests performed *in vivo* rather than *in vitro,* in eukaryotes rather than in prokaryotes and in mammalian species rather than in submammalian species. The most productive approach to screening for potential mutagens— carcinogens, would be to employ a 'Reversed Tier', that is to start with whole animal studies and use *in vitro* systems, where appropriate, to elucidate mechanisms of action. A deeper understanding of this mechanism of action is fundamental to any solution of the problems of hazardous materials wastes.

REFERENCES CITED

1. Berg, K. 1979 (Ed.) *Genetic damage in man caused by environmental agents.* Academic Press, New York, pp. 510.
2. Sharma, A. 1984. *Environmental chemical mutagenesis.* Persp. Rep. Ser. 6, Golden Jubilee Publ. Indian National Science Academy, New Delhi, pp. 53.
3. Sugimura, T., S. Kondo, and H. Takebe, 1982 (eds.) *Environmental mutagens and carcinogens.* Alan R. Liss Inc., New York.
4. Ray, V.A., L.D. Kier, K.L. Kannan, R.T. Haas, A.E. Auletta, J.S. Wassom, S. Nesnow, and M.D. Waters. 1987. An approach to identifying specialized batteries of bioassays for specific classes of chemicals: class analysis using mutagenicity and carcinogenicity relationships and phylogenetic concordance and discordance patterns 1. Composition and analysis of overall data base. *Mutat. Res.* 185:197-241.
5. Sharma, A.K. 1987. Genetic-Toxicology. *Indo-US Seminar on genetic toxicology, Calcutta,* Nov. 16-18, *The Nucleus.* 30: pp. 159-160.
6. Naismith, R.W. (ed.) 1987. Guidelines for minimal criteria of acceptability for selected short-term assays for genotoxicity. *Mutat. Res.* 189:81-183.
7. Sharma, A.K. and Sharma, A. 1980. *Chromosome Techniques—Theory and Practice,* 3rd ed. Butterworths, London.
8. Oftedal, P. and A. Brögger. 1986 (ed.) *Risk and reasons: risk assessment in relation to environmental mutagens and carcinogens.* Alan R. Liss Inc., New York, pp. 189.
9. Sharma, A.K. 1981. *Impact of the development of science and technology on environment.* Gen. Pres. Address, 66th Ind. Sci. Congress, Varanasi, 1-41.
10. Fishbein, L. 1984. Mutagens and carcinogens in the environment. *Genetics: New Frontiers* (eds. V.L. Chopra, B.C. Joshi, R.P. Sharma, and H.C. Bansal) III, 3-42, Oxford and IBH, New Delhi.
11. Sobels, F.H. 1987. *In vitro* testing as a step in the evaluation of *in vivo* toxicity. *Mutat. Res.* 18:7-10.
12. Zimmerman, F.K. and R.E. Taylor-Meyer. 1985. *Mutagenicity testing in environment pollution control.* Ellis Horwood Ltd. Chicester, U.K. pp. 47-68.
13. Sharma, A. 1986. Higher plants as cytogenetic monitors for chemical agents. *J. Ind. Bot. Soc.,* 64:9-16.
14. Sharma, A. and G. Talukder. 1987. Effects of metals on chromosomes of higher organisms. *Environmental Mutagenesis.* 9:191-226.
15. Ames, B.N., W.E. Durston, E. Yamasaki, and F.D. Lee. 1973. Carcinogens are mutagens: a simple test system combining liver homogenates for activation and bacteria for detection. *P.N.A.S. USA,* 70:2281-85.
16. Brusick, D.T., V.F. Simmon, H.S. Rosenkranz, V.A. Ray, and R.S. Stafford. 1980. An evaluation of the *Escherichia coli* WP2 and WP2 UVrA

reverse mutation assay. *Mutat. Res.* 76:169-190.
17. Moller, M., A. Bjorseth, and V.S. Houk. 1985. Chemical separation and *in situ* mutagenicity testing. In: *Mutagenicity testing in environmental pollution control* (eds. F. K. Zimmerman, and R.E. Taylor-Meyer). 47-68, Ellis Horwood Ltd. Chicester, U.K.
18. Zimmerman, F.K. and I. Scheel. 1984. Genetic effects of 5-azacytidine in *Saccharomyces cerevisiae. Mutat. Res.* 139:21-24.
19. Resnick, M.A., V.W. Mayer, and F.K. Zimmerman. 1986. The detection of chemically induced aneuploidy in *Saccharomyces cerevisiae. Mutat. Res.* 167:47-60.
20. Parry, J.M. 1982. Assay of the induction of mitotic crossing-over and aneuploidy in yeast by BC, 4-CMB and 4-HMB. *Mutat. Res.* 100:139-145.
21. Haynes, R.H. and B.A. Kuntz. 1981. DNA repair and mutagenesis in yeast. *Molecular biology of the yeast Saccharomyces Life cycle and inheritance.* Cold Spring Harbor, pp. 371-414.
22. Mukherjee, A. and A. Sharma. 1987. Effects of cadmium and zinc on cell division and chromosomal aberrations in *Allium sativum, Curr. Sci.* 56:1097-1100.
23. Heindorff, K., R. Rieger, I. Schubert, A. Michaelis, and O. Aurich. 1987. Clastogenic adaptation of plant cells— reduction of yield of clastogen—induced chromatid aberrations by various pretreatment procedures. *Mutat. Res.* 181:157-171.
24. Sprague, G.F., B. Brimhall, and R.M. Hixon. 1943. *J. Am. Soc. Agron.* 35:817-822 (Referred in M.J. Plewa, 1982).
25. Plewa, M.J. and J.M. Gentele. 1982a. The activation of chemicals into mutagens by green plants. In *Chemical mutagens principles and methods of detection.* Vol. 7. (eds. F.J. de Serres and A. Hollaender) Plenum, N.Y. pp. 401-420.
26. Plewa, M.J. 1985. Genetic assays of maize for environment pollution control. In: *Mutagenicity testing in environment pollution control* (eds. Zimmerman, F.K. and R.E. Taylor-Meyer) 47-68 Ellis Horwood Chicester, U.K.
27. Plewa, M.J., E.D. Wagner, G.J. Gentile, and J.M. Gentile. 1984. An evaluation of genotoxic properties of herbicides following plant and animal activation. *Mutat. Res.* 136:233-234.
28. Plewa, M.J. 1982b. Maize as a monitor for environmental mutagens. In: *Environmental mutagens and carcinogens* (eds. T. Sugimura, S. Kondo, and H. Takebe) 411-419, Univ. of Tokyo Press, Tokyo.
29. Degrassi, F. and M. Rizzoni. 1982. Micronucleus test in *Vicia faba* roottips to detect mutagen damage in freshwater pollution. *Mutat. Res.* 97:19-33.
30. Ma, T.H. and W.F. Grant. 1982. The Tradescantias — adventurous plants. *Herbarist.* 48:36-44.
31. Vant Hof, J. and L.A. Schaire. 1982. Tradescantia assay system for gaseous mutagens U: EPA Gentox Programme. *Mutat. Res.* 99:293-302.

32. Fujii, T. 1982. In: *Environmental Mutagens and Carcinogens (eds. T. Sugimura, et al)* Allen R. Liss Inc., New York, pp. 399-420.
33. Conger, A.D. 1953. Culture of pollen tubes for chromosomal analysis at the pollen tube division. *Stain Techn.* 28:289-293.
34. Kihlman, B.A. 1975. Root-tips of *Vicia faba* for the study of the induction of chromosomal aberrations. *Mutat. Res.* 31:401-412.
35. Nilan, R.A. 1978. Potential of plant genetic systems for monitoring and screening mutagens. *Environ. Health Perspect.* 27:181-196.
36. Sharma, A.K. and A. Sharma. 1960. Spontaneous and chemically induced chromosome breakage. *Internat. Rev. Cytol.* 10: 101-136.
37. Levan, A. 1949. The influence on chromosomes and mitosis of chemicals, as studied by the *Allium* test. 8th International Congr. Genetics *Hereditas* Suppl. vol. 325-337.
38. Grant, W.F. 1982. Chromosome aberration assays in *Allium*—a report of the US Environmental Agency Gene Tox Programme. *Mutat. Res.* 99:273-291.
39. Hsu, T.C. 1982 (ed.) *Cytogenetic assays of environmental mutagens,* Allanheld, Osmun, Totowa, N.J. pp. 430.
40. Vogel, E.W. 1985. The *Drosophila* somatic recombination and mutation assay using the *white-coral* somatic eye color system. In:*Progress in Mutation Research* 5:313-317, Elsevier, Amsterdam.
41. Prein, A.E., G.M. Thie, G.M. Alink, C.L.M. Poels, and J.H. Kosman. 1978. Cytogenetic changes in fish exposed to water in the river Rhine. *Sci. Total Environ.* 9:287-291.
42. Kligerman, A.D. 1980. The use of aquatic organisms to detect mutagens that cause cytogenetic damage. In: *Radiation effects on aquatic organisms* (ed. N. Egami) pp. 241-52, Japan Sc. Soc. Press, Tokyo.
43. Manna, G.K. 1986. Mouse bone marrow as a means of testing clastogenic agents. *The Nucleus.* 29(3):141-168.
44. Bateman, A.J. and S.S. Epstein. 1971. Dominant lethal mutations in mammals in A. Hollaender Ed. *Chemical mutagens, principles and methods for their detection.* Vol. 2 Plenum, N.Y. 541-68.
45. Manna, G.K. and T.K. Ghosh. 1983. Chromosome aberrations induced by *Staphylococcus* in partially hepatectomized mice and reduction by penicillin. *The Nucleus.* 26:153-159.
46. Green, S., A. Auletta, J. Fabricant, R. Kapp, M. Manandhar, Chin-In-Sheu, J. Springer, and B. Whitfield. 1985. Current Status of bioassays in genetic toxicology. *Mutat. Res.* 154:59-67.
47. Albanese, R. 1987. The cytonucleus test in the rat: a combined metaphase and micronucleus assay. *Mutat. Res.* 182:309-321.
48. Latt, S.A. 1974. Sister chromatid exchanges, indices of human chromosome damage and repair: detection by fluorescence and induction by mitomycin C. *Proc. Natl. Acad. Sci. USA* 71:3162-3166.

49. Wolff, S. 1977. Sister chromatid exchange. *Ann. Rev. Genet.* 11:183-201.
50. Solomon, E. and M. Bobrow. 1975. Sister chromatid exchange: a sensitive assay of agents damaging human chromosomes. *Mutat. Res.* 30:273-278.
51. Perry, P. and S. Wolff. 1974. New Giemsa method for differential staining of sister chromatids. *Nature.* 261:156-58.
52. Buckton, K.E. and H.J. Evans. 1982. Human peripheral blood lymphocyte cultures. In: *Cytogenetic assays of environmental mutagens* (ed. T.C. Hsu), Allanheld, Osmun, Totowa, N.J., pp. 183-202.
53. Schvartzman, J.B. 1987. Sister chromatid exchanges in higher plant cells: past and perspectives. *Mutat. Res.* 182: 127-145.
54. Sorsa, M., K. Hemminki, and H. Vainio. 1982. Biologic monitoring of exposure to chemical mutagens in the occupational environment. *Teratogenesis, Carcinogenesis and Mutagenesis2:137-150.*

Management of Hazardous Materials and Wastes: Treatment, Minimization and Environmental Impacts. Edited by S. K. Majumdar, E. W. Miller and R. F. Schmalz.

Chapter Twenty-Five

CURRENT TRENDS IN ENVIRONMENTAL EDUCATION

DEAN STEINHART

Director, Department of Education and Environmental Resources
Commonwealth of Pennsylvania
333 Market Street
Harrisburg, PA 17126

To understand the direction of environmental education in Pennsylvania today we must look at where we have been in our recent past. The records of the Department of Education show that the first evidenced concern for conservation of national resources occurred in the late 1930's when Bulletin 214, *Conservation Education*, was written by a Federal Writers Project and published by the Department of Education. It was not until the late 1940's that further concern was noted at which time correspondence indicated that the Department wished to withhold action on whether an interdepartmental committee on conservation was desirable.

In the mid 1950's a Department representative was named to the Conservation Education Laboratory Committee, and advisory committee to the staff of the Pennsylvania Teachers Conservation Laboratory, held each summer at the Pennsylvania State University. By 1958, the State Council of Education had employed a conservation education specialist to develop a teaching guide for natural resources conservation in Pennsylvania. From Bulletin 214, to a full time conservation educator took about 20 years.

The direction at that time was developmental and many programs evolved from these efforts. These programs included student resident experience at the Pennsylvania Sportsmen's Show, three day conservation education workshops for teachers, conservation and outdoor education roundtable discussions and the Pennsylvania Outdoor Education Conservation Association was formed during this time.

These efforts were mostly outdoor education and the publications of the day were: *Pennsylvania Teaching Guide to Natural Resources, Guidelines to Environmental Sensitivity* and *School Site Development for Conversation and Outdoor Education.*

CURRENT TRENDS

There are still remnants of these activities in environmental education, however, the focus has changed considerably in recent years. Since the early 1980's Pennsylvania has become a national leader in the field. By 1984 a new Master Plan was written for Pennsylvania and a very bold and an ambitious venture began again. Currently, the most important addition to environmental education is the new scope and sequence based on the Chapter V regulations of the State School Code. The scope and sequence covered grades K-12 and follows the six objectives set down for the quality goals for environment in Chapter V. This document is issue oriented and will serve as a guide for years to come. Hazardous and toxic wastes, resource recovery, environmental law and problem solving activities are all part of this guide.

Based on the scope and sequence several new curriculum programs have been developed for the state. In the Philadelphia area, REEP (Regional Environmental Education Program) has been prepared by teachers at the Schuylkill Valley Nature Center. This program has been produced so that it may be integrated into existing curricula and organized around ten hour time frames. In this way a school district would experience little difficulty in teaching the 30 hour requirement for environmental education.

The Lehigh Conservancy has been involved with curriculum development too. The focus for this effort has been in the Little Lehigh Watershed and the four school districts that are located in the watershed. This program, Aquatic Ecology, is also in the developmental stage and has gone through two years of writing, teaching, rewriting and teaching again. After the 3rd year, the materials will be ready for distribution and sale to schools and colleges in the Commonwealth.

The Pennsylvania Fish Commission is writing curriculum for the junior high/middle school years. These materials are related to the aquatic resources and are currently being piloted at rural, suburban and urban school districts. The Bureau of State Parks continues to be a leader in curriculum writing and development. Currently, the Bureau's efforts are focusing on the Chesapeake Bay problem. These materials are being piloted in several school districts. The Bureau of State Parks has also produced problem solving materials and water quality monitoring activities. Many other interagency programs are alive and well in Pennsylvania today. The Game Commission and the Department of Education have teamed up to sponsored Project "Wild" and the Bureau of Forestry is working with Education in sponsoring Project Learning Tree across the State. Literally thousands of teachers are being trained in these curriculum projects and the materials provided to teachers is of a very high quality. New programs like Nature Quest from the National Wildlife Federation are being evaluated for future utilization across the state. All of these programs have been used outside the formal classroom setting in girl and boy scout camp situations and with other conservation education groups.

EMPHASIS ON CURRENT ENVIRONMENTAL ISSUES

All of these programs are currently being used across the state with various commitments from schools and communities. The real action at the local level continues to be in the 4th, 5th and 6th grades. Many school districts continue to promote outdoor experiences at school sites, church camps and environmental centers while other remain classroom bound. Curriculum projects like "WILD" and "Learning Tree" are of great value to teachers tied to the school site or classroom. The National Wildlife Federation continues to publish materials for classroom use during National Wildlife Week and the Office of Environmental Education produces regular packets of materials for Environmental Education Week in October of each year. During 1988 the theme for Environmental Education Week will be Hazardous and Toxic Wastes. The Pennsylvania Alliance for Environmental Education will be writing the classroom activities for this publication of K-12 hands-on activities. The 1988 Environmental Education Week packet will be the 3rd collection of materials sent out to every school in Pennsylvania.

Mr. Robert Rodale, of Rodale Press, Inc., has become involved with environmental education in Pennsylvania and had produced classroom activities that explore his concepts of regeneration. The materials are very interesting and involve the individual at the personal responsibility level. The regenerative concept reaches to the heart of environmental education with its concern for positive personal input into our life styles and our communities. This approach places responsibility for the environmental with our individual decisions and how we interact with the earth as our home.

Perhaps, the most significant change in the teaching of environmental education has been related to the integration of environmental topics into the science or social studies curriculum of our public schools. Chapter V curriculum regulations require that every student in the secondary years receives a planned course in environmental education. This course may be integrated into other appropriate courses and if it is integrated, a minimum of 30 clock hours of instruction is required. The instruction is based on six objectives that are as follows:

1. Knowledge of Nature and Human Resources
2. Geographic Understanding, Local, Regional and Global
3. Interrelationships and Interdependence of Natural and Human Systems

4. Development of Personal Environmental Attitudes and Values
5. Environmental Problem Solving and Management Skills
6. Knowledge and Appropriate Uses of Energy

These objectives should set the basic standard for what is to be taught in environmental education in Pennsylvania even if the planned course is offered in life sciences or social studies. Many school districts have adopted the new Science, Technology and Society approach to these objectives and find that the program fits the requirement very well.

Whatever the issue, acid deposition, solid wastes, hazardous and toxic wastes, erosion and sedimentation, land use and planning, water quality, air quality or habitat loss and destruction the issue continues to be society based and technology oriented. This approach to environmental education is blessed with opportunity and well developed curriculum.

NETWORKING

With the growth of new materials and ever increasing numbers of issues, a formidable task begins to emerge. How do we expect teachers to stay abreast of this rapidly developing body of interdisciplinary information? A key concept for this concern has focused on the idea of networking and communication with the persons involved in the teaching of environmental education.

Currently, the networking within the environmental education community covers several formats. First, the general face to face communication that comes about at conferences and regional meetings of environmental groups. This is the tried and true method of history. More recently, the use of newsletters and other printed media has taken an active role in the networking process. The most talked about and hoped for system of networking is computer based and functions within the academic community. The schools and intermediate units of the Commonwealth are currently coming on line with Penn Link and all 501 school districts will be on line by 1990. Penn Link allows direct electronic communication with each school in the State and offers the most promising opportunity for rapid and effective contact at the local level.

A National Network for environmental education is also being proposed by the Alliance for Environmental Education, Inc. This National effort would tie together environmental education centers, colleges and universities and secondary schools throughout the country. These efforts are expected to raise the environmental literacy level in the United States. This network would increase the exchange and dissemination of information relative to many independent efforts designed to educate specific publics about a particular concern or resource area. This national effort will span ten years and be funded by the Alliance for Environmental Education, local business and industry, state and local education and resource agencies. It is expected that five centers will be established each year. The model for development of this network came from the Tennessee Valley Authority and is in use in the geographic region it serves. The TVA network involves 13 universities and has trained 20,000 teachers during its 10 years of operation. When this central mechanism is in place, the American public and professional environmental educators will have access to information that will assist them in making informed judgments on environmental issues.

THE FUTURE

No one can predict the future accurately but it is always interesting to try. The future seems to be tied to interagency cooperation in environmental education. In Pennsylvania a special interagency committee has been established to channel information and develop materials for classroom use. Each agency involved serves a distinct public and needs to educate adult and school age populations. Cooperative efforts have been ongoing and productive for the Commonwealth. A recent effort by the Pennsylvania Wildlife Federation has produced a publication called "First Environmental Quality Index" for Pennsylvania. This publication will be most useful for science teachers and environmental educators in the secondary years. In addition, this publication will become a basis for future analysis of Pennsylvania's Environmental Quality.

Many other States like Missouri, Wisconsin, Washington, Michigan and California have produced documents about environmental education with interagency help and cooperation. This trend is strengthened by the cooperative sponsorships of programs like Project Wild, Learning Tree and Nature Scope. Environmental education is closely tied to the initiatives of the Department of Environmental Resources too. Regular communications with Department officials are held to insure that environmental education is not pursuing its own objectives. Efforts have been tied to major programs like the Year of the Forest and this year we will focus on Hazardous and Toxic Wastes. Current legislation like Senate Bill 528 has placed the schools directly into the recycling issue. The future will find other legislation counting on students to educate their parents and other adults in the community.

Perhaps the brightest star in environmental education's future is the opportunity for our students to develop problem solving skills that focus on real problems. This unique opportunity will provide the perfect format for the teaching of higher order thinking skills such as analysis, synthesis, evaluation and communication. Materials are readily available and promote this type of effective schooling. A final hope for environmental education is that students will learn the action skills so necessary for successful citizenship. Helping students understand the system and the processes necessary to work within the existing framework to accomplish change is a very important part of our goal for an informed citizen. It is important for young people to become literate about their environment and have a conceptual notion about their impact on this fragile planet we call home. Being able to view, in a rational manner, complex issues like mass burn incineration and the relationship to their life style and consumerism will be our grandest hope for all of our futures in Pennsylvania, the Nation and the World. Picture a generation of young people with an awareness of consumerism, a feeling for conservation and a willingness to become good stewards of the earth and you can visualize a brighter future for all Pennsylvania. We can not afford to let this vision be controlled by the laws of chance. Each of us must be involved and seek to become constructive change agents. The environmental education community in Pennsylvania is committed to the tasks of educating all of our students, K-12 and all of the citizens within the Commonwealth. Our recent work with students majoring in elementary education has convinced us that our best chances for success lies with these young potential teachers. Each training session that is later used in a classroom situation will have a multiplying factor for the future. One teacher will touch 25-30 students per year in the elementary schools. If that teacher uses environmental education materials every year for 20 years the potential is obvious. Take the thousands of teachers being trained and multiply the student loads and we have begun a very important educational process. The final results of these teacher training programs should be found in a society that is environmentally literate. With all of the potential information that can interfere with environmental literacy it is important to start early and continue the process throughout life.

Management of Hazardous Materials and Wastes: Treatment, Minimization and Environmental Impacts. Edited by S. K. Majumdar, E. W. Miller and R. F. Schmalz.

Chapter Twenty-Six

SITING HAZARDOUS FACILITIES: PROBLEMS AND PROSPECTS

RICHARD J. BORD

Department of Sociology

The Pennsylvania State University

University Park, PA 16802

INTRODUCTION

Effective opposition to hazardous facilities has assumed the dimensions of a significant social problem. The growing familiarity of the neologisms "LULU," Locally Undesirable Land Use, and "NIMBY," Not In My Back Yard, reflect the realities facing those responsible for siting such facilities. While hazardous waste inventories have continued to increase, the corresponding increase in disposal facilities has not been forthcoming. Since 1980 there have been 81 applications for siting permits. Of these 81 applications only six are now operational while two more are under construction.[1] The evolution of the siting problem has witnessed the co-evolution of a number of proposed solutions. Experts of many persuasions have leaped into the fray in the hope of generating techniques which promote successful siting. This paper examines each of these proposed solutions in light of the network of problems presently characterizing hazardous facility siting. It concludes with a discussion of likely paths future siting attempts may take.

THE SEARCH FOR SOLUTIONS

Before discussing specifics, it is necessary to situate the general problem within its cultural context. There exists a matrix of common understandings, conditioned by our unique liberal democracy, which shapes the nature of the problem and the search for solutions.

Public participation is an integral part of our Jeffersonian heritage. The

Puritan's quest for religious freedom, American opposition to the British Monarchy, and the freedom made possible by the earlier frontier, resulted in the institutionalization of a general lack of respect for constituted authority and that tendency persists. It is understood that citizens should not be timid in fighting "the establishment." In the decade of the 1960's, public participation in decisions involving the siting of risky facilities was mandated by law. These legal innovations have made it possible for citizens, even small groups and individuals, to delay or stop the implementation of many risky technologies. From this perspective, opposition should be expected because the socio-political system both encourages and reinforces it. Furthermore, since many modern technologies are capital intensive, which translates into big business as operators and big government as regulators, it is easy to characterize the situation as a contest between the relatively powerless individual/community and the mobilized forces of corporate wealth and political power.

An understanding of this cultural context highlights the limitations of many attempts at solution presently in vogue. A problem which is inherently political in nature tends to be treated primarily as a problem in individual psychology, information transmission, or attitude change.[2] A basic argument of this paper is that approaches emphasizing individual psychology, information transmission, or attitude change are doomed to failure. Other approaches have higher probabilities of success but the magnitude of those probabilities is contingent on factors that are difficult to predict or control.

Experts are currently advocating three classes of solutions to the siting problem:

A. improved risk communication;
B. improved disposal or treatment technology; and
C. improved community participation programs.

Each of these proposed solutions, and their limitations, will be discussed in turn.

IMPROVED RISK COMMUNICATION AS A SOLUTION

Those favoring improved risk communication as a solution to public intransigence view the problem as one of correcting errors in perception and judgment. The assumption is that if the public only understood the problem as the experts understand it then fear and resistence would diminish. A noteworthy example of faith in improved risk communication is provided by EPA administrator, William Ruckelshaus in a speech delivered before the National Academy of Sciences at the beginning of his second term. Quoting Jefferson, Ruckelshaus stated:

> "If we think (the public) not enlightened enough to exercise their control with a wholesome discretion, the remedy is not to take it from them, but to inform their discretion."[3]

However, less than two years later, in the face of a number of failed attempts to site hazardous wastes, Ruckelshaus publically challenged the sentiments expressed in that quotation.

Two issues are essential in understanding the limitations of the improved risk communication approach:

1. the popularization and democratization of risk; and
2. problems associated with the risk information itself.

While there are excellent review articles discussing problems in risk communication,[4] they do not address this first issue.

The Popularization and Democratization of Risk

Concern with risk is not an invention of modern societies. It can be argued that most of what is viewed through the ages as significant examples of religion and law are concerned with risk control or aversion. All religions include rituals through which spirits can be invoked to protect person, family, and property. Law deals with the social control of risk and risky individuals and what is considered as the equitable distribution of risk in society. Also, support for science and technology is undoubtedly related to their success in reducing risks. The prestige and pecuniary rewards granted to medical professionals is a good index of the esteem granted to those who reduce the risks of living.

However, concern with risk in modern society has become fine-tuned. There are now hosts of professionals devoted to discovering, more accurately estimating, more effectively communicating, and more adequately controlling risks. Risk has become big business. In fact, risk discovery, estimation, communication, and control is becoming a somewhat distinct academic discipline embracing natural and social-behavioral scientists of many persuasions and even publishing its own professional journals.

While this growth of the risk establishment has much to commend it, there may be unexpected consequences which have not been adequately addressed. Consider that it is now virtually impossible to escape news of the latest "significant" risk on an almost daily basis. Every organ of the mass media provides constant reminders that life is one constant flirtation with disaster. While the cognitive and behavioral outcomes of constant attention to risk are debatable, there has been little research on that specific topic, there is no reason to believe that lowered anxiety or a willingness to entertain unwanted risks are among them. Research on public reactions to radiation risks indicates that the more that people know about the subject the more opposed they become.[5] This reaction may be true of other hazards as well.

Furthermore, general anxiety about risks may be diffusing throughout the population to a greater degree than previously. A noteworthy historical dimension of risk is that it is seldom distributed equitably. The poor usually bear greater burdens of risk than do the rich. Studies documenting that the poor breath fouler air, experience greater occupational hazards, and tend to live nearer to environmentally risky facilities should come as no surprise.[6] The more affluent have always had greater opportunity to relocate to areas of less risk, to purchase more adequate diets, and to assume safer occupations. However, contemporary". . . features of environmental risk seem so ubiquitous . . . that even the wealthy and powerful are becoming anxious."[7]

Risk experts have made it clear that environmental pollutants in the air, soil, water, and food chain do not discriminate by social status. Nor is it easy for most people to geographically relocate to escape the more common environmental hazards. Surveys of environmental attitudes, especially attitudes toward risky facilities, tend to demonstrate no consistent pattern for social class variables such as education, occupation, and income.[8] One reasonable interpretation of those results is that people from all social classes are anxious to some degree.

Paradoxically, while we may be the healthiest and longest-lived people to ever inhabit the planet, we may also be the most concerned with health and safety issues. It makes little sense to label this phenomenon a "phobia" or some other mental health aberration. There is a dense network of organizations whose major goals include keeping the public concerned about their health. That is how one garners contributions to health organizations, how numerous magazines and newsletters are promoted, how exercise equipment and health foods are marketed, how careers in various health related organizations are maintained, and how a plethora of health and beauty aids are sold. Promoting anxiety about health and well being is a very lucrative enterprise.

Problems Associated with Risk Information

Much has been made about the rationality, or absence of same, of the public's response to risk information.[9,10] However, judgments of rationality are clearly value based. The more pertinent question is exactly how do people process risk information and what is the probability that their judgments can be modified. Two issues are crucial:

1. how is risk information processed; and,
2. what information is available for processing.

Three interrelated aspects of risk information processing are central to this discussion: the importance of attention selectivity; the perceived voluntariness of the risk; and, the issue of generalizing risk.

"People's perceptions of risk . . . are influenced by the memorability of past events and the imaginability of future events. As a result, any factor that makes a hazard unusually memorable or imaginable, such as a recent disaster, heavy media coverage, or a vivid film, could seriously distort perception of risk."[11] Cognitive psychologists agree that attention selectivity is a key issue in information processing. The more vivid or dramatic the event the more likely it will garner attention. A review of how the television media covers risk concludes: "risk as calculated by scientists had little to do with the amount of coverage provided by the three networks' evening news broadcasts. Instead, the networks appear to be using the traditional journalistic determinants of news . . . plus the broadcast criterion of visual impact to determine the degree of coverage of risk issues."[12] In other words, the visual media, by which most people get their daily news, accentuates the human tendency to focus on that which is most vivid and compelling. Toxic spills and malfunctioning chemical disposal facilities are "news" and worthy of attention while smoking and automobile travel, which are far riskier but far less newsworthy, are given relatively little attention. It may be unrealistic to expect to influence public opinion with a dose of cold, calculating facts about risk.

Related to the issue of attention selectivity and biases in information availability is the fact that people seldom accurately interpret information based on probabilities.[13] While this can be used to fuel arguments of public irrationality, it may be more pertinent to ask exactly what kinds of public concerns risk probabilities address. People want to know whether a substance will or will not cause them health problems. News that something in drinking water increases the annual incidence of cancer by one in ten-thousand does not answer that question. Probability estimates answer the kinds of questions raised by experts, they do not answer the kinds of questions raised by those exposed to a risky substance. After all, people either get cancer or do not get cancer, people do not experience probabilities.

The perceived voluntariness of the assumed risk is close behind attention selectivity in importance. People routinely assume all sorts of risks on a voluntary basis. They do so because the risk taking provides benefits such as easier living, personal satisfaction, ego-enhancement, or simple pleasure. Comparison of this type of risk taking with living next to a waste disposal facility is futile. They are not comparable within any reasoned frame of reference. However, the involuntariness *per se* is not the critical issue. After all, many people are involuntarily exposed to radon gas and do not get too upset about it.[14] Involuntariness arouses public ire when it is imposed by some person or organization who is perceived as occupying a position of power. Recall the discussion of our historically based hostility toward authority viewed as arbitrary. In addition, there are enough examples of corporate, industrial, and governmental insensitivity resulting in negative outcomes to justify suspicion of, and hostility toward, those who would impose unwanted risks.

Equally as important as how people process risk information is the issue of the nature of information available for processing. Perhaps the single most vexing issue facing those who favor more objective risk information as a palliative to public opposition is the lack of certitude which plagues risk estimates.

> "It is unfortunately the case that in most instances the relation between the risks projected by the regulatory agencies and the actual risks associated with exposure to a particular agent are very poorly known. The basic problem is that scientists can now collect accurate risk information about enviromental agents only under exposure conditions that do not match—indeed, which are generally far removed from -- the exposure conditions experienced by the human population."[15]

In most cases information on potential environmental risks is highly problematic. Epidemiological studies are characterized by problems of the adequate identification and control of crucial variables and relevant populations. Laboratory studies on rodents and other animals demand assumptions and extrapolations that go far beyond the available data. Increased public awareness of the actual uncertainties involved in existing risk estimates cannot be expected to increase public confidence.

Finally, the current state of risk estimation is greatly deficient in one aspect of central concern to many people, the ability to estimate the cumulative effect of exposure to many environmental hazards. The person who is informed about the increased risk posed by a particular substance may wonder about how it interacts with all the other hazards being ingested, inhaled, or absorbed. This is especially likely in our highly charged risk information society. While EPA is presently considering a "total human exposure" methodology, useful information will not be forthcoming soon.[16] Furthermore, it should not be assumed that once this information is made available that it will diminish public anxiety. The accuracy of the information can always be challenged, its sources impugned, or it can result in increased fear and avoidance. A typical reaction to such news is, "My God, I didn't know things were so bad."

Summary: Improved Risk Communication as Solution

Improved education and communication about risks is unlikely to solve the problem of public opposition. Risk communication tends to be selectively perceived and filtered through lenses shaded darkly by a few dramatic failures. However, risk communication cannot be ignored. Recent additions to the Superfund law mandate citizens' "Right To Know." More effective risk communication is a worthy and necessary endeavor. It will have to be part of a package of approaches designed to deal with public opposition. However, it alone cannot be expected to make a significant dent in that problem.

IMPROVED DISPOSAL OR TREATMENT TECHNOLOGY AS A SOLUTION

The relative technical merits of alternative technologies is beyond the expertise and interest of this author. My focus is on probable reactions to alternative technologies by the concerned public.

Efforts continue to develop safer and more efficient hazardous waste incineration devices, treatment technologies which chemically alter the hazardous material to make it more safe and perhaps recyclable, and disposal methods which more effectively isolate the hazardous material from the biosphere. While all of these endeavors merit continued, and perhaps increased, support, their significant impact on public opposition is questionable for a number of reasons.

Convincing the public that a new technology significantly decreases their risk embodies all of the problems discussed above regarding risk communication in general. Moreover, alternate technologies carry risks of their own. For example, the association of incinerators with dioxin emissions has made it as difficult to obtain public support for incineration as it has to get permission to site new disposal facilities.

However, the key problem in the interaction between new technologies and the concern public is the increased complexity, and the accompanying uncertainty, of modern technology. Much of modern technology has developed well beyond the point at which even an educated public can understand its constituent elements and their interactions. Furthermore, beyond the issue of the scientific and mechanical integrity of a given technology is the question of management: the public increasingly questions the ability of industry and government to insure the safe operation of complex technologies. The Bhopal, India chemical disaster elicited the following comment:

> What truly grips us in these accounts is not so much the numbers as the spectacle of suddenly vanishing competence, of men utterly routed by technology, of fail-safe systems failing with a logic as inexorable as it was once...indeed, right up until that very moment—unforseeable. And the spectacle haunts us because it seems to carry allegorical import, like the whispering omen of a hovering future."[17]

News of mismanaged high tech is increasingly common. For example, it is well known that management decisions played a key role in the space shuttle disaster, The Three Mile Island, Chernobyl, and other nuclear accidents, and in toxic waste problems such as Love Canal. In five years of research on public attitudes toward radioactive waste treatment and disposal facilities, nuclear power plants, irradiated food, and toxic chemical disposal, one consistent and dramatic finding is that respondents frequently mention the high probability of human error as the major reason for their "anti" attitudes. Respondents do not believe that the experts know enough to judge whether the technology is safe or not, cannot insure the integrity and reliability of that technology, and cannot guarantee responsible and knowledgeable management. From their perspective, the only sensible option is to say "no."

Summary: Improved Technology as a Solution

Like improvements in risk communication, improvements in disposal and treatment technology are worthy goals but unlikely to solve the siting dilemma, at least in the short run. A crisis in public confidence in science, industry, and government regulators indicates that trust is the issue. Innovations in technology alone cannot solve problems of trust. This issue also interacts with the risk communication problems noted above: all disposal and treatment technologies involve risks which must be communicated to the public. These communications are as likely to foster higher anxiety as to encourage confidence, especially given the uncertainty characterizing risk estimates.

IMPROVED COMMUNITY PARTICIPATION PROGRAMS AS A SOLUTION

Public participation programs involving risky technologies have evolved considerably since first mandated by law in 1969. For the first decade "public participation" translated into the opportunity to comment on proposals and at least one public meeting at which concerns were aired. This process has been severely criticized from its inception for being mere tokenism. The concerned public is given no input on the actual design of the plan and is given no actual power to insure that its concerns are dealt with in a satisfactory manner.

More recent discussions of public involvement have focused on two quite separate innovations:

1. programs providing incentives/compensation; and
2. programs providing actual decision-making power to the concerned public.

Each of these will be dealt with in turn.

Incentives and power sharing options increasingly tend to be found in the same package offered communities but they are distinct entities and vary considerably in their degree of attractiveness to various communities and interest groups within communities. Incentives, or tangible material rewards, are generally viewed in the context of restoring equity. It is reasoned that a risky facility is a common good which distributes benefits widely but concentrates costs locally. The key to public cooperation is viewed as some sort of concession which raises the reward value of the facility and thus makes the costs less onerous. Some of the more common incentives are: a surcharge on the waste which is returned to the community to use as it sees fit; property tax relief; and, local buying and hiring agreements between the community and the site operator.

Material rewards appeal to pecuniary motives, are more attractive to financially strapped communities, and are subject to being labeled as crass payoffs by the opposition. They do not address issues of fear and distrust.

Power sharing options, on the other hand, are typically in the form of programs which provide the community with some means to protect itself and to check on information coming from government and industry experts. For example, Superfund legislation provides the option of a grant to be given to communities to hire their own independent experts. Other power sharing options being used or discussed include the following: providing trained community people with the opportunity to monitor on-site operations and/or to test groundwater on a regular basis; community input into who they trust as site operators; community input into which of several available technologies they feel most comfortable with; and, power granted to community decision makers to shut down the site should problems be detected.

One concession which bridges both incentives and power sharing is property value guarantees provided those who live near the site. Property value guarantees usually give the property owner the option of leaving or not leaving and protects that person's equity. It is not simply an incentive because it protects and provides options. Power sharing options do deal directly with issues of fear and distrust by providing community members some means to protect their own interests.

Incentive and power sharing packages are increasingly common in attempts to site all sorts of risky facilities, they are also used as lures to entice volunteer communities. They have been applied in hazardous waste siting attempts in a number of states with mixed results. Massachusetts has had some success with these programs. After almost a decade of failure to site toxic chemicals in Minnesota, the state adopted a program providing multiple community options; this program appears to be moving toward siting. Efforts to site low-level radioactive waste, including the State of Pennsylvania's plan, are incorporating incentive and power sharing option packages. It is still too early to evaluate their relative success in promoting siting.

There is a growing body of literature, from both Canada and the United States, which argues that incentives must be coupled with power sharing options to realistically increase the probability of successful siting. A 1985 Statewide survey of Pennsylvanians' attitudes towards various policy issues connected with low-level radioactive waste disposal siting produced some instructive results.[18] In this survey, respondents were asked to evaluate various material incentives and power sharing options. The incentives included local tax relief, a surcharge on waste to be returned to the community, and local hiring and buying agreements. The power sharing options included property value guarantees, community control over who operates the site, provisions to hire independent experts, the provision of resources to train locals to monitor the site, and local power to shut down the site should malfunctions occur.

The results indicated that, in general, the public felt that affected communities should receive both incentives and assurances. but, a solid majority expressed little confidence that the material incentives, by themselves, would promote local cooperation. Three power sharing options were selected by a majority of the respondents as factors which could significantly promote local cooperation. They were:

1. funding to train locals to monitor the site;
2. local power to shut down the site should malfunctions occur; and
3. guaranteed property values.

These options provide the community with substantial control over their own fates. They certainly deal directly with the fear and distrust issues. However, under existing legislation these option cannot be offered by those charged with siting hazardous chemical facilities.

The desire for local control is further illustrated by respondents' answer to the question of who they trust to represent the community to the siting agency. Somewhat surprisingly, county and township officials received relatively low votes of confidence, as did a special task force selected by the governor. High confidence was expressed in locals elected to a committee and the use of referenda.

The significance of these shared power options are even more dramatically highlighted when the data are broken down in terms of how various categories of people respond to the items noted above. Those people who can be expected to most vociferously oppose siting, environmental advocates and those who self-identify as "NIMBY's", overwhelmingly choose the community control options as the only factors which can be expected to promote siting while greatly underchoosing the incentive options. In contrast, civic and governmental leaders tend to overselect the incentive options while giving relatively little weight to the community control options.

This discrepancey in views between "NIMBY's" and civic and governmental leaders reflects fundamental differences in trust in agencies handling risky technologies. It also reflects a somewhat inaccurate perception, by the leaders, of what the dissenting public wants. During interview situations leaders often remarked that "money talks." While no community is likely to refuse material concessions, opposition elements are unlikely to be swayed. Money apparently talks best to those opposed least. Power to protect one's health and property talks best to those most likely to be involved in protesting a proposed facility.

Recent research from Canada has basically reached the same conclusions. Canada has been having ongoing problems siting both hazardous chemicals and low-level radioactive waste. In a recent influential siting document the authors argue:

> "Of all the factors accounting for community resistance to siting proposals, is the inequity and feeling of loss of control, that result from the imposition of a decision, which appear to be of primary significance in creating con-

flict in siting processes."[19]

They go on to argue that loss of control is the most important factor and propose that affected communities be involved in all site specific decision making, even to the extent of helping decide acceptable options.

Summary: Improved Public Participation Programs as a Solution

Standard public participation programs which provide opportunities for public comment and one or more public meetings are more likely to encourage opposition rather than cooperation in attempts to site hazardous facilities. A consensus appears to be emerging among those responsible for siting, and those studying the process, that programs which include both incentives and options which provide the community with the ability to protect itself are more likely to promote public cooperation.

CONCLUSIONS

Clearly, public opposition to hazardous facility siting is not amenable to simple solutions nor does it have highly predictable outcomes. While both improved risk communication and improved technology are valued objectives they are not likely to have a significant impact on the siting problem. Public participation programs which provide incentives and power sharing options hold more promise and have a better success record at present. However, even improved participation programs may not provide the needed facilities within a reasonable time frame. While protection of the public is not served by hastily and improperly developed sites, neither is it served by failing to properly manage the growing body of hazardous materials.

Other possibilities are being explored by those studying the siting problem. Treatment and disposal at the site of waste generation and treatment and disposal in remote areas are increasingly being considered as viable options.[20] Treatment and disposal at the site of waste generation appears to have some level of public support but is subject to cost, space, and environmental constraints. Remote treatment and disposal is a possible option but it runs headlong into the problem of transporting large quantities of hazardous materials along populated routes. These options, or combinations as solutions to the siting problem have yet to be adequately tested.

Should the various programs discussed above fail to solve the problem of safely dealing with hazardous materials more attention may shift to the political spectrum. State governments are increasingly pressed by both industry and environmental activists to provide legislation which promotes safe siting. However, the future of hazardous material disposal/treatment ultimately rests with the courts. The last decade has witnessed a court system which is highly sensitive to pleas of protection coming from small groups and individuals. This is exactly what the courts should do in a democratic society. However, the omnipresent dilemma of the rights of the few versus the rights of the many cannot be ignored. The paths down which this search for answers takes will have serious implications for the hazardous facility siting issue and for many other issues interfacing with the development and use of technology.

REFERENCES

1. New York State Legislative Commission on Toxic Substances and Hazardous Wastes. 1987. "Hazardous waste facility siting: a national survey." The Commission; Albany, NY. June, 21.
2. Douglas, M. 1985. *Risk Acceptability According to the Social Sciences.* Russell Sage Foundation, New York, NY.
3. Ruckelshaus, W.D. 1983. "Science, risk and public policy." *Science.* 221: 1026-1028.
4. Covello, V.T., P. Slovic, and D. von Winterfeldt. 1987. "Risk communication: a review of the literature." Unpublished Manuscript.
5. Fischhoff, B. 1985. "Managing risk perceptions." *Issues in Science and Technology.* 2: 83-96.
6. Epstein, S.S., M.D., L.O. Brown, and C. Pope. 1982. *Hazardous Waste in America.* Sierra Club Books, San Francisco, CA.
7. O'Riordan, T. 1983. "The cognitive and political dimensions of risk analysis." *Journal of Environmental Psychology.* 3: 345-354.
8. Douglas, Ibid.
9. Du Pont, R.L. 1981. "The nuclear power phobia." *Business Week.* Sept. 7: 14-16.
10. Mitchell, R.C. 1984. "Rationality and irrationality in the public's perception of nuclear power." In W.R. Freudenberg and E.A. Rosa (eds.). *Public Reactions to Nuclear Power: Are There Critical Masses?* Westview Press, Inc., Boulder. Co.
11. Covello, et. al., Ibid.
12. Greenberg, M.R., D.B. Sachsman, P.M. Sandman, and K.L. Salomone. 1988. "Network Evening News Coverage of Environmental Risk." Unpublished Manuscript. January.
13. Slovic, P., B. Fischhoff, and S. Lichtenstein. 1982. "Facts versus fears: understanding perceived risk." In D. Kahneman, P. Slovic, and A. Tversky, (eds.) *Judgment Under Uncertainty: Heuristics and Biases.* Cambridge University Press, Cambridge.
14. Smith, V.K., W.H. Desvousges, and A. Fisher. 1987. "Communicating radon risk effectively: a mid-course evaluation." EPA Cooperative Agreement No. CR-811075, July.

15. Environ 1986. "Elements of toxicology and chemical assessment." Environ Corp., Washington, D.C.
16. Yosie, T.F. 1987. "EPA's risk assessment culture." *Environment, Science, and Technology.* 21: 526-531.
17. Editorial. 1985. *The New Yorker.* Feb. 18: 25.
18. Bord, R.J. 1987. "Judgments of Policies designed to elicit local cooperation on LLRW disposal siting: comparing the public and decision makers." *Nuclear and Chemical Waste Management.* 7: 99-105.
19. Report of the Siting Process Task Force on Low Level Radioactive Waste Disposal. 1987. "Opting for cooperation." Energy, Mines and Resources Canada.
20. Zimmerman, R. 1988. "Enhancement of public acceptability of hazardous waste management through decentralized (on-site) treatment and isolated disposal." Paper presented at the 1988 Hazardous Materials Spills Conference, May 19, Chicago, Ill.

Management of Hazardous Materials and Wastes: Treatment, Minimization and Environmental Impacts. Edited by S. K. Majumdar, E. W. Miller and R. F. Schmalz. © 1989, The Pennsylvania Academy of Science.

Chapter Twenty-Seven

IMPLEMENTING INTERGOVERNMENTAL REGULATORY POLICY: THE CASE OF HAZARDOUS WASTES

JAMES P. LESTER

Colorado State University
Department of Political Science
Ft. Collins, CO 80523.

INTRODUCTION

Dramatic incidents such as kepone dumping in the James River, the "Valley of the Drums" in Kentucky, and the damage caused by the Love Canal chemical disposal sites pointed out the need for a mechanism to respond to emergency situations and to more long-term dangers created by hazardous substances. Consequently, the hazardous waste issue has become the "single most threatening environmental issue facing the country," or the "environmental problem of the century."[1]

The recognition of this environmental problem has provided a basis for the passage of such federal legislation as the Resource Conservation and Recovery Act of 1976 (RCRA) and the Comprehensive Environmental Response, Compensation, and Liability Act of 1980 (CERCLA or Superfund). The former was enacted in order to provide for "cradle to grave" management of hazardous waste, while the latter was enacted in order to provide for liability, compensation, cleanup, and emergency response for hazardous substances released into the environment and the cleanup of inactive hazardous disposal sites. Since their initial passage, both of these Acts have been amended to extend and enlarge the hazardous waste program.[2]

Yet, the implementation of these two federal programs has been difficult, to say the least. In fact, most analyses of the federal Superfund program reach a rather pessimistic conclusion: CERCLA implementation has made very meager progress, with very few cleanups completed.[3] RCRA implementation has proceeded slowly as well. In fact, it took EPA four years to develop the regulatory framework and it occurred only as a result of court ordered deadlines, imposed after the agency had been sued for delay by environmentalists and others.[4] Nevertheless, as of January, 1988, forty-one states have received delegation of authority (primacy) to run their own hazardous waste program under RCRA.[5]

In both cases, a number of perspectives have been offered to explain the slow pace of RCRA and Superfund implementation. These explanations generally focus on federal level (or subnational level) constraints as key determinants of state implementation. Yet each of these explanations, when considered separately, provides only a partial view of the implementation process. Moreover, these perspectives are generally confined to an explanation of RCRA and Superfund implementation at one point in time or at a single location. What is needed is an analysis of implementation that is genuinely intergovernmental and comparative (i.e., one that compares implementation across states).

The purpose of this chapter is to explain RCRA and Superfund implementation by using a conceptual framework that is both intergovernmental and comparative in nature. In this framework, constraints are identified at federal, state, and local levels that have adversely affected RCRA and CERCLA implementation in the fifty American states from 1980-1988. To accomplish this purpose, the paper is divided into three parts. First, an intergovernmental framework is presented to explain state implementation behavior. This model is used as a guide to probe and identify some key determinants of intergovernmental policy implementation in these two areas. Second, the current status of RCRA and Superfund implementation is described. In this section, the progress to date is discussed as far as Superfund cleanups and state authorization is concerned. Finally, on the basis of the following analysis, some prescriptions are offered for improving RCRA and Superfund implementation in an era of regulatory federalism.

EXPLAINING RCRA AND SUPERFUND IMPLEMENTATION: AN INTERGOVERNMENTAL FRAMEWORK

It is assumed that state implementation behavior is a function of inducements and constraints provided to or imposed on the states from elsewhere in the intergovernmental system—above or below—as well as of the state's own propensity to act and its capacity to effectuate its preferences.[6] A model of intergovernmental policy implementation must recognize the joint nature of decisions and actions of interdependent institutions at the subnational level of government, and the bargaining that occurs among them and between them and the national level of government. The model presented in Figure 1 takes this into account. In the subsections that follow, the components of the model are further explained and the particular cases of RCRA and Superfund implementation are analyzed using this model.

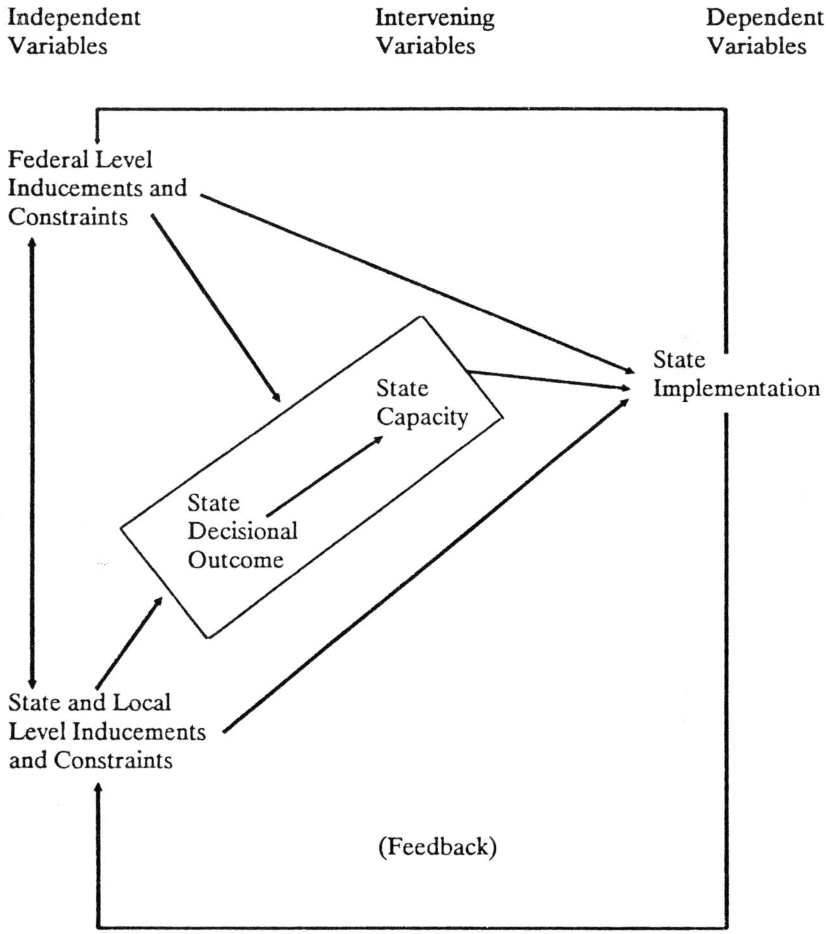

SOURCE: Goggin, Bowman, Lester, and O'Toole (1987).

FIGURE 1. A Conceptual Model of Intergovernmental Policy Implementation.

TABLE 1. National Priorities (Superfund) List: Final and Proposed Sites and Cleanups by State.

State	NPL Sites	Site Work Completed	Percent Completed (%)
AL	10	0	0%
AR	10	0	0
AK	0	0	0
AZ	9	1	11
CA	63	1	2
CO	15	0	0
CT	8	0	0
DE	18	0	0
FL	39	0	0
GA	8	1	13
HI	6	0	0
IA	13	1	8
ID	4	0	0
IL	27	1	4
IN	29	0	0
KS	8	1	13
KY	10	0	0
LA	8	0	0
MA	21	0	0
MD	10	2	20
ME	7	0	0
MI	70	2	3
MN	41	1	2
MO	19	0	0
MS	3	1	33
MT	9	0	0
NC	12	1	8
ND	1	0	0
NE	5	0	0
NH	13	1	8
NJ	101	2	2
NM	4	0	0
NV	0	0	0
NY	67	0	0
OH	31	2	6
OK	6	1	17
OR	6	0	0
PA	82	2	2
RI	8	0	0
SC	15	0	0
SD	1	0	0
TN	10	0	0
TX	26	2	8
UT	10	1	10
VA	19	1	5
VT	2	0	0
WA	28	1	4
WI	33	0	0
WV	6	0	0
WY	1	0	0
TOTAL	964	29	3%

SOURCE: U.S. Environmental Protection Agency, Office of Solid Waste and Emergency Response, September 30, 1987.

NOTE: The totals exclude American Samoa, Puerto Rico, Guam, and the Trust Territories.

The Dependent Variable: Superfund Cleanups, 1980-1987. Table 1 presents recent data (as of September 30, 1987) on Superfund cleanups by the fifty American states. Thus far, only 29 cleanups have been completed out of 964 National Priority List (NPL) sites (or 3% of the total sites).

At the same time, 494 sites (or 51%) are at the stage of "remedial investigations and feasibility studies," (or RI/FS). These studies are conducted at those sites identified through the NPL process as being priority candidates for remedial action. An RI/FS can be performed either by the EPA, the states, or responsible parties. A remedial investigation involves extensive sampling to determine the nature and extent of contamination at a site and to characterize a site in terms of topography, geology, and hydrology. Because of the comprehensive analysis required, an entire RI/FS requires, on the average, slightly more than one and one-half years to complete. State and federal agencies (other than EPA) can play significant roles during the RI/FS process and other stages of the remedial action. States can conduct fund-financed RI/FS and other remedial actions and must share in the costs of all remedial actions regardless of whether an action is conducted by a state or the EPA. In addition, 94 sites (or about 10%) are at the "remedial design" (RD) stage, and "remedial actions" (RA) have started at 89 sites (or 9% of the total NPL sites).

Remedial actions include the development of a detailed engineering design and the implementation of the selected permanent/long term remedial (cleanup) action of a hazardous waste site. As in the case for an RI/FS, remedial design and remedial action can be undertaken by the EPA or the states. Regardless of whether a remedial action is state or EPA-led, CERCLA requires that the state share the cost of the construction phase. If the state owns the site at the time of disposal, the required cost share is 50%; otherwise, the cost share is 10%. (The typical cost for the remedial design phase at one site is $440,000).

The Dependent Variable: RCRA Authorization, 1980-1988. Beginning with Delaware in 1984, a number of states have sought and received authorization from the EPA to run their hazardous waste programs under RCRA. Specifically, section 3006 of RCRA provides guidelines describing conditions under which states can be authorized to carry out their own programs.

The law provides for two stages in establishing state programs that would

TABLE 2. Status of Superfund Implementation.

Stage	1980-83	1984	1987
RI/FS	169	367	494
RD	17	33	94
RA	n.a.	n.a.	89

SOURCE: U.S. Environmental Protection Agency, Office of Solid Waste and Emergency Response, September 30, 1987.

n.a. = not available.

meet federal standards. The first is interim authorization under which states can operate their own less exacting programs for two years while upgrading them. The second is full, or final, authorization. Final authorization has three main criteria: (1) a state program must be "equivalent" to federal programs; (2) a state program must be "consistent" with federal or other state programs; and (3) a state program must provide for adequate enforcement. EPA has defined "equivalence" as requiring state programs to have the following elements: a manifest tracking system, criteria standards, adequate resources and legal authority, a designated lead agency and a public participation effort. It interprets "consistency" as preventing a state from imposing importation bans on hazardous waste from other states. Regulations elaborating upon the provisions for withdrawal of state approval involve the following: inadequate legal authority, unsatisfactory program operations or enforcement programs and modifications of a state program in such a way that it interferes with the free movement of hazardous wastes across state borders.

As of January, 1988, 41 states have received final authorization (primacy) from EPA for their hazardous waste programs. Of these, 11 received authorization in 1984, 19 in 1985, 10 in 1986, and one in 1987. Only Alaska, California, Connecticut, Hawaii, Idaho, Iowa, Maine, Ohio, and Wyoming have not yet received final authorization.[7]

Thus, as the preceding suggests, the federal government (especially the EPA) and the states are heavily involved in CERCLA and RCRA implementation, as well as site cleanup contractors, site operators, state environmental agencies, state legislatures, waste generators, local governments, and the courts.[8] Specifically, the major actors in RCRA and CERCLA implementation include three levels of government and the private sector. In order to adequately analyze the primary influences on RCRA and CERCLA implementation, a framework is necessary that incorporates all these actors. In the next section of this paper, an intergovernmental framework is utilized to describe the major constraints on RCRA and Superfund implementation.

Independent Variables: Federal Level Inducements and Constraints. The national decision that sets into motion an implementation process affects by its form and content the choices and behaviors of implementors. One set of variables has to do with the substance of the message sent by the national government to state implementors. For example, influential variables include the policy type, the ease with which the problem can be solved, the degree of certainty of effects, the provision for coercion, the financing arrangements, the anticipated effects on existing power structures, and the existence of a provision for citizen participation. Another set of potentially important variables includes the clarity of the provisions of the decision, the consistency of the decision with other policy objectives, and the flexibility of goals and procedures. A final set of federal level variables likely to affect state implementation includes the legitimacy, popularity, and credibility of regional officials and agencies.[9]

A recent survey by The Association of State and Territorial Solid Waste Management Officials (ASTSWMO) suggests that CERCLA delays are being caused in the site assessment and remedial action process by a lack of clear standards and guidance from the EPA, especially in the areas of determination of cleanup levels, setting priorities among sites, and demonstrating environmental damage. Moreover, EPA's Superfund process and coordination procedures were considered "too long, too bureaucratic, inefficient, and cost-ineffective."[10]

Other factors at the federal level causing delays in cleanup activities include the difficulties involved in assessing the magnitude of the problem, Congressional politics, partisan politics and a change in administration from Carter to Reagan in 1981. For example, "conducting the arduous research necessary to produce the data required for comparisons (of relative risks of individual sites) brings with it the great political liability of delay."[11] In addition, Superfund reauthorization was stalled in Congress because of the inability of the House and Senate to reconcile their differences about what sort of taxing scheme to use to finance it.[12] Moreover, an action suggesting strong partisan bias in Superfund implementation was the use of political criteria, rather than technical and statutory standards, for determining how to award federal grants to clean up priority dump sites under Superfund. EPA administrators were frequently accused of selectively withholding awards to clean up dangerous waste sites in the hopes of influencing the 1982 gubernatorial and congressional elections.[13] Finally, one of the factors that delayed implementation of Superfund was the change in administrations from Jimmy Carter to Ronald Reagan. It is clear that a hazardous waste program was given higher priority during the Carter administration than during the first two years of the Reagan presidency.[14] The original Superfund proposal and the legislation enacted were strong, federally-oriented programs. When the Reagan administration came into power, priorities shifted and the EPA was instructed to quickly delegate as much of Superfund to the states as possible. This shift in emphasis had the effect of leaving unplanned much of the fund-financed cleanup programs.[15] Moreover, by seeking a delegation of responsibility for Superfund to the states, the Reagan administration maximized the role and influence of state and local inducements and constraints.

In the case of RCRA implementation, recent survey results suggest that state implementors perceive the RCRA "message" from the federal government to be unclear, inconsistent, and inflexible.[16] For example, one state implementor commented that "the state authorization process is seemingly neither designed nor implemented to enable states to gain primacy for the program. The flexibility of RCRA is non-existent, thereby precluding the states from using their knowledge of the hazardous waste universe in their state, and being able to adjust resources and priorities as necessary."[17] Moreover, the nature of federal-state relations in this area was also perceived by state implementors to be a significant constraint on RCRA implementation. One state implementor commented that "state and federal relationships are supposed to be cooperative (but) . . . EPA makes them anything but that."[18] On the other hand, help from EPA's regional staff was perceived by state implementors as an inducement to RCRA implementation.[19]

Independent Variables: State and Local Inducements and Constraints. State and local politics, that is, the organization of interests within a state, also influence state implementation.[20] Three interacting organizational clusters—interest groups, state and local political institutions, and the focal state agency—generate the receptivity climate that the federal directive encounters. Thus, the propensity of a state to comply is influenced by the preferences and the relative strength of "advocacy coalitions."[21] Strength is a function of three characteristics of the coalitions: skill, cohesiveness, and level of resources. If the prevailing sentiment of the advocacy coalitions is oriented toward the preferences of the federal level actors, compliance is much more likely.

In general, groups representing business, and the chemical industry in particular, opposed the original Superfund legislation while those speaking for environmental interests tended to be supportive.[22] Beyond this general orientation of interest groups, however, there are a number of intergovernmental and intersectoral tensions that affected Superfund implementation at the state and local levels. In a careful analysis of Superfund implementation in South Carolina, Ann Bowman describes intersectoral tensions between the court system and the site operators and between the state environmental agency and the site operators as factors that affected Superfund cleanups in one site. In another site, intergovernmental tensions between the EPA and the state environmental agency (i.e., the South Carolina Department of Health and Environmental Control or DHEC) slowed Superfund implementation. She concluded that such tensions "served to delay goal achievement—the cleanup of uncontrolled hazardous waste sites."[23] Implementation of environmental legislation (such as Superfund) that blends intergovernmental operational responsibilities has vast tension producing potential. Moreover, an ASTSWMO survey in 1987 noted that there was a lack of public/political acceptance of remedial actions at NPL sites which contributed rather heavily to Superfund cleanup delays.[24]

In the case of RCRA implementation, however, a supportive mix of implementing officials made up of governors, state legislators, state administrators, judges, and local officials were a facilitating factor as far as implementation was concerned. States that received primacy early in the process were characterized by an "advocacy coalition" in support of RCRA implementation, while those that received primacy much later were characterized by an absence of supportive officials.[25]

Intervening Variables: State Decisional Outcomes and State Capacity. The model positions state decisionmakers in a pivotal role. They are the receivers and evaluators of streams of information flowing from the federal, state and local levels. Even under optimal conditions in which there is perfect congruence between information from the "top" and the "bottom," a decision to proceed or to implement must be made. The decision to implement is a function of the manner in which decisionmakers interpret the information that they are receiving. Interpretation is related to characteristics internal to the decisionmakers (e.g., psychological predispositions) and external to them (e.g., interpersonal relations with politicians and representatives of organized interests). This is a little recognized but powerful step in the implementation process.

The decision to implement is a necessary precondition for the next part of the process: the state's capacity to act. State actions regarding compliance are constrained not only by federal level and state/local level conditions, but also by the organizational and ecological capacity of the state to act on those preferences. Organizational capacity is related to the structure, personnel, and financial resources of a state agency. For example, unified structures reduce coordination costs and ease the transmission of information, thus enhancing state capacity to act; conversely, organizational fragmentation inhibits capacity.[26] Similarly, when relevant organizational actors are sufficiently numerous, qualified, and disposed to accept and undertake implementation, the capacity to act is further increased. Additionally, agency capacity requires possession of adequate financial resources to manage the program. Finally, ecological capacity is determined by socioeconomic and political conditions in the state where the policy is to be implemented. These conditions include state wealth, policy liberalism, general economic conditions, political culture, the salience of the problem to the state, and the media attention given to the issue for which the policy is to be addressed.

State capacity to implement Superfund was recently assessed by an ASTSWMO survey in 1987.[27] The survey found that there was a lack of funds for site assessment and cleanup efforts, including a shortage or an uncertainty of CERCLA funding and a shortage of contractor funds in thirty-five out of the forty-five responding states. Moreover, there was a lack of staff for site assessment and cleanup activities, including the lack of technical and project management staff and a lack of training and trained staff.[28] Specifically, there was an average of $8 million per state for total appropriation and eleven states reported no available funds for contractors. In addition, 41 responding states indicated that there was an average of 26 state staff per state and six states reported no staff available for Superfund activities. State staff turnover, including hiring delays, was also a factor constraining Superfund implementation.[29]

In addition, a number of states do not have a separate state funded Superfund program, beyond working with the EPA on the federal Superfund program. Many states have Superfund programs that are in their early stages and are still being developed. They have not gone beyond the preliminary assessment or site investigation stage for most all of their sites and may not have completed any remedial actions.[30] Moreover, a number of states indicated that they did not have a state statute or statutes authorizing them to conduct assessments and remedial actions at hazardous waste sites.[31] Thus, another factor delaying Superfund implementation is the lack of state Superfund legislation, authority and/or general legislative support.

The concern over the capability of the states to implement RCRA prompted EPA to initiate an assessment of the states' program capability as a key component of the final authorization process. In early 1984, regional administrators were instructed to work with the states to evaluate their capabilities and come to an agreement on whether action was required to strengthen those capabilities before authorization was recommended. In these "state capability assessments," particular attention was directed to the availability of resources in the states, the training of staff, the skill mix of the staff, organization of the state agency, and the salary structure of the state agencies involved in RCRA implementation. From 1984 until the present time, EPA has collected information on these areas of concern with a view toward helping the states develop their capabilities in these areas prior to authorization.

The results of a recent survey, however, suggest that state capabilities were not important determinants of the pace of RCRA implementation during 1984-1988.[32] Rather, these findings emphasize the nature of the RCRA "message" and the nature of relationships between regional EPA offices and state agencies charged with implementation as key determinants of authorization. In addition, the support given to the RCRA mandate by state implementors also influenced the speed of authorization. That is, if the state implementors were supportive of this federal directive and believed that it would make a significant difference in the responsible management of hazardous waste, then this attitude facilitated prompt compliance.[33] Thus, in the case of RCRA implementation, state capabilities were not important determinants of implementation compared to the nature of federal-state relations and intergovernmental communication.

In summary, a number of factors have influenced the rate of RCRA and Superfund implementation, including federal level constraints (such as the nature of the federal message), state/local constraints (such as intersectoral and intergovernmental tensions), and state capacity constraints (such as staffing and financial resources). In the final section of this chapter, suggestions are made that might alleviate the conditions that create delays in the implementation of RCRA and Superfund.

CONCLUSION: SOME PRESCRIPTIONS FOR IMPROVING RCRA AND SUPERFUND IMPLEMENTATION

In this chapter, it has been suggested that a number of factors have influenced the rate of RCRA and CERCLA implementation. In the area of Superfund implementation, major constraints center around state/local constraints and the issue of state capability, while RCRA implementation is constrained by intergovernmental relations and communication. On the basis of the previous analysis, some suggestions may be made for improving the implementation process in these two areas.

One of the central questions in redesigning a policy that enhances Superfund implementation is where and how to focus the authority for clean up activities: at the federal level or at the subnational level. That is, should the EPA or the states themselves have primary responsibility for implementing Superfund? Traditional arguments for centralization of authority typically originate with a criticism of decentralization. The criticisms that Henig summarizes are often used to justify centralization: (1) state and local governments lack the expertise of national governments; (2) state and local officials tend toward parochialism; (3) state and local governments are more unable or unwilling to raise money to meet service demands than national governments; (4) state and local governments are dominated by a conservative, business-oriented elite; and (5) state and local governments are structurally incapable of engaging in redistributive policies.[34] In the area of environmental problems specifically, it is often argued that many types of pollution problems—air and surface water pollution in particular—require national regulation because state control is not technically feasible. Rivers and winds refuse to respect state borders. Since the pollution they carry spills over state lines, national action is necessary. Moreover, research on the hazardous waste problem benefits all those who are afflicted regardless of where they live. It would prove both difficult and counterproductive to try to prevent those states which did not invest in research from obtaining the information produced by those who did. National coordination is necessary to prevent expensive duplication of effort and to ensure that the most effective deployment of the limited pool of properly trained researchers. In addition, it has been argued that even if state cleanup control is technically possible, the national government must intervene to protect the rights of all Americans, regardless of where they live. The "Mississippi syndrome" recognizes that some states have the capacity to administer technically complex programs, but others do not. For every California, there is a Mississippi. Therefore, an intrusive and restrictive federal presence is necessary to prevent the least capable states from misbehaving, even if this has the effect of stifling initiative and creativity among the most capable states.[35] Under this approach, uniform national standards would be mandated by the federal government. The essential problem with this approach is that states vary in terms of their need for implementation of a given policy.

The decentralization argument has often been invoked to varying degrees by students of state government generally and state environmental policy specifically and supported by at least two studies of state regulation of hazardous waste.[36] The essential arguments are that: (1) rulemakers in state agencies are in closer touch with field-oriented enforcement staffs, giving them a greater familiarity with the variety of circumstances to which proposed rules will be applied; (2) state regulators are more approachable for local groups and firms and therefore could respond more rapidly to rulemaking errors and changing conditions; (3) small enforcement staffs at the state level allow more flexible methods of enforcement than are possible for federal agencies; (4) the diversity of initiatives and programs from different states enhances innovation and enables states to learn from the experiences of pioneering states before committing the nation's resources to a single solution; and (5) states are more able than the national government to formulate approaches uniquely suited to their individual conditions and preferences.[37] Under this approach, the states, not the federal government, would serve as the ultimate arbiter of which sites receive priority attention and how much is spent on each site. Each state would make those decisions itself. The essential problem with this approach is that states vary in terms of their capability to implement a given policy.

A Mixed Approach: The Case for Selective Decentralization. One of the strongest implications of these arguments is that both have merit, depending on the conditions in the states themselves. Some of the states are quite capable of implementing Superfund (e.g., California) while others are not (e.g., Alabama or Mississippi). Nonetheless, according to Marc Landy, "changes in the design of the policymaking apparatus would serve to encourage greater responsibility and a fuller consideration of the need to integrate cleanup goals with other worthy policy objectives."[38] Such a policy could be accomplished by restructuring Superfund to a federal revenue sharing program. Under this policy, the national government would allocate monies to the states on the basis of a formula which attempted to measure their relative need and capability to implement CERCLA. Such a formula could be based, for example, on the number of NPL sites in the states, the resources available to the individual states to deal with cleanups, and the available staff. States with a more severe problem and few resources would receive more funds than those states with a less severe problem or greater resources. Moreover, "by creating a ceiling on the amount of funds available to a given state, the revenue sharing approach encourages the exercise of political responsibility by state chief executives."[39] In addition, this approach may help to remedy some of the tensions apparent in Superfund implementation that were discussed above. For example, tensions between EPA and the state environmental agencies might be relieved by such an approach which provides substantial funds (as well as regulations) for administering cleanups at the state level. States have often complained that they receive far too much in the way of federal advice but far too little in the way of federal financial support for administering what are essentially federal programs.

This approach is similar to one advocated by recent scholars of federalism.

Peterson, Rabe, and Wong argue that it is desirable to decentralize some federal programs but not others. They suggest a strategy of "selective decentralization," whereby redistributive programs are centralized while distributive programs are decentralized.[40] Thus, the programs are responsive to the conditions and preferences in the states themselves. By recognizing that a single approach (e.g., centralization or decentralization) will not work for all fifty states, a policy may be redesigned that is responsive to the differences inherent to the states themselves and thus reflects one of federalism's greatest virtues.

Finally, in the case of RCRA, the extant research suggests that EPA might be better advised to try to improve the nature of communication between headquarters staff and state implementors rather than to concentrate the bulk of their attention on assessing states' capability to implement RCRA. In addition, regional EPA staff should be encouraged to provide greater support to state implementors since, by doing so, they affect the style of implementation adopted by the states and the eventual pace of implementation. State implementors should also do what they can to enlist the support of governors, state legislators, and judicial officials in the authorization process. By doing so, a supportive mix of implementing officials may be attained which, in turn, affects the speed of implementation in a positive way.

It is entirely possible that changes such as those suggested above may facilitate RCRA and Superfund implementation in the future. Indeed, redesigning public policies will be a major challenge in the future as we increase our experiences with regulatory federalism. Hopefully, as we gain more experience with such programs, we will be able to redesign them so as to make them work rather than to turn implementation over to the private sector or to accept policy failure as the inevitable outcome of an intergovernmental system.

The views expressed in this paper are those of the author alone and do not necessarily represent the views of The Council of State Governments or The Center for the Environment and Natural Resources (CENR).

ACKNOWLEDGEMENTS

This research was supported by The Council of State Governments/Martin School of Public Administration Visiting Scholars Program at the University of Kentucky. The author is grateful to the Director of the Martin School, Phillip W. Roeder, for his generous support and encouragement throughout the project. In addition, I am especially indebted to The Council of State Governemnt's Center for the Environment and Natural Resources (CENR) and its Director, R. Steven Brown, for his support and helpful criticism.

REFERENCES

1. Epstein, Samuel S., Lester O. Brown, and Carl Pope. 1982. *Hazardous Waste in America.* San Francisco, CA: Sierra Club Books. See also James P. Lester and Ann O'M. Bowman, eds. 1983. *The Politics of Hazardous Waste Management.* Durham, NC: Duke University Press.
2. The Superfund Amendments and Reauthorization Act of 1986 (SARA) increased the funding level for the federal share of cleanup from $1.6 billion (1980-1985) to $8.5 billion (1986-1991). It also established a new $500 million cleanup program for leaking underground storage tanks and a new community-right-to-know program. Among the most important changes are provisions designed to increase significantly the role of the states in implementing the Superfund program. SARA makes changes in six key areas that have significant ramifications for states and local governments. These changes affect state obligations to pay for cleanup; requires states to site new disposal facilities or risk losing their Superfund monies; expands the states' role in selecting cleanup remedies; and gives the states major new responsibilities for implementing new aspects of the Superfund program.

   The Hazardous and Solid Waste Amendment Act of 1984 (HSWA) amended RCRA to bring those producing 100 kilograms or more of hazardous waste/month into the regulatory framework.
3. Bowman, Ann O'M. 1984. "Intergovernmental and Intersectoral Tensions in Environmental Policy Implementation: The Case of Hazardous Waste." *Policy Studies Review* 4:230-244. See also, Bowman, Ann O'M. 1985. "Hazardous Waste Cleanup and Superfund Implementation in the Southeast." *Policy Studies Journal* 14:100-110; and Bowman, Ann O'M. 1988. "Superfund Implementation: Five Years and How Many Cleanups." in Charles E. Davis and James P. Lester, eds. *Dimensions of Hazardous Waste Politics and Policy.* Westport, CT.: Greenwood Press.
4. Riley, Richard. 1983. "Toxic Substances, Hazardous Wastes, and Public Policy: Problems in Implementation." In: James P. Lester and Ann O'M. Bowman eds. *The Politics of Hazardous Waste Management.* Durham, NC: Duke University Press.
5. U.S. Environmental Protection Agency, Office of Solid Waste and Emergency Response, State Programs Branch, January 1, 1988.
6. Goggin, Malcolm L., Ann O'M. Bowman. James P. Lester and Laurence J. O'Toole, Jr. 1987. Third Generation Implementation Research: A Comparative State Diachronic Analysis. A Proposal Submitted to the National Science Foundation.
7. Lester, James P. 1987. "RCRA and Regulatory Federalism." *State Government News* 30:12-13.
8. Bowman, 1984. *op. cit.*

9. Crotty, Patricia M. 1988. "Assessing the Role of Federal Administrative Regions: An Exploratory Analysis." *Public Administration Review* 48:642-648.
10. Association of State and Territorial Solid Waste Management Officials. 1987. *State Programs for Hazardous Waste Site Assessments and Remedial Actions.* Washington, D.C.: ASTSWMO.
11. Landy, Marc. 1986. "Cleaning Up Superfund." *The Public Interest* 85:58-71.
12. Ibid.
13. Rosenbaum, Walter A. 1985. *Environmental Politics and Policy.* Washington, D.C.: Congressional Quarterly Press.
14. Cohen, Steven. 1984. "Defusing the Toxic Time Bomb: Federal Hazardous Waste Programs." In Norman J. Vig and Michael E. Kraft, eds., *Environmental Policy in the 1980s: Reagan's New Agenda.* Washington, D.C.: Congressional Quarterly Press.
15. Cohen, Steven and Mark Tipermas. 1983. "Superfund: Preimplementation Planning and Bureaucratic Politics." In James P. Lester and Ann O'M. Bowman, eds., *The Politics of Hazardous Waste Management.* Durham, NC: Duke University Press.
16. Lester, James P. 1988. "Implementing Intergovernmental Regulatory Policy: A Comparative State Analysis. " Lexington, KY: The Council of State Governments.
17. Ibid.
18. Ibid.
19. Ibid.
20. Goggin, et al., *op. cit.*
21. Sabatier, Paul A. 1987. "Knowledge, Policy-Oriented Learning, and Policy Change: An Advocacy Coalition Framework." *Knowledge: Creation, Diffusion, Utilization* 8:649-692.
22. Bowman, Ann O'M. 1984. *op.cit.*
23. Ibid.
24. Association of State and Territorial Solid Waste Management Officials. 1987. *op. cit.*
25. Lester, James P. 1988. *op. cit.*
26. Garnett, James L. 1987. "Operationalizing the Constitution Via Administrative Reorganization: Oilcans, trends, and Proverbs." *Public Administration Review* 47:35-44.
27. Association of State and Territorial Solid Waste Management Officials. 1987. *op. cit.*
28. Ibid.
29. Ibid.
30. Ibid.
31. Ibid.
32. Lester, James P. 1988. *op. cit.*

33. Ibid.
34. Henig, Jeffrey. 1985. *Public Policy and Federalism.* New York, NY: St. Martin's Press.
35. Landy, Mark. 1986. *op. cit.*
36. Lester, James P. et al. 1983. "Hazardous Wastes, Politics, and Public Policy: A Comparative State Analysis." *Western Political Quarterly* 36:258-285; and Bowman, Ann O'M. 1984. "Explaining State Response to the Hazardous Waste Problem." *Hazardous Waste and Hazardous Materials* 1:301-308.
37. Scholz, John. 1981. "State Regulatory Reform and Federal Regulation." *Policy Studies Review* 1:347-359.
38. Landy, Mark. 1986. "Cleaning Up Superfund." *The Public Interest* 85:58-71.
39. Ibid.
40. Peterson, Paul E., Barry Rabe, and Kenneth Wong. 1986. *When Federalism Works.* Washington, D.C.: The Brookings Institution.

Management of Hazardous Materials and Wastes: Treatment, Minimization and Environmental Impacts. Edited by S. K. Majumdar, E. W. Miller and R. F. Schmalz. © 1989, The Pennsylvania Academy of Science.

Chapter Twenty-Eight

RESURRECTING THE TERRITORIAL DIMENSION OF POLITICS: PUBLIC OPINION AND INTERGOVERNMENTAL RESPONSIBILITY FOR HAZARDOUS WASTE MANAGEMENT

MICHAEL R. FITZGERALD, WILLIAM LYONS and PATRICIA K. FREEMAN

Department of Political Science
University of Tennessee
Knoxville, TN 37996

INTRODUCTION

From Cooperative to Contentious Federalism

The evolution of the American federal system from "dual federalism," in which the various levels of government were strictly separate (and largely competitive) in dealing with governmental responsibilities, to "intergovernmental relations" in which the various levels of government are inextricably bound (and largely cooperative) in dealing with social problems, has been well documented.[1] The "discovery" of cooperative federalism by scholars, with its concomitant focus upon intergovernmental relations, recognized the dramatic change in the nature of federal and state relations brought by the New Deal.[2] Elazar has argued that the conventional wisdom regarding 19th and early 20th Century federal-state relations greatly exaggerated the separation and competition between levels of government. From the inception of the Republic, contends Elazar, the dominant pattern of relations between the national and state governments has been consistently cooperative.[3] By the 1970s the dominant paradigm of federalism, both in theory and in practice, was the functional integration of governmental activities based on national mandates, constraints, and domination of fiscal and regulatory activities.[4]

By the close of the 1970s, however, there were portents that cooperative federalism was evolving into a more disputatious mode. Walker's chronicle of the "collapse of key constraints in the system," led to his conclusion that intergovernmental relations had become overloaded and dysfunctional, and his call for a rearrangement of responsibilities along more decentralized lines that recognize "the limits of sharing."[5]

In addition to burgeoning scholarly disquiet over the evolution of intergovernmental relations, there was the growing sense among the electorate that in domestic affairs the national authority had at last been pushed beyond the bounds of "good" government. Thus, the putative Reagan Reformation in federalism emphasized a reduction in fiscal federalism, deregulation of key industries, and a presidential commitment to restoring authority and responsibility to subnational governments. The resurgence of the states as "viable and progressive" units of government has significantly bolstered the movement toward reshuffling of functions in a more decentralized federal arrangement.[6]

Clearly, the opening of the 1980s signaled a time of transition in American federalism, and the determined effort of the Reagan Administration to guide the transition, was bound to foster conflict. Less apparent, but surely no less significant in contributing to the rise of "contentious" federalism, has been the development of intergovernmental conflict over hazardous waste management; a dispute that, some argue has "bred more federal-state conflict than any other since the battle over desegregation in the South."[7] The roots of this burgeoning conflict have been attributed to the "residual effect of national government policy failures in the area,"[8] the failure of federal enforcement and neglect by individual states of their responsibilities,[9] federal agency disregard for the "potential power of state and local officials whose opinions reflect the consensus of their constituency on matters of health and safety,"[10] and continuing state experience that "federal promises are . . . unreliable."[11]

Federalism and the Environmental Agenda of the 1980s

Significant shifts in, and additions to, the national agenda for environmental regulation and protection have developed since the early seventies. Perhaps the most far reaching, critical problem area to be added to that agenda is the regulation of hazardous waste substances.[12] By the early 1980s, in fact, hazardous wastes had become such an issue that even a national administration otherwise dedicated to the curtailment of environmental regulation found itself under pressure to do something to defuse the "toxic time bomb":

> Hazardous waste policy posed a particularly vexing problem to the Reagan Administration. The administration was ideologically committed to scaling down environmental programs, but it was clear that the public wanted action on hazardous waste, a newly discovered environmental danger that required immediate attention.[13]

Unfortunately, the well-publicized, ineptness of the Environmental Protection

Agency's efforts to implement federal hazardous waste policy proved an acute political embarrassment verging on a national scandal, as well as an administrative disaster for the Reagan administration. The EPA imbroglio rekindled concern for, and reinforces the volatile political nature of, hazardous waste management policy.[14]

Hazardous wastes—radioactive, toxic, corrosive, ignitable, or chemically reactive materials posing a threat to humans and the ecosystem—are the by-products of our industrial processes and have been accumulating in geometrically increasing levels and effects. In fact, hazardous waste management and disposal has been designated by some as the "major environmental issue of the 1980s,"[15] and by others as "the environmental problem of the century."[16] Environmental policy making, implementation, and oversight is—despite an intrinsically scientific and technological nature that requires a prominent role for the scientific community—inevitably a political affair, and in the hazardous waste management area seems acutely so.[17] It has been noted that the hazardous waste management and disposal issue "exemplifies the transference of a technological problem with economic overtones into the political arena."[18] That is, policy decisions and their implementation in this area are intrinsically bound with the kind of public health, social, and economic consequences that are almost certain to engender significant political dispute.

Virtually all areas of environmental regulation possess an acutely intergovernmental character, but the effective management of hazardous waste problems intensifies the need for coherent federal, state, and local policy implementation.[19] Hazardous and toxic wastes are produced throughout the nation, and have been disposed by public and private producers in sites (many yet to be located) of varying size and safety in virtually every American community. The failure of government to effectively deal with the problem will affect the short and long-term prospects of virtually every American—especially those in future generations.

Consistent with the nature of the problem and its consequences, the fundamental framework of hazardous and toxic substance control, as it has emerged in federal law since 1976, requires extensive state and local activity. Indeed, Bowman has posited that national "hazardous waste policy is designed to produce an expanded role for state and local governments while that of the national government contracts."[20] Moreover, the degree of intergovernmental activity in the entire waste management field shows signs of demanding some fundamental redefinitions of federalism.[21] Early indications are that intergovernmental relations in this emerging era will be increasingly contentious as the implications of who does what, when, and at whose expense becomes more apparent to the American public.[22] There is ample reason to believe that this particular environmental domain will prove the most difficult of all because of the politically volatile, and technologically uncertain, question of waste disposal facility siting.[23] In this public policy area, which one scholar has called "bile" as compared to "pork barrel" politics,[24] the intragovernmental distribution of authority and responsibility for site construction and regulation is critical. As late as the mid-1970s most Americans assumed that the issues and technology related to hazardous waste management were beyond the understanding of the average citizen. Thus, influence in the decision making process was largely reserved for a small elite of experts, mostly employed in private industry. But, the occurrence of such well, and justifiably, publicized incidents as Times Beach in Missouri, and the Love Canal in New York, created a broadening public awareness of the great costs of the failure to provide adequate treatment and storage methods. This awareness eventually produced strong demands for more active government regulation of hazardous wastes, as well as a more open decision making process involving "a larger number of citizens and interest groups."[25] Zinberg's perspective on nuclear facility siting aptly applies to the full spectrum of waste management issues—nuclear and non-nuclear alike:

> Civic unease has been aggravated by the publicized assessments of scientists whose disagreements with industry and government provide such a range of views that by selection, any position can be supported. Seeing no consensus among experts, the public has begun to insist not only that it be educated, but also that it be allowed to participate directly in decisions. . . .[26]

Nowhere is the demand for an open decision making process more apparent than in the area of hazardous waste facility siting. When facility siting was essentially the purview of the scientific and industry experts, the calculation of risks associated with a given site could be based largely on what Rayner has called, "rival estimates of the probability and magnitude of unlikely events."[27] But the deepening public awareness, often fear, of the real and potential societal costs of locating these facilities—especially to those in the vicinity of the site—has rendered risk assessment and acceptance a critical element of virtually each and every decision. This shifts the decision toward "the explicit basis of social issues of trust and equity."[28] When hazardous waste management moves onto the policy agenda, government officials, at all levels, become in a very real sense public "risk managers" who are responsible for determining "what levels of perceived risk are publicly acceptable in situations where individuals are unable to negotiate a satisfactory solution for themselves."[29] Whether or not, however, this determination is acceptable, and capable of implementation, inevitably depends on three factors: First, the trust citizens generally have in the decision maker(s); second, recognition that the specific decision rendered falls within the "proper" authority of the decision maker(s); third, the "institutional credibility" of those specific agencies upon which implementation responsibility falls.[30]

Risk assessment and management decisions constitute only a part, and perhaps ultimately not even the most difficult part, of the hazardous waste management domain. There is the emerging problem of social equity (the fair and impartial distribution of costs and benefits) inherent in all public policy. As one of the primary purposes of the positive administrative state, social equity inevitably has become a major component of evolving intergovernmental efforts to struggle with hazardous waste management. Difficult equity issues pervade public choices across the policy spectrum. In this particular policy domain "undoubtedly the most visible and volatile equity problem currently is the geographical separation of beneficiaries and those who will bear burdens within this generation."[31] But, as Kates and Braine so succinctly put it, "despite the number of locations and the complexity of interest, the waste eventually ends up in someone's backyard."[32] The issue of social equity enters the picture because:

> Deciding whose "backyard" shall be chosen raises questions of whether some people should bear the risks for others, of which people should bear the risks and if and how they should be compensated by the beneficiaries, and of how accountability can be achieved in societal decisionmaking. Inadequate attention to these issues could produce highly inequitable policies.[33]

Inadequate attention to equity issues, we contend, is likely to produce not only inequitable policies, but policies that are virtually impossible to implement in an era of contentious federalism; citizens today can expect to find some level, or institution, of government that is both willing and able to support their resistance to perceived inequity. Again, the factors of trust, recognition of legitimacy, and institutional creditability, are bound to significantly shape the contours of intergovernmental contention over hazardous waste management and social equity issues. The scope of conflict is bound to widen because "there is almost surely a mismatch between the diverse locational interests and those institutions (at different levels in the federal system) that purport to represent or speak for them."[34]

The opening of the decision making process, and the intrinsically intergovernmental nature of the problem, in conjunction with proliferating public concern with risk management and social equity, clearly assures a very wide scope of conflict over hazardous waste management — especially with respect to disposal facility siting. The extent to which citizens trust those who make key decisions in this significant policy area is a major factor affecting the prospects of successful conflict resolution, and effective implementation. One concerned observer has summarized the situation by noting that:

> There is a growing trend of opposition in the United States to big government, large-scale technology, and the centralization of authority, and there is general skepticism about the government's ability to deal with the problem of society. Yet these same elements must play a central role in the resolution of societal problems. It is urgent that the tension between the federal government and the individual citizen be resolved through the intermediary organizations of state and local governments, through ad-

vocacy groups, and through direct participation.[35]

Such considerations, of course, go to the very heart of the federal system. But which set of "risk managers" and "arbiters of social equity," and at what level of government do citizens want to become involved in making hazardous waste facility siting decisions? This paper provides a preliminary answer to these important questions based on public opinions in one state.

RESEARCH DESIGN

Basic Research Question

One of the major features of whether or not a hazardous waste facility is going to be successfully sited and kept in operation is public trust in those who are directly involved in the siting decision. Thus, the basic research question framing this research is: Whom, and at what level of government, does the public generally trust, and what level does the public prefer to be involved in regarding hazardous waste facility siting decisions? Our research task was that of estimating the degree to which federal, as opposed to state and local institutional actors were seen as preferable in making siting decisions. In addition, we sought to model these preferences in the context of opinion in one state. Our research was framed around two premises. The first was that a preference for both state and federal involvement in siting was a function of a more generic notion of governmental trust. However, given the lack of proximity and familiarity with federal actors, there should be a more direct link between trust and siting involvement preference for that level, as opposed to same link for state and local siting preference. The greater role of the national government in policies related to income redistribution and in environmental issues more relevant to women leads us to our second expectation: one's being lower income and/or female to be more related to a preference for federal siting involvement than for state and local involvement.

Data and Methods

The Survey. This research is based on a public opinion survey conducted in the state of Tennessee. The survey was administered between February 16, and February 23, 1986. Calls were made Monday through Friday evenings from 6:30-9:30, and on Saturday and Sunday from 11:00 a.m. to 7:30 p.m. Households were randomly chosen from the universe of all Tennessee residences having listed telephone numbers, with respondents randomly chosen from the members of the selected household who were eighteen years old or over. Three attempts were made in order to contact the person selected, and the response rate was 63%. The sample size was 588 yielding an error margin of + or − four percent.

Citizen Trust and Involvement Preference. Citizens were asked a series of questions to determine their attitudes and opinions regarding hazardous waste management and disposal. Respondents were initially asked to rate a variety of government institutions and officials (federal, state, and local) according to their ability to make decisions on behalf of the public.[36] To determine the degree to which the public discriminates in its preferences for whom, at what level, should make siting decisions depending on the distinctive types of hazardous waste (low-level nuclear; high-level nuclear; toxic and chemical), citizens were asked separate questions for each type.[37] Finally, a series of questions regarding public preferences for which level should handle hazardous waste management of government, national or subnational, were asked.[38] Thus, based on responses to these questions, we are able to assess who Tennesseans prefer to make decisions on their behalf regarding hazardous waste management in general, and to locate hazardous waste disposal facilities in particular.

Modeling Preference. We are also interested in trying to understand why citizens vary in their preferences in this key area of intergovernmental responsibility. Accordingly, the survey also tapped additional attitudes regarding hazardous waste as well as the background characteristics of respondents. Three of these are used as independent variables in the analysis that follows: (1) Community size; (2) Family income; (3) Gender of respondent. The fourth independent variable used in the multivariate analysis is trust in governmental institutions. These four independent variables are used in regression analysis to predict which level of government, federal or state and local, citizens prefer to (a) make hazardous waste decisions in general, and (b) handle the siting of hazardous waste facilities. Table 1 provides the operational definition of each of these variables.

FINDINGS

The foundation of effective hazardous waste management and facility siting in an open decision making system is public confidence in those decision makers and institutions that must act as risk managers and equity arbiters. The comparative degree of general trust apparent in Tennessee public opinion is reported in Table 2. The most trusted of the rated institutions are the chief executives, president and governor—roughly 9 of 10 report having at least "some" trust in these officials. The least trusted are local government, (with about 1 in 4 expressing no trust), the EPA, and the Congress. The potential for executive leadership based on general citizen confidence is readily apparent. Moreover, although there is substantial variation about the mean, Tennesseans on average tend toward the more trusting end of the General Trust Scale.

Does general citizen confidence extend to preferences for hazardous waste management? The answer, as Table 3 shows is mixed. Tennesseans clearly prefer

TABLE 1

Operational Definitions for Variables

INDEPENDENT VARIABLES

1. RURAL: Size of community in which respondent lives (1 = over 100,000; 2 = 10-100,000; 3 = under 10,000; 4 = rural)
2. INCOME: Income of respondent's family (1 = under $10,000; 2 = $10-20,000; 3 = $20-30,000; 4 = $30-50,000; 5 = Above
3. SEX: Gender of respondent (1 = male; 2 = female)
4. TRUST: General trust in government institutions (Responses for citizen trust in the president; Department of Energy; EPA; Congress; Supreme Court; state governor; legislature; local government were first factor analyzed. The result was a one-factor solution indicating a single dimension to trust. Thus, we summed responses across all the trust items to create a General Trust Scale ranging from 9 (no trust) to 16 (great trust in all)

DEPENDENT VARIABLES

5. FEDERAL INVOLVEMENT: An index of the extent to which respondent preferred that the Federal government made decisions regarding hazardous waste. (Responses for whether respondents wanted DOE, EPA, & Congress to make decisions about high-level nuclear, low-level nuclear, & chemical-toxic waste were summed resulting in a Federal Involvement Scale ranging from 0 to 9)
6. STATE/LOCAL INVOLVEMENT: An index of the extent to which respondent preferred that state and local governments make decisions regarding hazardous waste. (Responses for whether respondents wanted the governor, legislature, & local government officials to make decisions about high-level nuclear, low-level nuclear, & chemical-toxic waste were summed resulting in a State & Local Involvement Scale ranging from 0 to 9).
7. FEDERAL SITING: A measure of the extent to which respondent preferred that the Federal government decide the location of hazardous waste disposal facilities ranging from 0 = strongly disapprove to 3 = strongly approve.
8. STATE/LOCAL SITING: An index of the extent to which respondent preferred that state or local government decide the location of hazardous waste disposal facilities ranging from 0 = strongly disapprove to 6 = strongly approve.

to have these substances regulated by the EPA—about 8 of 10 wanted it to manage them. The Congress, regardless of the waste type involved, is consistently rejected as a management locus by about half of Tennesseans. It appears that citizens clearly prefer to have management of hazardous waste exercised primarily by federal agencies—especially the EPA. This is reflected in the relative means for the respective involvement indices; the federal average is higher and there is less variation from the mean than is the case for state/local involvement. Tennesseans on average prefer federal, rather than state/local involvement, in managing hazardous waste, and are more in agreement regarding the former's involvement than of the latter. These results suggest that citizens tend to view level involvement in hazardous waste management as a package rather than differentiating by type of materials; Tennesseans do not clearly separate distinctive roles for a given level based on the type of substance, nuclear or non-nuclear, involved.

Regulation, it seems is one thing, facility placement is another. As is shown in Table 4, citizens in Tennessee tend to prefer subnational, rather than national decisions on siting disposal facilities. Only about one-third want the federal government to make siting decisions, while one-half prefer the state government, and just under half the local government. This preference for subnational institutions is reflected in the higher average state and local siting involvement index (2.4) as compared to the federal mean (2.1). Tennesseans clearly differentiate among levels of government depending on whether regulation or facility

TABLE 2

Degree of Trust in Institutions & Levels (N=588)

Institution or Level	Degree of Trust		
	No Trust	Some Trust	Great Deal of Trust
The President	11.1%	44.7%	44.2%
U.S. Department of Energy	17.8	69.7	12.5
U.S. EPA	21.6	61.1	17.3
The Congress	18.7	64.6	16.8
The Supreme Court	16.7	54.3	29.0
The Governor	12.1	49.7	38.1
State Legislature	12.5	71.6	15.8
Local Government	0.0	83.4	16.6
	Mean	Standard Deviation	
General Trust Scale	8.6	2.9	

TABLE 3

Institution or Level Citizen Would Like to Have Manage Different Types of Hazardous and Toxic Waste (N=588)

	Low Level Nuclear		High Level Nuclear		Chemical & Toxic	
	Yes	No	Yes	No	Yes	No
U.S. DOE	70%	30%	65%	35%	58%	32%
U.S. EPA	80	20	78	22	80	20
Congress	48	52	52	48	49	51
Governor	59	41	59	41	59	41
Legislature	56	44	56	44	47	43
Local Government	57	43	56	44	60	40

	Mean	Standard Deviation
Federal Involvement Index	5.75	2.63
State/Local Involvement Index	5.08	3.41

siting is at stake. They do not entirely reject a role in either for any level, but appear to prefer stronger subnational involvement when their "territorial interest" is more directly at stake in affecting the placement of facilities perceived as involving great potential risks.

Model Results

Citizen preference for whether national or subnational units make facility siting decisions, we contend, is the product of background, trust, and regulatory-locus factors. Figures 1 and 2 empirically test this proposition for federal and state/local siting involvement preferences using multiple regression in a causal model of the process. The path coefficients (Beta, or standardized regression coefficients) demonstrate the relative importance of each factor as a predictor of citizens preferences for, first, hazardous waste management involvement, and, second, siting involvement.

The model presented in Figure 1 statistically explains 11% of the variance in federal site involvement preferences. Clearly, this model is just the beginning of an explanation of variance in citizen preferences, but given the limited number of predictors, and the difficulty in social science of accounting for attitudinal differences on most issues, we consider the results quite satisfactory. The best predictor (B=.25) of management involvement in this model is general trust; those people with the highest trust levels tend to prefer greater levels of federal involvement. Women, and those citizens with relatively lower income, also tend to prefer federal involvement, but the relationships are only about half as strong (B=.10 and .13 respectively). The type of community in which Tennesseans reside is not significantly related to their management preferences. As expected, the direct relationship between background variables and federal site involvement preference disappears. Insofar as these factors have impact, it is through general management preferences. But general trust continues to have an independent effect; those people with the highest levels of overall trust in government tend also to have the strongest preference for federal siting decisions. Trust, does

TABLE 4

How/Who Should Decide Hazardous Waste Facility Location (N=588)

	Strongly Disapprove	Disapprove	Approve	Strongly Approve
Federal Government	30	37	26	07
State Government	22	29	38	12
Local Government	29	27	29	15
			Mean	*Standard Deviation*
Federal Siting Involvement Index			*2.10*	*.91*
State and Local Siting Involvement Index			*2.35*	*.88*

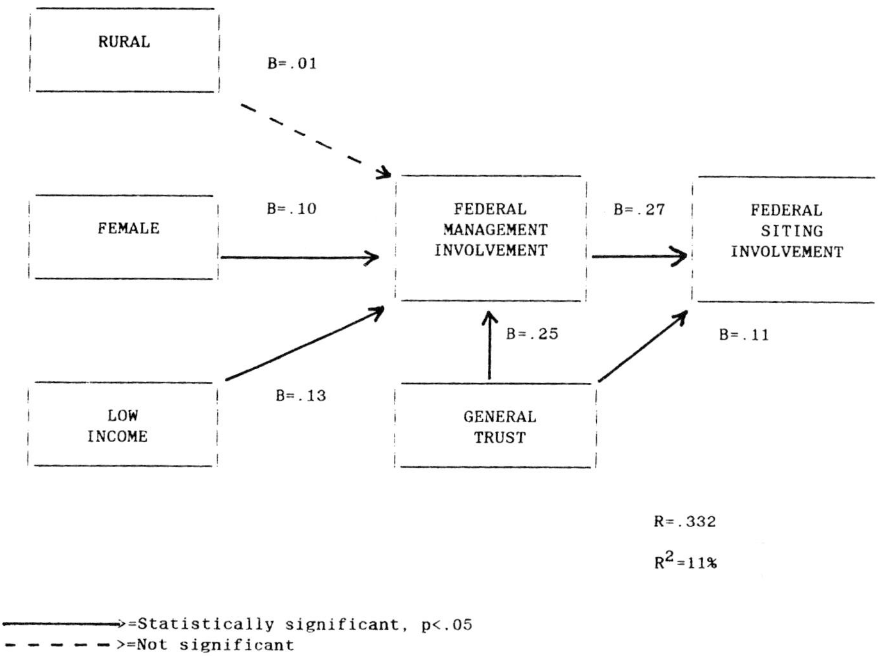

FIGURE 1. Regression Model of Federal Involvement in Hazardous Waste Regulation and Facility Siting.

prove something of a foundation for the willingness to have federal decisions about the siting of hazardous waste facilities. Moreover, among those groups for which federal regulation, as compared to state and local operation, has proven the most salutary in recent years (women and low income) their preference for national management in this issue domain is quite clear.

The state/local model reveals a different pattern of relationships. The model in Figure 2 is more successful in statistically explaining state/local siting involvement preferences, accounting for 17% of the variance. Gender and income are not significant predictors of state/local preference. Trust again is a significant predictor, but carries less weight and does not have an independent effect upon siting involvement preference. Of essentially the same predictive power (B=.17) is the size of the community in which respondents live. Historically, it has been rural Americans who have borne the brunt of unwanted, or relatively unattractive, public facility sitings. Further, the less populated regions of the U.S. have also traditionally been better able to protect themselves at the state, rather than the national level from urban electoral majorities. This likely accounts for the significant relationship between preference for state/local regulation and community size: rural Tennesseans, as compared to urban

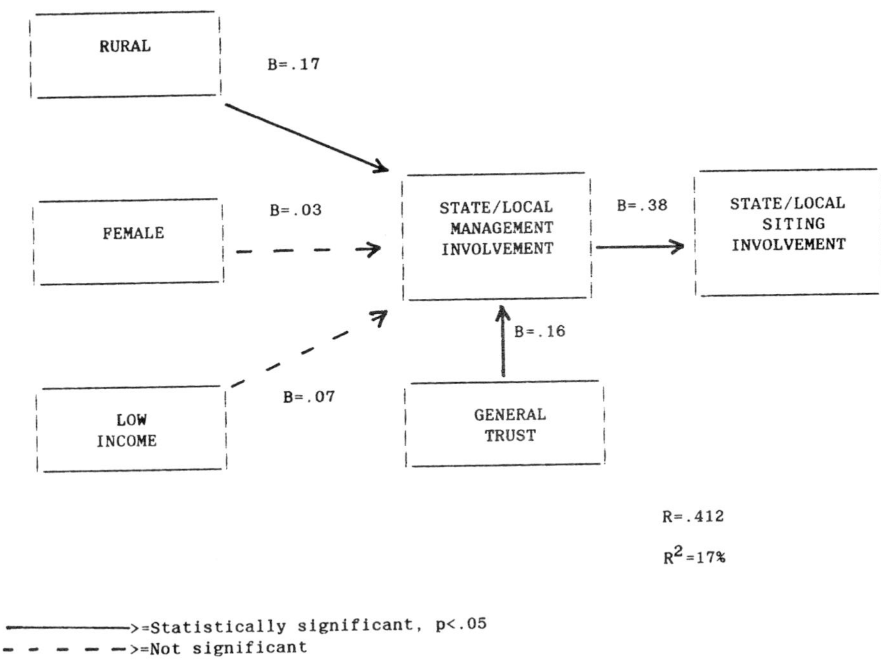

FIGURE 2. Regression Model of State/Local Involvement in Hazardous Waste Regulation and Facility Siting.

counterparts want the problem regulated by the level of government that has proven more responsive to their interests. The model demonstrates clearly that, among those citizens who most desire that state and local government regulate the problem, there is the strong tendency to want these units to control siting decisions as well. The most striking findings apparent in Figures 2 and 3 are that there are significant differences between Tennesseans who tend to prefer national or subnational decisions on hazardous waste siting regulation and siting; most especially, higher levels of general confidence in government accrues to the national government's benefit.

SUMMARY AND CONCLUSIONS

One of the major contributing factors to the decline of "cooperative" and rise of "contentious" federalism is the ascendancy of the ubiquitous problem of hazardous waste management on the national environmental policy agenda. We contend that the intrinsic nature of the waste disposal facility siting problem, in conjunction with what have thus far proven, and are likely to continue, to be intractable issues of risk assessment, risk management, and social equity management, guarantees deep and abiding intergovernmental conflict. As we move toward an "open" decision making system in this area, the fundamental basis for the successful resolution of conflict that must precede effective policy making and implementation is popular trust in institutions and actors at different governmental levels.

Based on a survey of public opinion in Tennessee, this study sought to establish the empirical parameters of this trust. We found significant variation among citizens regarding whom, at what level, they wanted to serve as their "risk managers" and "arbiters of equity" in locating hazardous waste disposal facilities. We then tested and evaluated a model that attempted to explain this variance. What conclusions can we draw based on these findings?

First, popular trust is a necessary but not sufficient condition regarding hazardous waste facility siting. It is the starting point, but a good start is no guarantor of success. Still, public confidence is one of the prerequisites for political leadership. If Bowman's call for "effective political leadership" to overcome "desultory governmental behavior" on hazardous waste problems is to be met, this research indicates that it will have to be from the executive, rather than the legislative branch, that leaders are drawn; public confidence is so much greater for the executive branch at both the state and federal levels.[39] "Local public opposition," observes Portney, has virtually rendered the siting process unworkable in most areas of the country."[40] While state preemption of local authority, and economic compensation to localities, have been offered as methods of breaking what has become a siting deadlock,[41] there is growing evidence that these approaches simply are not working.[42] Prescriptions, however, rooted in a better understanding of risk perception, and attention to provisions for local control, show some promise.[43] Major changes in how we've handled risk will have to be made. Informed consent of communities,[44] redefinition of property rights to accommodate binding community referenda on proposed sites,[45] and the restoration of "moral leadership,"[46] are suggested strategies. But, one way or the other, citizens want better control over the risk managers. Nevertheless, some feel it may already be too late, that there may well not be a sufficiently broad base of public and institutional support to sustain effective policy making and implementation in this area.[47]

Second, nuclear and non-nuclear hazardous waste issues are inextricably intertwined. Our results show that while the public does discriminate between these wastes, they find all of it distasteful and want it all stored somewhere other than their backyard. The finding that citizens tend to view the different types of hazardous waste as a package is consistent with previous research, and confirms Kasperson's view that nuclear and non-nuclear waste problems exacerbate one another.[48] Public fears regarding these dangerous substances are bundled together, Three Mile Island, Cherynobl, Love Canal, Times Beach—such incidents have had a mutual reinforcing effect. As a result, the prospects for siting one type of waste facility rather than another have become essentially the same.[49]

Third, the burgeoning, if hitherto quiet, national crisis in hazardous waste facility siting problems may well, in conjunction with other energy, environmental, and economic problems signal the resurrection of the "territorial dimension of politics" that shaped the original federal system into its dual spheres.[50] This new federal era, we suggest, may well reflect the kind of politics that fostered duel federalism in the first place; i.e., "the old view that the federal constitutional division of functions is the way the national government is kept in check."[51] The key to understanding the evolution of modern intergovernmental relations, according to Morton Grodzins, was the realization that "sharing" rather than separation of federal and state functions is the tradition, that governmental functions cannot readily be distinguished as local or national, and expanding national activities did not threaten excessive centralization of power so long as informal independent centers of political power continued to exist.[52] The hazardous waste management issue in general, and disposal facility siting in particular, is no longer viewed as entirely satisfactory: the resurrection of formal state and local institutions of politics and government, with a mandate to protect territorially based interests. In the closing decades of this Century it appears that the reduction of state and local officials to intergovernmental administrators of national programs is not enough. That is, Americans once more are more directly thinking about themselves, their politics, and government, in terms of "the territorial interest group—an aggregate of individuals and groups who share not only common experiences, values, fears, and purposes but also an awareness of the territorial dimensions of their collective interests and actions."[53] The results of our model testing provide some evidence of this. Certainly, the geographic locus inherent in the facility siting question, lends itself to the restoration of "territorial interests" as a prominant feature of our political life.

This is the more significant because, if valid, it suggests a revitalized role for what Elazar calls the "civil community" (the complex of governments and quasi-governments maintained locally to serve a given locality) in our political life. That is, we are perhaps seeing, because citizens are increasingly recognizing the need once again to use their local governments and leaders as bulwarks against "centralized" threats to community health and safety, the restoration of those political institutions "designed to serve the immediate communal ends of people who live in more or less immediate proximity to one another." The purpose of the civil community is to "serve local needs in light of the local value system."[54] Intergovernmental relations for all of its contributions to the improvement of our national life, rendered much of the civil community indistinguishable to the average citizen from national and state governments. Contentious federalism renders local independence and autonomy relevant once more—especially as communities seek, according to local needs and values, to handle "threats" or "opportunities" for becoming sites for hazardous waste disposal facilities. As Morell and Magorian have observed:

> While new hazardous waste management facilities undoubtedly will be needed in the United States in the years ahead, it is highly rational for communities and their residents to oppose waste facilities proposed for their immediate locale, and to oppose them as strenuously as possible. This inherent conflict between the majority's need for sites and facilities and the minorities right to resist being exploited in this manner lies at the heart of the politics of the siting dilemma.[55]

Finally, whether or not the system can manage the hazardous waste problem remains, and is likely for a long time to remain, an open question.[56] We are in for a long struggle, hopefully "contentious federalism" and the restoration of our "civil communities" will provide an effective framework for the resolution of conflict over hazardous waste disposal facility sting.

FOOTNOTES

1. See William Anderson, *The Nations and the States, Rivals or Partners?* (Minneapolis: University of Minnesota Press, 1955); Morton Grodzins, *The American System* (ed.) Daniel J. Elazar (Chicago: Rand McNally, 1966); William H. Riker, *Federalism: Origin, Operation, and Significance* (Boston: Little, Brown, 1964); W. Brooke Graves, *American Intergovernmental Relations* (N.Y.: Scribner's, 1964); and Daniel J. Elazar et al. (eds.) *Cooperation and Conflict: Readings In American Federalism* (Itasca, Ill.: F.E. Peacock, 1969).
2. See Richard H. Leach, *American Federalism* (N.Y.: W.W. Norton, 1976), and Daniel J. Elazar, *American Federalism: A View From the States* (N.Y.: Crowell, 1966).
3. Daniel J. Elazar, *The American Partnership: Intergovernmental Cooperation in the Nineteenth-Century United States* (Chicago: University of Chicago Press, 1962).
4. Michael D. Reagan, *The New Federalism,* (N.Y.: Oxford University Press, 1982); and Deil S. Wright, *Understanding Intergovernmental Relations* (North Scituate, Mass.: Duxbury, 1978).
5. David B. Walker, *Toward a Functioning Federalism* (Cambridge, Mass.: Winthrop, 1981): chapters 1 and 9.
6. Ann O'M. Bowman and Richard C. Kearney, *The Resurgence of the States* (Englewood Cliffs, N.J.: 1986).
7. This, at least, is the perspective of Richard C. Kearney and Robert B. Garey respecting intergovernmental relations in the nuclear waste management area. See "American Federalism and the Management of Radioactive Wastes, "*Public Administration Review* 42 (January/February 1982); 15.

These scholars argue that conflict in this policy area, "in conjunction with other energy and environmental policy areas, may force a reconceptualization of the system in terms of "contentious federalism'." Intergovernmental conflict is considered by Roger E. Kasperson as one of the "institutional uncertainties" that significantly compromises the prospects for the development of effective nuclear waste management. "Social Issues in Radioactive Waste Management," in Roger E. kasperson (ed.) *Equity Issues in Radioactive Waste Management* (Cambridge, Mass.: Oelgeschlager, Gunn and Hain, 1983): 41-48. See also Sheldon Kamieniecki, Robert O'Brien, and Michael Clark, "Intergovernmental Cooperation in Environmental Policy Making: Six Case Studies of Toxic Waste Sites," delivered at the Annual Meeting of the American Political Science Association, 1985; John A. Worthley and Richard Torkelson, "Managing the Toxic Waste Problem: Lessons From the Love Canal," *Administration and Society* 13 (August 1981): 145-160; and "Intergovernmental and Public-Private Waste Sector Relations in Hazardous Waste Management: The New York Example," in Lester and Bowman (eds.). *The Politics of Hazardous Waste Management* (1983): 102-11.

8. Ann O'M. Bowman, "Hazardous Waste Management: An Emerging Policy Area Within an Emerging Federalism," *Publius* 15 (Winter 1985): 131.
9. Richard W. Riley, "Foreword: Nuclear Waste and Governance," in E. William Colglazier, Jr. (ed.) *The Politics of Nuclear Waste* (N.Y.: Pergamon Press, 1982): viii-ix.
10. Richard G. Hewlett as quoted in Colglazier (ed.) *The Politics of Nuclear Waste* (1982): xiv.
11. E. William Colglazier and Paul Doty, "Preface," in Colglazier (ed.) *The Politics of Nuclear Waste* (1982): xiv.
12. Walter A. Rosenbaum, *Environmental Politics and Policy* (Washington, D.C.: CQ Press, 1985): 179-215; Stahrl W.w. Edmunds, "Environmental Policy: Bounded Rationality Applied to Unbounded Ecological Problems," in Dean E. Mann (ed.) *Environmental Policy Formation* (Lexington, Mass.: Lexington Books, 1981): 196-98; James P. Lester and Ann O'M. Bowman (eds.), *The Politics of Hazardous Waste Management* (Durham, N.C.: Duke University Press, 1983); Stephen Cohen, "Defusing the Toxic Time Bomb: Federal Hazardous Waste Programs," in Norman J. Vig and Michael E. Kraft (eds.) *Environmental Policy in the 1980s: Reagan's New Agenda* (Washington, D.C.: CQ Press, 1984): 273-91; and Bruce A. Williams and Albert R. Matheny, "Testing Theories of Social Regulation: Hazardous Waste Regulation in the American States," *Journal of Politics* 46 (May 1984): 435-438.
13. Cohen, "Defusing the Toxic Waste Bomb," (1984): 273-4.
14. Cohen, "Defusing the Toxic Waste Bomb," (1984): 279-90; Ann O'M. Bowman, "Epilogue," in James P. Lester and Ann O'M. Bowman (eds.)

The Politics of Hazardous Waste Management (1983): 252-56.

15. Bowman, "Hazardous Waste Management," (1985): 131.
16. Samuel S. Epstein, Lester O. Brown, and Carl Pope, *Hazardous Waste in America* (San Francisco: Sierra Club, 1982): 37.
17. Walter A. Rosenbaum, *Environmental Politics and Policy,* (1985): chapter 2, 179-180; Ann O'M. Bowman, "The Politics of Hazardous Waste Regulation: Theoretical and Practical Implications," in James P. Lester and Ann O'M. Bowman (eds.), *The Politics of Hazardous Waste Management* (1983): 243-245; Cohen, "Defusing the Toxic Waste Bomb," (1984): 276; David L. Morell and Christopher Mangorian, *Siting Hazardous Waste Facilities: Local Opposition and the Myth of Preemption* (Cambridge, Mass.: Ballinger, 1982: 85.
18. Bowman, "Hazardous Waste Management an Emerging Policy Area," (1985): 131.
19. Charles Davis, "Perceptions of Hazardous Waste Policy Issues Among Public and Private Sector Administrators," *Western Political Quarterly* 38 (September 1985): 448. Morell and Mangorian, *Siting Hazardous Waste Facilities* (1982): 21; Roger E. Kasperson, "Rethinking the Siting of Hazardous Waste Facilities," paper delivered at the Conference on Transport, Storage, and Disposal of Hazardous Materials, Institute for Applied Systems Analysis, Vienna, Austria, 1985: 10-11; Richard Riley, "Toxic Substances, Hazardous Wastes, and Public Policy: Problems in Implementation," in Lester and Bowman (eds.), *The Politics of Hazardous Waste Management,* (1983): 41; James P. Lester, "The Process of Hazardous Waste Regulation: Severity, Complexity, and Uncertainty," in Lester and Bowman (eds.) *Politics of Hazardous Waste Management,* (1983): 16-18.
20. Bowman, "Hazardous Waste and Emerging Policy Area," (1985): 137.
21. Bowman, "Hazardous Waste Management an Emerging Policy Area," (1985); Stephen A. Graham, "The Nuclear Waste Policy Act of 1982: A Case Study in American Federalism," *State Government* 57 (Winter 1984): 7-12; Yearn Hong Choi, "Issues of New Federalism in Low-Level Radioactive Waste Management: Cooperation or Confusion?" *State Government* 57 (Winter 1984): 13-20; and Kasperson, "Rethinking the Siting of Hazardous Waste Facilities," (1985): 28-9.
22. Bowman, "Hazardous Waste Management an Emerging Policy Area," (1985): 135; Kearney & Garey, "American Federalism and the Management of Radioactive Wastes," (1985): 210.
23. Michael O'Hare, Lawrence Bacow, and Debra Sanderson, *Facility Siting and Public Opposition* (N.Y.: Van Nostrand Reinhold, 1983); David L. Morell, "Technological Policies and Hazardous Waste Politics in California," in Lester and Bowman (eds.) *The Politics of Hazardous Waste Management* (1984): 139-175. John E. Seley and Julian Wolpert, "Equity and Location," in Kasperson (ed.) *Equity Issues* (1983): 69-93.

24. John J. Pitney, Jr., "Bile Barrel Politics: Siting Unwanted Facilities," *Journal of Policy Analysis and Management* 3 (Spring 1984): 446-448.
25. Robert M. O'Brien, Michael Clarke, and Sheldon Kamieniecki, "Open and Closed Systems of Decision Making: The Case of Toxic Waste Management," *Public Administration Review* 44 (July/August 1984): 334.
26. Dorothy S. Zinberg, "Public Participation: U.S. and European Perspectives," in Colglazier (ed.) *The Politics of Nuclear Waste* (1982): 160.
27. Steve Rayner, "Learning From the Blind Men and the Elephant, or Seeing Things Whole in Risk Management," paper delivered at the Annual Meeting of the Society for Risk Analysis, Knoxville, Tennessee, 1984: 11. See also Robin Cantor, Steve Rayner, and Bob Braid, "The Role of Liability Preferences in Societal Technology Choices: Results of a Pilot Study," paper delivered at the Annual Meeting of the Society for Risk Analysis, Washington, D.C., 1985.
28. Rayner, "Learning From the Blind Men," 1984: 11.
29. Steve Rayner, "Disagreeing About Risk: The Institutional Cultures of Risk Management and Planning for Future Generations," in Susan G. Hadden (ed.) *Risk Analysis, Institutions, and Public Policy* (Port Washington, N.Y.: Associated Faculty Press, 1984): 151. Risk assessment and acceptance as a feature of environmental policy making has developed to such an extent that some scholars now write of the emergence of an environmental risk establishment consisting of "individuals whose professional activities center on the identification, assessment, and regulation of environmental risks." James L. Regens and Thomas M. Dietz, "Risk Assessment and Environmental Regulation: The Role of Technical Analysis in Policy Choice," paper delivered at the Annual Meeting of the American Political Science Association, New Orleans, LA., 1985: 1.
30. Roger E. Kasperson, "Social Issues in Radioactive Waste Management," in Kasperson (ed.) *Equity Issues* (1983): 48-9.
31. Roger E. Kasperson, Patrick Derr, and Robert W. Kates, "Confronting Equity in Radioactive Waste Management: Modest Proposals for a Socially Just and Acceptable Program," in Kasperson (Ed.) *Equity Issues* (1983): 341.
32. Robert W. Kates and Bonnie Braine, "Locus, Equity, and the West Valley Nuclear Wastes," in Kasperson (ed.) *Equity Issues* (1983): 115.
33. Kasperson, *Equity Issues* (1983): x.
34. Kates and Braine, "Locus, Equity, and the West Valley," in Kasperson (ed.) *Equity Issues* (1983): 116.
35. Dorothy S. Zinberg, "Public Participation," in Colglazier (ed.) *The Politics of Nuclear Waste* (1982): 160.
36. The question was: I'm going to ask you how you feel about government institutions. For each I read, tell me whether you feel no trust, some trust, or a great deal of trust in their ability to make decisions on your behalf.
37. Each type of waste was defined for the respondent prior to seeking their

response to the following question: I am going to read you a list of government officials and institutions. For each, please tell me whether you would, or would not, prefer for them to make decisions regarding low level nuclear waste [or, in order, chemical and toxic waste, high level nuclear waste].

38. The question was: Decisions concerning the location of hazardous waste facilities can be made in many ways. I am going to read you a list of ways which have been suggested for making this decision. For each, tell me whether you approve, or disapprove of making the decision on that basis?
39. Ann O'M Bowman, "The Politics of Hazardous Waste Regulation," in Lester and Bowman (eds.) *The Politics of Hazardous Waste* (1983): 251.
40. Kent E. Portney, *Risk Perception and Opposition to Hazardous Waste Treatment Facility Siting: Some Evidence and Implications for Hazardous Waste Management and Policy From Survey Research"* (Medford, Mass.: Lincoln Filene Center for Citizenship and Public Affairs, Tufts University, 1986): 1.
41. Lawrence S. Bacow and James R. Milkey, "Overcoming Local Opposition to Hazardous Waste Facilities: the Massachusetts Approach," *Harvard Environmental Law Review* 6 (1982): 265-305; O'Hare, Bacow, and Sanderson, *Facility Siting and Public Opposition* (1983): 166-181; Morell and Mangorian, *Siting Hazardous Waste Facilities* (1982): chapters 5 and 6.
42. See Kent E. Portney, *Citizens Attitudes Toward Hazardous Waste Facility Siting: Public Opinion in Five Massachusetts Communities* (Medford, Mass.: Lincoln Filene Center for Citizenship and Public Affairs, Tufts University, 1983); "Allaying the NIMBY Syndrom: The Potential for Compensation in Hazardous Waste Facility Siting," *Hazardous Waste* 1 (Fall 1984): 411-21; "The Potential of the Theory of Compensation for Mitigating Public Opposition to Hazardous Waste Treatment Facility Siting," *Policy Studies Journal* 14 September 1984): 81-89; Mario Ristoratore, "Siting Toxic Waste Disposal Facilities in Canada and the United States: Problems and Prospects," *Policy Studies Journal* 14 (September 1985): 140-48; Richard N.L. Andrews and Terrence K. Pierson, "Local Control or State Override: Experiences and Lessons to Date," *Policy Studies Journal* 14 (September 1985): 90-99.
43. William Lyons, Patricia K. Freeman, and Michael R. Fitzgerald, "Public Opinion and the Legislative Response to the Hazardous Waste Challenge," delivered at the Annual Meeting of the American Political Science Association, Washington, D.C., 1986; Portney, "Risk Perception and Opposition to Hazardous Waste Treatment Facility Siting" (1986).
44. Roger E. Kasperson and Barry L. Rubin, "Siting a Radioactive Waste Repository: What Role for Equity?" in Kasperson (ed.) *Equity Issues* (1983): 143-4.
45. Robert Cameron Mitchell and Richard T. Carson, "Siting of Hazardous Facilities: Property Rights, Protest, and the Siting of Hazardous Waste

Facilities," *AEA Papers and Proceedings* 76 (May 1986): 285-90.

46. John J. Pitney, Jr., "Bile Barrel Politics: Siting Unwanted Facilities," *Journal of Policy Analysis and Management* 3 (Spring 1984): 448.
47. Roger E. Kasperson, "Social Issues in Radio Active Waste Management," in Kasperson, *Equity Issues* (1983): 25, 60.
48. Kasperson "Social Issues in Radio Active Waste Management," in Kasperson, *Equity Issues* (1983): 60.
49. Kasperson "Social Issues in Radio Active Waste Management," in Kasperson, *Equity Issues* (1983): 60.
50. Ivo D. Duchacek, *Comparative Federalism: The Territorial Dimension of Politics* (N.Y.: Holt, Rinehart and Winston, 1970).
51. Martin Diamond, "On the Relationship of Federalism and Decentralization," in Elazar et al. (eds.) *Cooperation and Conflict* (1969): 72.
52. As paraphrased in Martin Diamond, "On the Relationship of Federalism and Decentralization," in Elazar et al. (eds.) *Cooperation and Conflict* (1969): 72-3.
53. Ivo Duchacek, *Comparative Federalism* (1970): ix.
54. Daniel J. Elazar, *American Federalism* (1972): 84.
55. Morell and Mangorian, *Siting Hazardous Waste Facilities* (1982): 85.
56. Roger E. Kasperson, "Social Issues in Radio Active Waste Management," in Kasperson, *Equity Issues* (1983): 25.

Management of Hazardous Materials and Wastes: Treatment, Minimization and Environmental Impacts. Edited by S. K. Majumdar, E. W. Miller and R. F. Schmalz.

Chapter Twenty-Nine

NEIGHBORHOOD RESPONSE TOWARD AN EXISTING HAZARDOUS WASTE FACILITY

OWEN J. FURUSETH

Department of Geography and Earth Sciences
University of North Carolina at Charlotte
Charlotte, NC 28223

While technological and scientific advances are a cornerstone of the United States' economic growth and development, recent opinion polls have shown that the American public has become increasingly concerned about the cost of unbridled technological growth.[16] The news media's coverage of the events surrounding Three Mile Island, Love Canal, and Times Beach in the U.S. as well as the international tragedies at Bhopal and Chernobyl have sensitized a significant portion of the public to the risks of technological hazards at the community and neighborhood level. Among these risks, one which has been increasingly visible, but is as yet unresolved, is the production, disposal and reprocessing of hazardous waste.

Hazardous chemical wastes are a normal by-product of many industrial operations. Until recently, the transportation, recycling, or disposal of these materials was ignored by public policy-makers. The public was equally unconcerned. During the past five years, however, the location of existing waste disposal facilities and storage sites and the approval of new facilities have become controversial and divisive public policy decisions. Once a site is designated as a possibility for hazardous waste disposal, the response is usually the same: NIMBY—Not in my Backyard.

More critically, the locational conflict surrounding the siting of hazardous waste facilities can be expected to become an even more visible and widespread public policy issue. The full implementation of the Federal government's Hazardous and Solid Waste RCRA Amendments of 1984 (Public Law 94-580), along with the development of "mini RCRAs" by state governments, are expected to increase the need for additional waste management facilities at a time when these plants are already among the premier LULU's ("Locally Unwanted Land Use") in the United States.[14,10] Not surprisingly, public opposition to siting facilities is considered by some analysts to be the most critical problem in developing new hazardous waste management facilities.[8]

Generally, the announcement of a decision to locate a new waste treatment or disposal facility in a community is marked by public apprehension and opposition. The possibility that a facility which handles or stores hazardous materials will become a neighbor is not viewed as a positive land use change. Local elected officials and planners are often unable to handle the technical and safety issues surrounding these facilities nor deal with the political consequences. The level and intensity of public opposition to hazardous waste facilities is rarely matched by any other land use controversy.

While most research surrounding community attitudes toward hazardous waste facilities comes from the experiences of locating new facilities, little attention has been directed at existing facilities. Despite the fact that there are over 1600 licensed plants storing or treating hazardous materials on premises in the US, our understanding of how neighbors or community residents respond to these facilities and why they hold their particular attitudes and perceptions is unclear. One obvious, and potentially important, question that has little research is the degree of public awareness of existing hazardous waste facilities. Do neighbors know there is a hazardous waste facility situated nearby and how is this land use perceived?

Awareness or knowledge of a facility's existence is a precursor to forming opinions about specific impacts of a particular land use. If residents are aware of a specific facility it may be a "community stressor", producing concern and anxiety among populations living or working near the activity. Alternatively, residents may not perceive it as a noxious land use. In either case, opinions surrounding such a facility are potentially strong and likely to be conflicting. Much of the existing literature has dealt with situations such as the former, where owing to accidents or unintentional releases of hazardous materials, plants have contaminated surrounding areas, resulting in widespread and often chronic public distress and severe economic and political disruption.

An alternative scenario, is that neighboring residents may be unaware of a facility or its hazardous waste function. If residents fall into this category it suggests that the mental stress and physical externalities created by the facility are minimal. Whether these circumstances are realistic is unknown. If news accounts and popular literature are believable virtually all hazardous waste plants are LULU's.

The research reported in this chapter is part of a larger study of neighborhood attitudes and response toward an existing hazardous waste facility located in Charlotte, North Carolina. The facility is a manufacturing and research plant owned and operated by a large international chemical company. It produces, processes and stores over 30 hazardous substances. Under RCRA and state rules it is classified as a major generator.

In spite of its operational function, the plant is non-traditional in several ways. First, is its appearance. It does not fit the traditional image of a chemical plant. It is a low rise facility with a "clean" appearance. This impression is enhanced by good site planning. The plant is situated on a very large parcel with heavy landscaping around the perimeter. This site design screens negative visual externalities and reinforces the campus-like impression intended by the plant operators.

One explanation for the elaborate site design is the location of the plant. The facility is located in a predominately middle income residential area. It is the only industrial facility in a six mile radius. This type of location is not traditionally associated with chemical facilities. The plant and the residential character of the neighboring land use both evolved about the same time 18 years ago. At that juncture, this area was on the urban fringe and the zoning in the area was much less restrictive. In the intervening years the plant has been surrounded by the city of Charlotte and residential zoning paved the way for the development of a stable, predominately single family neighborhood.

RESEARCH METHODS

The primary source of data was a structured survey administered during July and August of 1987. Personal interviews were conducted with 135 randomly selected residents living in the neighborhood surrounding the facility. The refusal rate was extremely low, less than 5 percent. The survey instrument queried the respondents on a number of issues related to hazardous waste facilities and their management, including perception of the risk associated with hazardous waste facilities, the probable response to the siting of a facility nearby, as well as knowledge of the present facility situated in the neighborhood and its impacts on the community.

The sampling frame and interviewing process were stratified to include approximately an equal number of male and female respondents and to select respondents living a varying distance from the facility. Approximately, one quarter of the respondents lived within one-quarter of a mile, one-half of a mile, three-quarters of a mile, and a one mile radius of the plant.

AWARENESS OF THE FACILITY

Among the most unexpected findings of the survey, was the low level of community awareness of the "neighborhood" hazardous waste facility. As seen on Table 1, when asked whether there was any neighborhood facilities producing, storing, or handling hazardous waste less than one in five respondents (19 percent) answered affirmatively, and over 60 percent responded negatively. The remaining 19 percent of the sample were unsure or didn't know.

The small proportion of residents who were knowledgeable about the plant differs from popular image as well as past scientific research. News accounts of community protests against hazardous wastes facilities infer that the location of these activities is an important land use issue on which there is widespread spatial knowledge. This impression is supported by previous research. For example, Bachrack and Zautra found, that only 8.6 percent of rural residents living in a community near Phoenix Arizona proposed as the site for a hazardous waste treatment facility were unaware of the plant.[1]

One potential explanation for the low level of awareness is that the respondents may exhibit positive attitudes toward hazardous waste disposal technology and, therefore, the presence of this plant is not a salient personal or community concern. In such an environment, where hazardous waste facilities are perceived as relatively non-threatening, the location of a facility nearby would be acceptable and may be ignored by residents. It becomes an insignificant element on the local landscape.

This explanation was not supported by the survey results. The findings displayed on Table 2 show that the overwhelming majority of respondents were concerned about hazardous waste facilities. Seventy-seven percent were either concerned or very concerned about hazardous waste and facilities which handle it. More critically, the respondents indicated that they would be even more

TABLE 1

"As far as you know are there any factories or plants in this area (neighborhood) which produce, store, or handle hazardous chemical waste?"

Yes	19%
No	62%
Don't Know	19%
N = 135	

Source: The Author

TABLE 2

"How concerned are you about hazardous chemical waste and facilities that produce or store it?"

Very Concerned	47%
Concerned	30%
Somewhat Concerned	13%
Not Very Concerned	7%
Not Concerned at All	4%
N = 135	

Source: The Author

concerned if faced with the prospect that a hazardous waste facility was proposed for their neighborhood. (Table 3) Over 60 percent answered that they would be "very concerned", with an additional 30 percent responding that they would be "concerned" at the location of a hazardous waste facility nearby. Conversely no respondents indicated a lack of any concern.

This strong degree of suspicion and anxiety parallels earlier attitudinal research carried out in North Carolina and nationally. Although not directly comparable, statewide research by Richardson and county scale research by Runge and the Urban Institute at the University of North Carolina at Charlotte found strong opposition among North Carolinians at the prospect of hazardous waste producing manufacturing plants locating in an individual's home county.[11,12,13] Table 4. In a 1981 statewide survey Richardson found that 62 percent of North Carolinians would oppose a manufacturing plant producing hazardous waste by-products moving into their county. Two years later, the same question asked to survey respondents in Guilford and Warren Counties, jurisdictions with controversial hazardous waste disposal proposals pending, elicited even stronger opposition. More recently, a very similarly phrased question was asked in a survey in Mecklenburg County, which encompasses Charlotte. The results found that over three-quarters of those surveyed would oppose this type of new industrial establishment.

One measure of the implied strength of public opposition to hazardous waste

TABLE 3

"How would you react if you read or heard that a facility producing or storing hazardous waste was going to be located in this neighborhood?" Would you be:

Very Concerned	61%
Concerned	30%
Somewhat Concerned	8%
Not Concerned at All	0%
Don't Know	1%
N = 135	

Source: The Author

TABLE 4

"In general, would you ________ (a plant producing large quantities of by-product hazardous wastes) moving somewhere into your county?"

N.C. Statewide	%	Guilford County	%	Warren County	%	Mecklenburg County	%
Favor	26.4	Favor	11.6	Favor	11.8	Strongly Favor	1.4
Against	62.7	Against	74.8	Against	83.0	Favor	6.6
Indifferent	7.3	Indifferent	12.2	Indifferent	3.3	Against	32.6
						Strongly Against	45.2
						Indifferent	11.6
N = 682		N = 147		N = 153		N = 828	

Sources: After Richardson 1981, Runge 1983, Urban Institute, of North Carolina at Charlotte, 1987.

facilities measured by these questions is the geographic scale of response. In each case the question asked about the location of a hazardous waste source within the respondent's home county not his or her neighborhood or community. The emphatic opposition at the larger scale suggests even stronger concern at the personal scale.

The critical importance of protecting ones neighborhood from the risk of hazardous activities was as important to the Charlotte survey respondents as

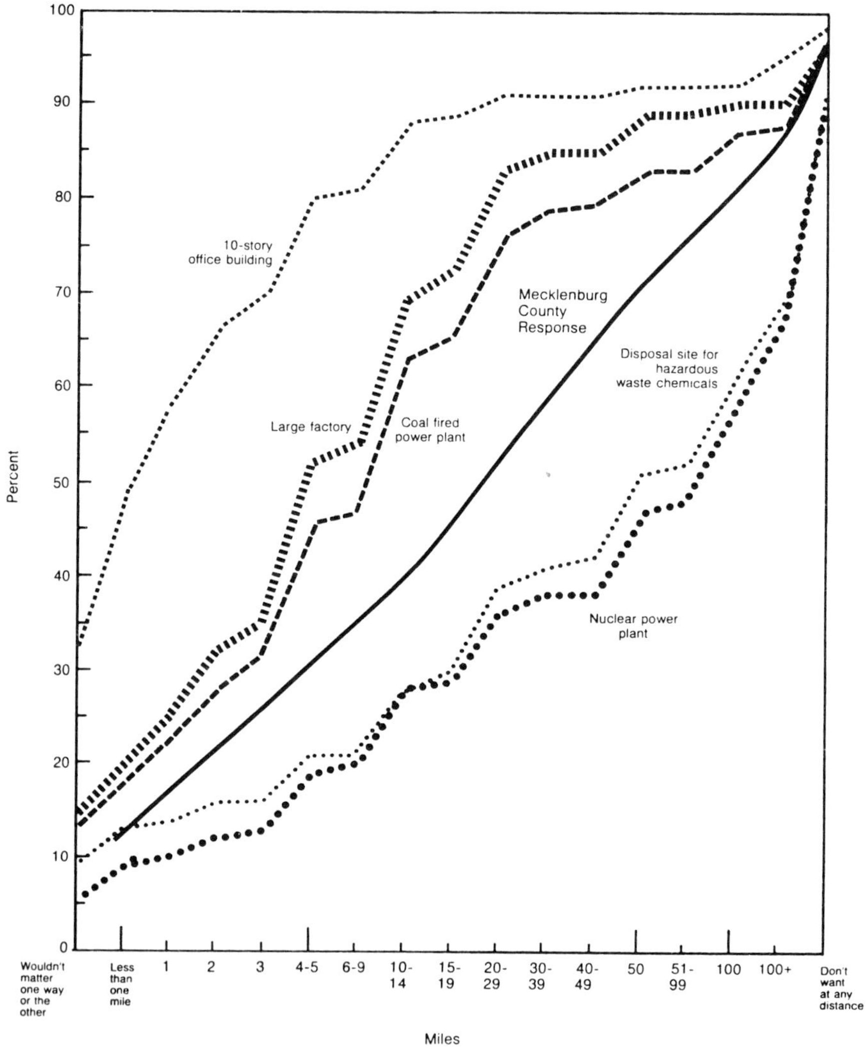

FIGURE 1. Cumulative Percentage of Respondents willing to accept varying Land Uses near their homes.

it was to respondents in a 1979 national survey sample. Figure 1. When the Charlotte respondents were asked how far they would be willing to live from a hazardous waste producing or storing plant, their cummulative response mirrored the findings of a nationwide survey by the Council on Environmental Quality which ask the same questions.[14]

ANALYSIS OF AWARENESS

Having established that the respondents were concerned about the safety of hazardous waste facilities but that neighborhood awareness was low, the research turned to examining socio-economic and behavioral attributes which might help explain the pattern of response. This analysis was carried out by cross-tabulating the survey question measuring individual knowledge of the hazardous waste facility with a number of predictor variables. The measurement of association between the predictor variables and individual awareness was carried out using the Theta (θ) and Lambda (λ) coefficients of association. The θ coefficient is a non-parametric statistic which measures the strength of association between ordinally scaled variables. The λ coefficient measures the relationship between an ordinal and a nominally scaled variable. Significant testing was carried out using the Kruskal-Wallis (K) and the Chi square (χ^2) tests.

Based on a review of the literature dealing with community response and attitudes toward hazardous waste facilities and other technological hazards, a set of eight variables were hypothesized as affecting community awareness. While the variables selected had not previously been applied to an analysis involving neighborhood response to an existing hazardous waste facility, most had been linked to individual response in other technological risk investigations.

All of the study variables were collected on the survey questionnaire. Table 5 contains these factors, their hypothesized operational affect, and the source of research hypothesis. Owing to restrictions in the questionnaire format and the composition of the sample, several potentially valuable variables cited in past research as important affectors of attitudes or response to technological risks, such as ethnic and racial characteristics and environmental group membership, could not be included in the analysis.

A review of the predictor variables examined shows that they can be arranged into four broad groupings: residential character, socio-economic status, demographic character, and technological orientation. For the residential character variables, DISTANCE and LENGTH, it was expected that respondents living at closer distance to the hazardous waste facility and those living in the neighborhood longer would have a greater knowledge of the facility.

Socio-economic status and demographic characteristics have been widely used to assess public attitudes toward nuclear power plants and other technological risks, however, the results have been mixed. In some studies selected socio-

TABLE 5

Explanatory Variables

Description	Label	Hypothesized Relationship	Source
Household distance from the facility	DISTANCE	Inverse—shorter distance greater awareness	Johnson, 1986; Vleeming, 1985
Length of residence in the neighborhood	LENGTH	Positive—longer residency greater awareness	
Total Family Income	INCOME	Positive—higher income greater awareness	Bachrach and Zautra 1985, Richardson, 1981
Total years of schooling completed	SCHOOL	Positive—greater educational attainment greater awareness	Bachrach and Zautra 1985, Johnson 1986, Richardson, 1981
Respondent age	AGE	Negative—greater age lower awareness among adults	Hamilton, 1985; Johnson 1986
Gender	GENDER	Greater awareness among females	Ottum, 1984; Hamilton, 1985; Bachrach and Zautra, 1985; Vleeming, 1985
Presence of children under 18 in household	KIDS	Greater awareness among households with children in home	Hamilton, 1985; Johnson 1986
Positive attitude toward hazardous waste disposal technology	TECHNOLOGY	Negative—lower awareness among supporters of waste technology	Baum, Fleming, and Davidson, 1983

economic and demographic variables have been very effective predictors of individual attitudes or perceptions, while in others they were not significant. In many cases, variables have operated in a contradicting fashion from earlier research, that is, they were statistically significant in opposite directions from previous results. Despite these problems, two of the strongest and most reliable socio-economic factors were included in this analysis. Both SCHOOL and INCOME were hypothesized to operate in a positive direction, indicating that an individual's awareness of the nearby facility would increase with the amount of formal education and income level.

Similarly, three demographic variables were analyzed. AGE was expected to be correlated in a negative direction, with younger residents being more aware of the facility. GENDER and KIDS were shown to be powerful predictors in earlier research, with women and parents of small children exhibiting greater concern about hazardous waste. In this research it was anticipated that individual awareness would be similarly effected.

The final variable TECHNOLOGY reflected the respondent's confidence in the capability of existing technology to mitigate the risk of hazardous waste. Previous attitudinal research has found that technocentric individuals, that is, persons who have confidence in scientific and technologically based solutions to modern problems exhibit stronger support for higher risk technology. In this investigation technology was measured by several likert-type questions on the survey instrument. Individual responses were subsequently aggregated to derive a technological confidence index score. For this analysis it was hypothesized that a high score would correlate with more confidence in hazardous waste technology, less anxiety about such a facility and therefore less awareness of the nearby plant.

The analysis of the crosstabulated data, presented in Table 6, shows that most predictor variables were only weakly associated with neighborhood awareness.

TABLE 6
Analysis of Crosstabulation Results

	Degree of Association		Test Significance	
	θ	λ	K	χ^2
Distance	.538		*	
Length	.126			
Income	.093			
School	.213		**	
Age	.047			
Technology	.037			
Gender		.137		
Kids		.020		

* Significant Beyond .99 Level
**Significant Beyond .95 Level

Source: Calculations by the Author

For example, less than 10 percent of the variation in an individual's knowledge of the facility was explained knowing the respondents income, age, whether or not there was a child present in the household, or the attitude toward hazardous waste technology. Gender and length of residence in the neighborhood were slightly stronger predictors accounting for almost 14 and 13 percent of the variation, respectively. Neither variable performed as well as in previous investigations. None of these variables were statistically significant.

The only statistically significant factors were the amount of education and the location of respondent households. The θ coefficient of .213 for SCHOOL indicates that 21 percent of the variation in a respondent awareness was related to educational background. A review of these data finds, as expected, that the association was positive. This indicates that knowledge of the facility increased with educational attainment.

Without doubt, the most impressive association was related to respondent location from the hazardous waste facility (DISTANCE). Nearly 54 percent of the variation in individual awareness of the facility was accounted for by the locational criterion. As expected, the closer one lived to the facility the more aware of it they were. For geographers, this distance-decay relationship comes as no surprise. Past work has found community awareness and opposition to the LULU's, including sanitary landfills, groups homes, and nuclear facilities, is strongly linked to the actual or perceived distance from the noxious land use.[3,6,16,4] These data extend the previous findings one step further and indicate that knowledge or awareness of a noxious facility is similarly affected by distance.

CONCLUSIONS AND IMPLICATIONS

For Charlotte residents living around the subject hazardous waste facility, the location and operation of this plant represent an insignificant risk. Most were unaware of the risk. While they, like the members of the public, have serious concerns about living near a hazardous waste facility, their anxiety is not reflected in a complete knowledge about their local environment. Only those residents living immediately around the plant or those with higher levels of educational attainment exhibit any knowledge about the hazardous plant sited in their midst.

Although these data are limited to a single neighborhood and did not specifically query residents about what they thought was occurring in the nearby factory or what was being manufactured at the plant, one potential explanation for the low level of awareness is the appearance of the facility and the operational management of it. The campus-like setting, the large expanse of open area separating the manufacturing structure from the neighborhood, and the landscaping barrier around the plant all tend to "de-industrialize" its appearance. Similarly, a security gate and absence of noxious externalities, such as smell and smoke, tend to control casual or unintentional knowledge of the plant. Basing their perceptions on images gained from the news media (the most important source of information about hazardous waste according to respondents) or past experience regarding the appearance "chemical plants", most respondents probably did not associate this facility with hazardous waste.

These data are from a single neighborhood sample in only one community, they do suggest that appearance may be very important to reducing public awareness and, therefore anxiety, regarding hazardous waste facilities. While it is infeasible and unethical to disguise all hazardous waste plants so as to make them look like suburban business parks, design and aesthetic considerations may help minimize the negative public impression of hazardous waste facilities. If increased attention can be given helping make a hazardous waste facility an attractive or unobtrusive component of the local scene, it may help reduce anxiety and the resulting public conflict. Obviously, this is no panacea. Nonetheless any strategy to help resolve the mounting problem of siting new hazardous facilities warrants consideration.

Finally, a caveat. It should be remembered that the study facility is 18 years old. How would the community react if this facility were proposed today? Certainly, information about the operation and the production of hazardous waste would be more widely disseminated in the community. Awareness would be much higher and feelings stronger. Whether respondents would act upon their stated opposition to a hazardous waste facility is an interesting prospect. A hazardous waste facility, even one that looks like a office park, might exceed neighborhood tolerance. Again, the relationship between facility appearance and public anxiety is one which warrants further study.

Short of restructuring our current economic and social fabric, we are going to produce increasing amounts of hazardous waste and have to safely deal with the disposal of those materials. This will mean new facilities to treat these substances. Efforts to understand citizens attitude's toward these facilities and ways to reduce public anxiety are critical to making a hazardous waste disposal system work. Hopefully, this research has made a small contribution in this cause. Failure to address these social dimensions of the problem will only create serious long term problems.

REFERENCES

1. Bachrach, K.M. and A.J. Zautra. 1985. "Coping with a Community Stressor: The Threat of a Hazardous Waste Facility," *Journal of Health and Social Behavior,* 26:127-141.
2. Baum, A., R. Fleming and L.M. Davidson. 1983. "Natural Disaster and Technological Catastrophe" *Environment and Behavior,* 15:333-354.
3. Dear, M. and S.M. Taylor. 1977. *Not on Our Streets, Community Attitudes to Mental Health Care.* London: Pion Ltd., 180 pp.

4. Furuseth, O.J. and M.S. Johnson. 1988. "Neighborhood Attitudes Toward a Sanitary Landfill: A North Carolina Study," *Applied Geography* (forthcoming).
5. Hamilton, L.C. 1985. "Concern About Toxic Waste, Three Demographic Predictors," *Sociological Perspectives,* 28:463-486.
6. Johnson, Jr. J.H. 1986. "A Model of Evacuation-Decision Making in a Nuclear Reactor Emergency," *The Geographical Review,* 76:405-418.
7. Ottum, M. 1984. "Local Residents' Perception of the Williamstown Pollution Problem: Telephone Survey-February 1, 1984: (unpublished) Department of Environmental and Health Sciences, Johnson State College, Vermont.
8. Popper, F.J. 1983. "LP/HC and LULUS: The Political Uses of Risk Analysis in Land-Use Planning," *Risk Analysis"* 3:255-263.
9. Portney, K.E. 1986. "The Perception of Health Risk and Opposition to Hazardous Waste Treatment Facility Siting: Implications for Hazardous Waste Management and Policy from Survey Research," *In:* J.W. Frazier, B.J. Epstein, and J.F. Langowski (Eds.) *Papers and Proceedings of the Applied Geography Conferences,* 9:114-123.
10. Regens, J.L. 1983. "Siting Hazardous Waste Management Facilities," *In:* G.A. Daneke, M.W. Garcia, and J.D. Priscoli (Eds.) *Public Involvement and Social Impact Assessment,* Boulder: Westview Press, 121-128.
11. Richardson, C.J. 1981. *Hazardous Substances in North Carolina: An Analysis of Public Perceptions.* Durham: School of Forestry and Environmental Studies, Duke University, 70 pp.
12. Runge, C.F. 1983. *Local Attitudes Towards Hazardous Waste Siting in Guilford and Warren Counties, North Carolina.* Raleigh: North Carolina Board of Science and Technology, 44 pp.
13. Urban Institute, University of North Carolina at Charlotte. 1988. "Results of the 1987 Omnibus Citizens Survey of Charlotte-Mecklenburg Residents" (unpublished) Charlotte, N.C.
14. U.S. Council on Environmental Quality. 1980. *Public Opinion on Environmental Issues, Results of a National Public Opinion Survey.* Washington, D.C.: U.S. Government Printing Office, 37 pp.
15. Vleeming, R.G. 1985. "Factors Affecting Attitudes toward Nuclear Power in the Netherlands," *The Journal of Social Psychology,* 125:119-125.
16. Zeigler, D., J.H. Johnson, Jr., and S. Brunn. 1983. *Technological Hazards.* Washington, D.C.: Association of American Geographers Resource Paper, 99 pp.

Management of Hazardous Materials and Wastes: Treatment, Minimization and Environmental Impacts. Edited by S. K. Majumdar, E. W. Miller and R. F. Schmalz. © 1989, The Pennsylvania Academy of Science.

Chapter Thirty

A MULTIOBJECTIVE DECISION METHODOLOGY FOR SITING NOXIOUS PUBLIC FACILITIES

BENJAMIN G. MAIDEN[1] and HENRY L. HUNKER[2]

Vice President, Environmental Sciences
Battelle Columbus Division
Columbus, OH 43201-2693
and
[2]Professor of Geography
The Ohio State University
Columbus, OH 43201-2693

The siting of noxious or locally undesirable public facilities is one of the more intractible location problems in contemporary America. Such facilities include hazardous chemical waste, low-level nuclear waste, high-level nuclear waste, and sanitary landfill disposal operations, among others. In general, it is agreed that society must effectively address the safe disposal of its waste products. When a disposal facility is proposed in a specific state or locale, however, those who may be impacted inevitably mobilize against the proposal. This chapter explores the unique characteristics of the noxious public facility location decision, and outlines a proposed methodology for effectively balancing the multiple objectives involved in making such decisions.

The location of a high-level nuclear waste repository is examined to illustrate the dimensions of the noxious public facility location problem, and the potential utility of the proposed methodology to support such location decisions. More specifically, the chapter considers the following topics: (1) the nature and character of the physical requirements of a geologic repository for high-level nuclear waste disposal, (2) the specific location factors that are significant to repository siting, (3) the principal differences between public and private sector location decisions, and (4) a description of a proposed methodology for structuring and supporting noxious public facility location decisions.

REPOSITORY DESCRIPTION

Nuclear power constitutes a significant, and controversial, share of the nation's total energy budget. The U.S. Department of Energy indicates that, in 1984, domestic nuclear power plants generated in excess of 13% of the total energy produced in the United States.[1] With the benefits of nuclear power comes the social obligation to safely dispose of the waste products generated by the nuclear option. Through 1984, approximately 10,000 metric tons of uranium waste were discharged by commercial reactors in the U.S.[2] Projections are that 50,000 metric tons of waste will be produced by the year 2000, and 130,000 metric tons by 2020. Clearly, the safe and effective disposal of high-level nuclear waste is a task with major social implications for the continued use of nuclear energy.

The ultimate disposal of high-level nuclear waste is designed to limit or prevent the release of highly radioactive materials into the accessible environment for literally thousands of years. No disposal facility has been licensed anywhere in the world, but the U.S. has been a leader in pursuing questions relating to the feasibility of locating such a facility including physical, economic, social, and political constraints. To this end, a wide range of physical disposal options were examined initially, including ice sheet disposal, sub-seabed disposal, outer space disposal, transmutation, and geologic disposal. Geologic disposal was selected by the federal government as the preferred option. This concept relies upon the emplacement of packaged nuclear wastes in stable rock formations 1000 to 4000 feet below the earth's surface. Figure 1 illustrates an artist's concept of such a repository. This facility would resemble a conventional deep min-

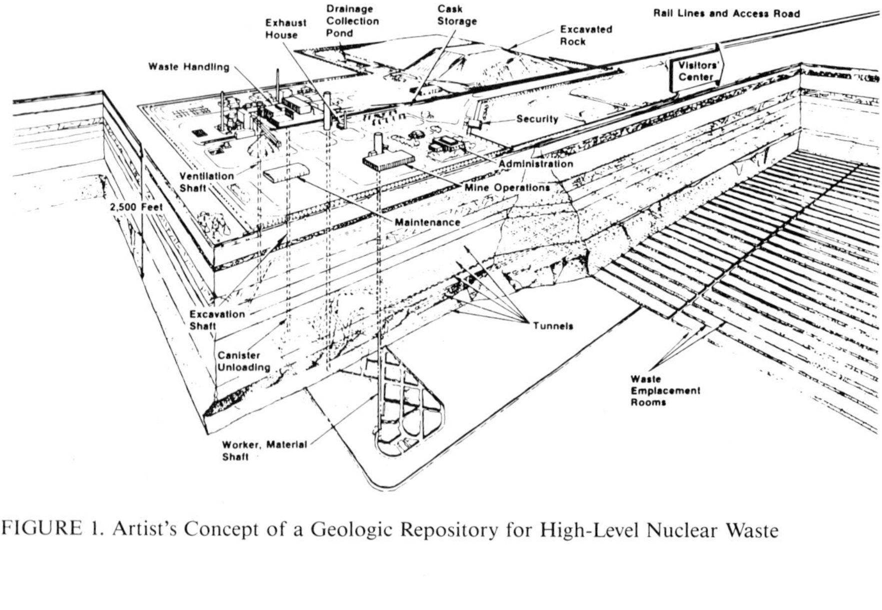

FIGURE 1. Artist's Concept of a Geologic Repository for High-Level Nuclear Waste

ing operation in many respects. It would rely upon both natural and engineered barriers to ensure waste isolation for long time periods (i.e., in excess of 10,000 years).

REPOSITORY LOCATION FACTORS

Location factors involved in respository siting include a complex array of considerations, some of which are not ordinarily relevant to a private sector location decision. Included are factors related to: (1) health and safety, (2) environmental impact, (3) socioeconomic impact, (4) transportation "costs", in the broadest sense, and (5) the ease and cost of construction.

As might be expected, the issue of paramount importance to repository siting is the protection of public health and safety at a high level of confidence. Safety must be maintained during repository construction, during facility operation, and long after facility closure or long-term sealing off of the facility. Key factors that must be considered over the long-term include geohydrology (because water is the principal vehicle for waste migration), geochemistry (due to possible attenuating effects of the geologic medium), specific rock characteristics (porosity, density, etc), possible climatic change (potential long-term impact of temperature changes), erosion, dissolution (salt), tectonic changes (faulting, earthquakes), and those related to human interference (particularly, mining operations).

The generalized objective related to environmental quality is to protect the environment adequately by avoiding impacts where possible, and by mitigating those impacts where they are unavoidable. Air quality, water quality, mine waste, and land use issues are, therefore, of principal concern.

The very nature of the siting problem—i.e., safeguarding the human population and the physical environment from adverse impacts—places emphasis upon areas of relatively low population density. Even within this geographic context, repository construction and operation has the potential to result in a broad range of social and economic impacts. Employment impacts, public finance and public service effects, gross regional product impacts, and community disruption are representative considerations.

Efficient and effective transportation of the nuclear wastes from the power plant sources to the repository for final disposition is another key concern. Accessibility, population risk, safety, and transportation costs are factors to be evaluated in repository siting. Certainly, there is a growing awareness in our society of the environmental risks and "costs" associated with the transportation of nuclear wastes as well as other noxious and toxic materials. This introduces a "cost" element in the consideration of the transportation factor not associated, typically, with more traditional location decisions.

The ease and cost of constructing the repository is the *least important category by Federal law,* but is still required to be considered. The ease- and cost-related factors include consideration of surface characteristics, rock characteristics, hydrology, and tectonics. Here, again, how these costs are evaluated relative to the overall objectives of the siting decision and to the total set of costs is distinctively different from how they would be treated in a routine plant location decision.

This quick review of repository location factors demonstrates that the siting of such a facility is a complex undertaking involving multiple objectives. The balance of this chapter will examine a suggested framework for structuring such complex, multi-objective, public facility siting decisions.

PUBLIC VERSUS PRIVATE LOCATION DECISIONS

Most of the attention in the literature has been directed toward the explanation of private sector industrial location decisions. The predominant driving forces in such decisions have been economic—for example, cost minimization, profit maximization, and/or revenue maximization. Typically, such decisions rely upon an optimization framework using the dollar as the common metric.

In contrast, the public sector location decision is customarily driven by a different set of objectives. These objectives may include service maximization, risk minimization, and/or adverse impact minimization. Economic forces, such as cost minimization, are usually subordinate to these other objectives in the public facility location decision. In addition, the public sector location decision is influenced by a complex array of "authority constraints", public policy considerations, and technical factors, both monetary and non-monetary.[3] Of particular importance are the set of "authority constraints" which include: (1) legal and/or regulatory constraints, such as the major public policy instrument that directs the high-level waste program, the Nuclear Waste Policy Act of 1982 (as amended), and relevant Federal, state, and local area legal and regulatory influences, (2) institutional constraints, such as the roles, responsibilities, authority, and accountability of major Federal, state, local, and judicial actors, and other interested parties, such as Indian tribal councils, and (3) political constraints, such as the political environment in which noxious public facility location decisions are to be made and the related political pressures. Historically, the basic problem of the final disposition of high-level nuclear wastes has suffered from the absence of a functional public policy, in part a reflection of these authority constraints.

The consideration and evaluation of these complex forces requires a multi-objective model capable of measuring on a reasonably common metric, and capable of making variable trade-offs explicit in decision making. It also requires a satisficing rather than an optimizing model of the location decision. The *satisficing* model allows for the treatment of complex, multiple objectives in a manner that recognizes inverse relationships between objectives, and that explicitly characterizes the trade-offs between the objectives. It is in response to these characteristics of the public sector location decision that the model described here has been developed.

THE NOXIOUS PUBLIC FACILITY SITING MODEL

Given the characteristics of the noxious facility location problem, a weighted summation model is preferred for initial candidate site identification. This method relies upon the development of standard measures for each location factor and upon the assignment of weights to each. These weights represent the relative importance of each location factor in the location decision. In the weighted summation method, aggregate or composite suitability for each unit of analysis (e.g., the square mile grid cell) is estimated as follows:

$$\text{Suitability} = \sum_{i=1}^{n} \text{Wi Vi (Xi)}$$

where Wi is the weight, and Vi ( ) the value function of location factor Xi.

This method offers several important advantages as a model of the multi-objective noxious public facility siting problem. First, it is readily adaptable to geographic expression on maps. In addition, it explicitly recognizes the need to evaluate trade-offs among multiple objectives that otherwise would be obscured, including trade-offs between economic and non-economic objectives. Further, it is a model oriented to satisficing, rather than to optimizing. It can also accommodate the various kinds of legal and/or regulatory, institutional, and political authority constraints that materially affect such location decisions.

These results are accomplished through the selection, definition, and measurement of appropriate location factors, through the use of factor weights that are broadly representative of a range of value systems, and through the use of maps that can express geographically the implications of alternative value systems for the location decision.

In the following, a four-step model for supporting noxious public facility siting is described. Illustrative examples are drawn from the experience of applying the model to the siting of a geologic repository for high-level nuclear waste in the eastern United States.

Step 1: Apply Relevant Disqualifying Conditions

This step applies to those location factors that can be defined in clear-cut terms of unacceptability and acceptability (i.e., in black/white terms). Unacceptable areas are *eliminated* from further consideration. This step recognizes that there are features that, if present, are enough to disqualify an area from further consideration. These features may be defined by law or regulation, as in the repository case, or may be defined by the public agency with the siting authority. In the latter case, this is commonly done with public input. Disqualifying factors are typically a relatively small subset of the total list of location factors. Steps 2 and 3 are designed to address the other factors. Figure 2 illustrates a sample disqualifer map.

Step 2: Apply Potentially Adverse and Favorable Factors

Step 2 is applied to the areas that emerge after the completion of Step 1. This Step recognizes there are many location factors that cannot be defined in strictly unacceptable/acceptable or black/white terms. These factors may be defined in terms of degrees of favorability. For a given variable or location factor, those areas which exhibit favorable characteristics are favored over areas which display potentially adverse characteristics. When applying a large number of location factors, the objective is to select areas for further examination that, in the aggregate, exhibit favorable over adverse characteristics.

The degree of adverse conditions/favorable conditions present for each Step 2 variable are numerically portrayed by a standardized 1 through 5 scale, with 1 representing the adverse extreme and 5 the favorable extreme. Scaling is the process by which physical conditions for each screening variable (potentially adverse or favorable) are translated into a numerical value that can be used to evaluate the aggregate suitability of the remaining study area after the application of Step 1. Figure 3 illustrates representative scales that were developed as part of the repository siting process.

Each of the scale values for *all* of the Step 2 variables or location factors is assigned a standard shade of grey to depict that scale value assignment on

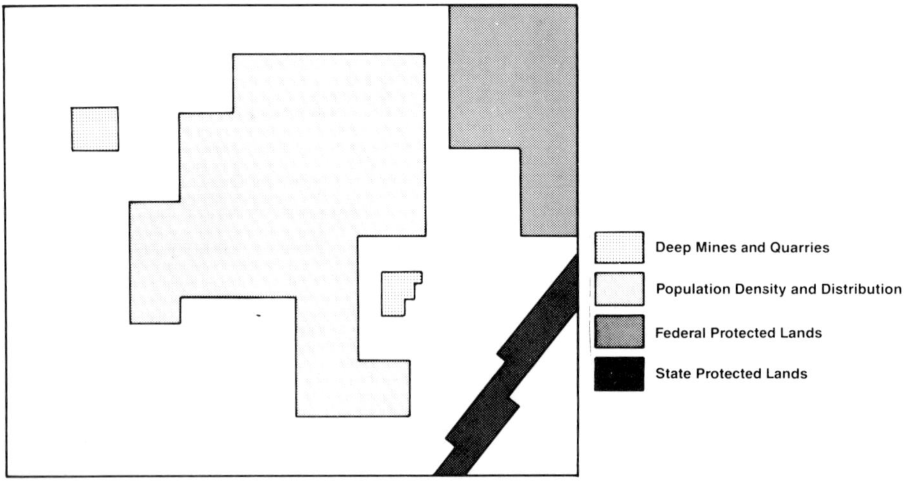

FIGURE 2. Sample Map of Disqualifying Factors in Repository Siting

variable favorability maps. Generally, the convention used in the preparation of these maps is the darker the grey, the more adverse the condition being portrayed. Figure 4 provides an example of a variable favorability map. Such maps provide a key building block for the estimation of aggregate or composite favorability using the weighted summation model. They do so by, in effect, assigning a single scale value for *each* variable to each square mile grid cell in the study area.

1. SEISMICITY

Maximum Probable Ground Acceleration (%g)

>70	>50-70	>30-50	>10-30	<10
1	2	3	4	5
More Adverse		Scale Value		More Favorable

2. POSTEMPLACEMENT FAULTING

Distance From Fault (miles)

0-3	>3-4	>4-5	>5-6	>6
1	2	3	4	5
More Adverse		Scale Value		More Favorable

3. POPULATION DENSITY

Population Density (persons per square mile).

800-999	. 600-799	400-599	200-399	0-199
1	2	3	4	5
More Adverse		Scale Value		More Favorable

4. PROXIMITY TO STATE-PROTECTED LANDS

Distance From Boundary (miles)

0-3	>3-4	>4-5	>5-6	>6
1	2	3	4	5
More Adverse		Scale Value		More Favorable

FIGURE 3. Step 2 Variable Scales
Source: Ibid, DOE, *Region-to-Area Screening Methodology for the Crystalline Repository Project*, pp. 88-117.

Variable Weighting. While variable scales depict a physical condition for an individual location factor, variable weights are developed to depict value judgments concerning the relative importance of the complete set of location factors. These value judgments may be influenced by technical inputs, by variable scales, and by other considerations. However, the assignment of weights is, in essence, an exercise in expressing human value judgments or preferences. Because of this, and particularly for public facility location decisions, it is important to capture the implications of a wide range of value judgments or preferences in the location decision. Such weights can be effectively derived through the use of nominal group technique at weighting workshops. Such workshops should involve a broad spectrum of participants representing the various interest groups affected by the location decision. Initial weighting exercises completed by individual participants can be statistically evaluated to place individuals in small groups (e.g., up to 10 people) with similar weights. A group process facilitated by a neutral outside party can then be used to derive a set of weights that are representative of that group's perspective on the relative importance of the location factors. Multiple groups will yield a range of weight sets that can be used to derive composite favorability maps.

Composite Maps. After location factor scales and weights have been developed, the weighted summation approach can be utilized to estimate the aggregate favorability of each square mile grid cell in the study area of interest. This was done for the regional phase of the repository siting study using the following formula:

$$AF = \sum_{i=1}^{n} Wi\ Vi\ (Xi)$$

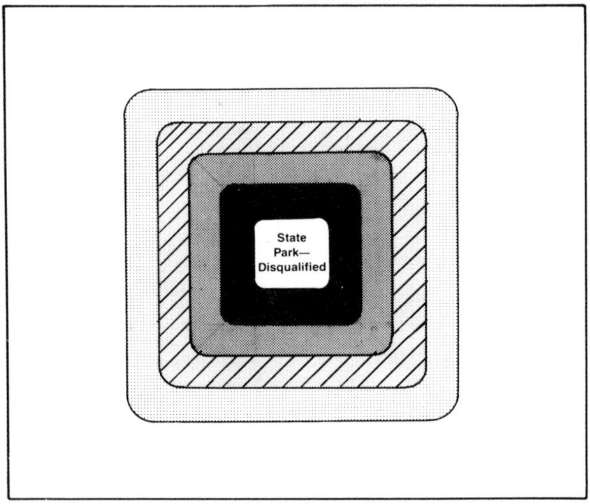

FIGURE 4. Sample Screening Variable Favorability Map—State Protected Lands

	AF	= estimate of aggregate favorability
where:	n	= number of variables or location factors
	$\sum_{i=1}^{n}$	= summation over n items
	Wi	= weight of variable i
	Si	= scale value for variable i
	1000	= total number of weighting points allocated.

The mathematical calculation outlined above was performed for each square mile grid cell in the repository study area. The result was thousands of estimates, on a grid cell basis, of the aggregate favorability for a repository within the constraints of a regional investigation. When these results were assigned shades of grey according to standardized ranges of aggregate favorability estimates (e.g., 4.5 - 5.00 is the lightest grey), the product was a composite favorability map that graphically illustrated the geographic distribution of the most favorable areas, as defined by a given set of scales and weights.

In essence, the composite map displays the geographic expression of values or preferences regarding the relative importance of location factors to the location decision. By preparing composite favorability maps for multiple sets of weights, it is possible to examine the spatial implications of different value systems or preference judgments for the location decision. Figure 5 illustrates the composite map concept for the equally weighted case applied to a portion of the repository study area. An examination of such composite maps can illustrate clearly the distinct geographic differences in the most favorable portions of the study areas.

Step 3: Sensitivity Analysis

The third step in the proposed screening model for noxious public facilities allows for the opportunity to perform sensitivity analysis. Sensitivity analysis is structured to examine the spatial effects on the location decision of the following:

(1) Modifying selected variable scales, and

(2) Developing and evaluating summary composite maps.

Scale Modification. The concept of variable scaling for application in the weighted summation model was described previously. Each variable scale value represents a physical condition or range of conditions. Recognizing that there may be varying opinions on scale definition, Step 3, Sensitivity Analysis, allows for the evaluation of alternative scales and their impact upon the geographic distribution of favorable areas.

The Summary Composite Map. Given that multiple sets of weights are used in the proposed location decision process, and given the various permutations

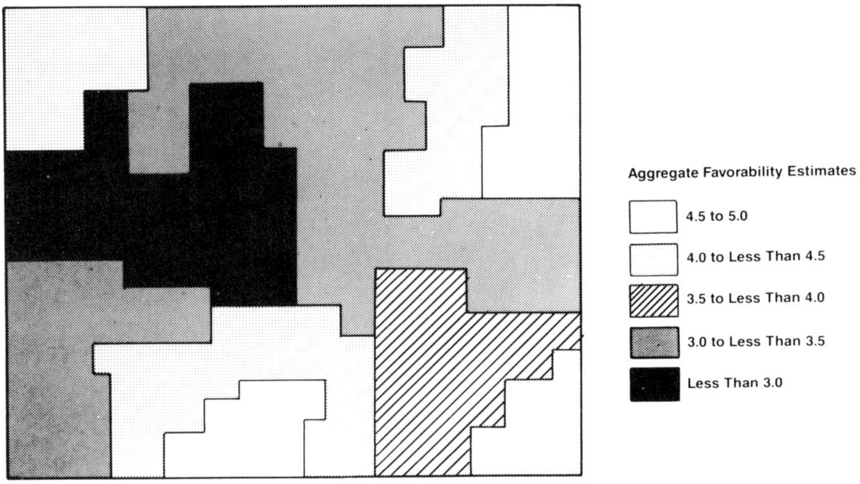

FIGURE 5. Sample Composite Favorability Map

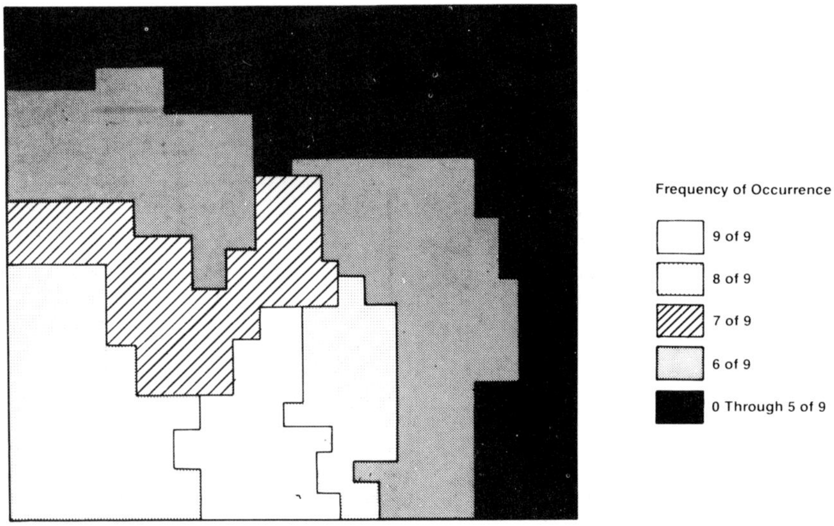

FIGURE 6. Sample Summary Composite Map

and combinations of sensitivity analysis that are possible, an approach is needed to simplify and clarify the location decision. The summary composite map concept effectively meets this need. The summary composite map is a synthesis of geographic data extracted from a related series of composite maps. It is designed to identify similarities or overlaps in the definition of the most favorable areas on sets of related composite maps.

The summary composite map utilized by the repository siting example was developed on a frequency of occurrence-best candidate area basis. To develop such a summary composite map in the repository example, all nine available weight sets/composite maps were used.

Every undisqualified grid cell was examined (via computer software) to determine the number of times (out of 9) that the cell had an aggregate favorability score higher than an established benchmark for *each* of the nine sets of weights. The results of this examination are geographically expressed on the summary composite map as a frequency of occurrence (e.g., 9 out of 9, 8 out of 9, etc.). It is assumed that the higher the frequency of occurrence for a given grid cell or cluster of grid cells, the more broadly that area is perceived as worthy of further location analysis.

Figure 6 illustrates the concept of the summary composite map outlined above for a portion of the repository study area. This map illustrates, in effect, the degree of consensus on the top-rated areas in this subarea, taking into consideration widely varying value systems (represented by nine sets of weights). As the map shows, there are areas which show up at the 9 out of 9, 8 out of 9, and the 7 out of 9 frequency of occurrence levels. Thus, the summary composite map is capable of illustrating the spatial distribution of areas that are highly regarded by a broad spectrum of interests. This can be of substantial significance to both the technical and political defensibility of the location decision.

Step 4: Selecting Candidate Areas/Identifying Potentially Acceptable Sites

The final step of the screening model involves the review of the results from the previous three steps. This is done to confirm the acceptability of the candidate areas that emerge as most favorable. This final step is subject to the influences of the full range of legal/regulatory, institutional, and political authority constraints described earlier in this chapter.

SUMMARY

The siting of noxious public facilities is a difficult task in contemporary America. Such location studies are multiobjective undertakings that typically must address a broad array of factors, including health and safety, environmental impact, engineering, transportation, and economic considerations. The principal objective of such studies is to examine systematically location factor trade-offs in a satisficing rather than an optimization process.

The weighted summation method provides a useful tool for supporting noxious public facility location decisions. Using exclusionary location factors, certain areas can be disqualified early from further consideration. Remaining areas can be evaluated in terms of their aggregate or composite favorability by using location factor scales and weights. The results of such evaluations can be geographically expressed on composite maps that represent the location preferences of alternative value systems. The use of summary composite maps can, in turn, identify the geographic coincidence between the most favored candidate areas on related sets of composite maps. Doing so may be taken to represent a consensus among variant views in the location decision process. This ability enhances both the technical and political defensibility of the location decision. Hopefully, the use of such tools in an open process will help to break the intractible nature of noxious public facility location decisions.

REFERENCES

1. U.S. Department of Energy, "General Guidelines for the Recommendation of Sites for the Nuclear Waste Repositories; Final Siting Guidelines", 10CFR Part 960, December 6, 1984.
2. U.S. Department of Energy, Office of Civilian Radioactive Waste Management, *Mission Plan for the Civilian Radioactive Waste Management Program,* Vol. I, DOE/RW-0005, June 1985, p. 3.
3. Cox, Kevin R., "The Behavioral Revolution in Geography: Definition and Evaluation", *Geographical Perspectives,* No. 33 (Spring 1974), pp. 41-50.

Management of Hazardous Materials and Wastes: Treatment, Minimization and Environmental Impacts. Edited by S. K. Majumdar, E. W. Miller and R. F. Schmalz. © 1989, The Pennsylvania Academy of Science.

Chapter Thirty-One

THE INSURANCE DIMENSIONS OF HAZARDOUS WASTE LIABILITY

J.D. HAMMOND
William Elliott Professor of Insurance
The Pennsylvania State University
University Park, PA 16802

The general awareness level about pollution dangers was given strong impetus by Rachael Carson's *Silent Spring*, published in 1962.[1] Although hazardous waste production and management were not the principal focus of the work, its impact extended generally into the problems of chemical releases, accidental and otherwise. Concerns over the quality of the environment and the potential for carcinogenic harm were heightened, even though clear-cut causal connections between environmental impairment and personal injury were and remain difficult to establish. Nonetheless, negligence law was about to become a major vehicle for redress of those exposed to such harm and liability insurers could not yet foresee the contractual and claims difficulties which were to come.

Man-made toxins were not the only contributor to increased environmental concern and awareness. Oil spills and their graphic portrayal on television were another. In 1967, the oil tanker Torrey Canyon ran aground off Britain. Just over a decade later another tanker, the Amoco Cadiz, ran aground in the English Channel, spilling tons of oil over the Brittany Coast.[2] If environmental liability exposures were ever thought of as routine, that status had certainly ended by the mid 1970's. Those exposures, together with expanding problems from products and medical malpractice liability had combined to create difficult, perhaps unprecedented insurance underwriting and pricing problems. The claims potential for such low-probability, high severity events was surely fully realized when hazardous waste storage became a major and highly' visible problem with the Love Canal episode. That combined with asbestosis, agent orange, and D.E.S. made it clear the liability insurers were entering a contentious and litigous era. Though causal connections to personal injury may have been tenuous, the public showed little tolerance for risks, actual or perceived, imposed on them by others and essentially beyond their control.

For most major risks, firms seek insurance. In most instances, an active and competitive market brings insurers and buyers together. It is hardly unprecedented, however, for some insurance coverages to be unavailable or available only at a very high price. If a risk is insurable, insurance suppliers will be forthcoming. If it is not, transfer of a risk to an insurer is no longer a viable risk management alternative.

The insurance technique for redistributing losses is increasingly called upon to allocate the costs of major and complex claims. Those associated with the environment and pollution liability are but one such category. The economic losses associated with AIDS are another prominent and recent example.[3] Whether a particular loss exposure is insurable becomes important in determining how major economic losses are to be met. If they are uninsurable, there is no alternative but for them to be managed in other ways, perhaps borne by individual persons or firms, perhaps by the public through government programs of one kind or another, or by increased expenditures on loss prevention.

Objective criteria about insurability will be addressed in this paper. Because the insurance contract sets the terms and conditions of coverage, and especially because some provisions may be disputed, contractual language and its interpretation by the courts are a key part of any review of the insurance dimension of hazardous waste management. That will follow the discussion of insurability. Finally, the marketplace is the ultimate arbiter of whether a risk is insurable and whether the price and terms of risk transfer will bring underwriters and insurance buyers together. The status of the current market for liability coverage for the hazardous waste exposure and its development over the past decade or so complete the basis for a set of concluding observations.

Although hazardous waste events can encompass a variety of economic losses, the focus of the paper is on legal liability, both under tort law and by statutory imposition.

THE HAZARDOUS WASTE EXPOSURE

Hazardous wastes and the danger which may come from exposure to them can be associated with their manufacture, transport, storage, and disposal. A toxic release can occur at any point in the chain. The release may be sudden, such as the rupture of a container in transport, or it may be gradual from the accumulated effects of slow releases over a long period of time. The Environmental Protection Agency estimates the presence of some 30,000 to 50,000 hazardous waste dumpsites, with perhaps as many as 2,000 of them presenting imminent hazards to public health or the environment. The principal industries involved appear to be chemical, textiles, rubber and plastics, electronic components, petroleum refining, leather and tanning operations, electroplating and primary metals.[4]

Estimates of the dollar magnitude of the problem vary, but the U.S. Congress Office of Technology reports the cost of the waste-site problem to be in excess of $100 billion. The cost would presumably include all of those expenditures required to clean up a particular site and to restore the affected environment to its normal condition. Those costs would not include damages paid for personal injury under the tort system. It appears that every one of the chemical dump-sites across the country is a potential source of liability. The amount of that liability is extremely difficult, if not impossible, to estimate, though it can easily be of catastrophic proportions; particularly of exposure to toxins takes place in a highly-populated area.

Liability costs are imposed from two sources. The most recent are the statutory requirements of federal and state legislation while others are imposed through the operation of negligence law which, as will be shown, has evolved in favor of hazardous-waste plaintiffs.

Federal legislation. Pollution and hazardous waste problems are logical candidates for for regulation. Safety standards in hazardous waste management and the handling of claims costs seem to fall naturally within the public interest. While market forces and negligence law serve as partial controls on careless behavior, they would not likely provide the consistency or reflect all of the public concerns as would carefully framed legislation. Federal laws governing the disposal of waste or pollutants are numerous, but one of the early important laws was the Water Pollution Control Act of 1972, aimed at deterring those who would dump waste solvents into sewers. The major acts dealing with hazardous waste management requirements came later. The Resource Conservation and Recovery Act (RCRA) and the Comprehensive Environmental Response, Compensation and Liability Act (CERCLA) are the most important.

The Resource Conservation and Recovery Act, passed in 1976, established requirements for the management of potentially hazardous materials, beginning with their origin and continuing to their ultimate disposal—sometimes referred to as a "cradle to grave" system of managing such materials. The Environmental Protection Agency was to develop regulations and standards for essentially any organization involved in any aspect of hazardous waste management. Specific procedures and rules were left to the EPA. Among these were requirements that hazardous waste facility operators be financially capable of stopping seepage. Financial responsibility under the rules was to be shown by surety bonds, trust fund arrangements or liability insurance.[5] Reclamation and recycling were strongly encouraged not only by the law but by the economic pressures imposed by the OPEC oil embargo. Petroleum-based products had become expensive.

The Comprehensive Environmental Response, Compensation and Liability Act of 1980 created a "superfund" to pay for cleanup expenses associated with hazardous waste dumpsites. The imposed liability measure of the law was severe. Dumpsite owners, generators, and transporters are all subject to strict liability on a joint and several basis. Once leakage is established, all parties who may have been in any way connected with the site are liable. The fund is collected from taxes on petroleum, chemical feedstocks or business generally and will pay for cleanup when potentially liable persons or organizations cannot be identified. Contributions to the fund are made with little or no relationship to the separate risks of each of the contributors. Court cases involving liabilities under the act leave little doubt about the principle of "the polluter pays." The government need not even prove direct or proximate cause in cases against current or former owners and operators.[6]

The potential financial obligations of those associated with dump sites are extensive. Even banks which hold such property through foreclosure proceedings are not immune. In short, someone will pay. Insurance relief for mandated clean-up costs is not a part of the Commercial General Liability contract but insureds have still sought coverage through litigation. Other contracts, when available, do provide some relief for site clean-up costs.

Negligence law. Under traditional applications of negligence law, an individual or organization is expected to act with reasonable care. If, as a result of the failure to act with reasonable care, an individual is injured or property is damaged, the victim is entitled to recoveries for economic loss and for the general damage of pain and suffering. Under traditional standards, there must have been a direct causal connection between the failure to use reasonable care and the resulting injury. Other such standards have developed over time, but the general social intent has been to shift the economic loss of victim to the wrong-doer, provided that fault of the wrong-doer could be established.

The basic standards of negligence law seem reasonable enough,but their interpretation and application are not always clear. The determination of a causal connection can be difficult and where several defendants are involved, assigning fault to only one or equitable allocating it among several, can be troublesome. Complicating everything is an evolving social issue of whether all accident victims should be compensated for economic loss instead of just those where fault could be established.[7] Some courts have indicated a willingness to adapt common law and statutory principles for humanitarian purposes and, in the case of hazardous waste injuries, to provide remedies to those exposed to hazardous materials, to those who become ill from them, and even to those who are at increased risk from having been exposed, even though actual harm has not and may not materialize.[8] The latter is a significant extension of negligence law. Plaintiffs have been able to secure remedies more readily through such extensions. The adaptations are usually stimulated by initial actions in the courts of leading states, such as New Jersey, California, Michigan, and Massachusetts. The result has been to reduce recovery barriers traditionally imposed by the fault or negligence system.[9]

The hazardous waste exposure presents one of the most complex hazards to be addressed by either negligence law or insurance underwriting. An incident may not be well-defined at all, but rather take place over a period of several years. Causality and harm may be difficult to prove. Questions easily arise as to when an event or an injury occurred. Does the injury take place at the time it manifests itself with the victim or does it take place at the time when the victim was exposed to the toxic elements? If the victim does not show actual illness, is an increased probability of illness compensable? If so, then statistical correlation estimates replace actual harm and observable causality as evidence. Moreover, the number of sites appears to be so numerous and the presence of the problem so permanent that the problems of injury compensation are likely to remain for some time.

The complexities of the exposure together with an apparent desire to compensate victims—that the "polluter should pay"—have produced major changes in the legal environment in which insurers operate. Statutes of limitations have been changed so that the law begins to run at the time the illness or harm was discovered and not at the time of the initial exposure. Courts have supported strict or absolute liability so that harm is the only prerequisite for compensation. Under the doctrine of joint and several liability, any and all defendants may be held liable for damages from hazardous waste exposures and that liability need not in any way be proportionate to the relative fault involved. Even subsequent owner of a hazardous waste site can become liable, even though no hazardous waste materials were contributed to the site.

The legal changes summarized are of relatively recent origin, with the Love Canal episode serving as a major catalyst. Prior to that, insurers and the law were more concerned with incidents, such as oil spills, that were relatively fixed in time and easy to define. Those insurance and legal concerns were later followed by a variety of pollution claims, some sudden and some gradual. At least in the early days of such exposures, the insurance market was positively responsive. That condition was not to remain.

CONCEPTS OF INSURABILITY

Even casual observation indicates that insurance is not available for all risks. Flood insurance on fixed location property is not available in the commercial market; only from a federal program and not without limitations even then. Life insurance is generally unavailable for those with terminal diseases or at advanced ages. War risk insurance is unavailable when hostilities reach some intolerable point. In short, there may sometimes be no market since sellers do not come forward. In other instances,a market may be described as "tight" when the number of sellers is limited or if prices are higher than what many buyers are willing to pay. Still, a few insurers may be willing to offer coverage while the majority prefer otherwise. For the latter group, a particular risk is considered uninsurable even though the marketplace has not made that final judgement.

The fact that some risks are uninsurable and private insurance therefore unavailable, suggests that criteria are available to help make that assessment; that it is not arbitrary or capricious. Although virtually every standard insurance text presents a discussion of the requisites of an insurable risk, the most detailed effort to date has been by Baruch Berliner.[10] His work is useful to all inquiries addressing issues of insurability and questions about insurance availability, including the insurance dimensions of hazardous waste management. The discussion of insurability here summarizes the extended criteria of Berliner.

Randomness. Ideally, underwriters would prefer that risks be perfectly accidental, beyond the control of the insured, and independent from other losses. Such perfection does not always exist, but the normative standard is a useful benchmark. The perfect antithesis of randomness is complete predictability. If losses were completely predictable, there would be no need for insurance and the commercial insurance market would not exist. Random losses within a homogeneous rating classification, however, would fall randomly across the members of the class, each facing essentially the same probability of loss and each paying the same price for insurance.

There is little to suggest that the liability exposure from the management of hazardous waste meets the randomness criterion. Such losses are not easily predictable and exposure to hazardous waste is nearly tantamount to a legal judgement. Liability claims are always under some influence of the court system, but damages for hazardous waste exposures have become increasingly severe with the severity and probability of a claim apparently increasing significantly in correspondence to damage awards under the tort system. Moreover, hazardous waste releases appear to be within the insured's control and some may actually be intentional.

Maximum possible loss. Underwriters need to know the worst out come possible. With liability contracts, that can easily be the liability limits of the coverage for one or for a series of separate occurrences. For real property, it is the replacement cost. Under older liability contracts, even though no longer in force, coverage can still exist by virtue of court interpretation of contract language (i.e, exposure to injury can occur at one point in time and actual injury or manifestation of injury, at a later point).

The maximum possible loss for hazardous waste liability is virtually impossible to estimate. Certainly at the time many liability contracts were underwritten, there was no expectation of claim payments several years hence after the expiration of coverage. Hazardous waste exposures and other risks with the potential of catastrophic loss can be underwritten, but only if the catastrophe potential is minimal and measurable to a reasonable degree. Even so, high rates and coverage restrictions are always possible. At the extreme, insurance becomes unavailable—the risk uninsurable.

Average loss amount and average time between two loss occurrences. Insurance premiums are the product of the distributions of loss severity and loss frequency. Adverse underwriting results can occur from a single large claim or a series of lesser claims. Generally, risks which develop a high frequency of claims tend to produce a relatively low average claim size and, at the extreme, are not suitable for insurance. Deductibles, for example, are a common way of eliminating small and frequent claims, thereby lowering the required premium. An expectation of repeated loss occurrences within the coverage interval would likely cause underwriters to reject the risk or, if coverage were already in place, to cancel the contract or not renew it.

Liability claims are repeatable and a single insured could present more than one claim within the coverage period. Large liability losses within a given period for a given policyholder can also result from multiple claims generated by a single occurrence. That is characteristic of hazardous waste claims since most result from a single or continuing incident; one not easily repeated. Courts may also view a collection of separate claims from a single event as separate occurrences, with liability coverage limits applying separately to each. Large claims, filed with increasing frequency, increase the likelihood of uninsurability.

Insurance premium. The pure premium for an insurance contract (that amount required to pay only for losses with no margin for expenses) is the expected value of the annual loss plus a contingency loading to provide for adverse fluctuations in claims results. The greater the uncertainty in pricing, the greater the contingency margin should be. The contingency margin should increase as the variation in the pure premium distribution increases, the latter being the product of the frequency and severity distributions noted above. At some point, the premium becomes too high in relation to the value being insured and the risk becomes uninsurable.

A high premium relative to the value being insured also prompts another underwriting phenomenon, adverse selection. In the large, adverse selection describes a condition where asymmetrical information exists between the insured and the insurer and where the insured has little or no incentive to reveal information that would allow estimation of loss probabilities.[11] The insured has less incentive to reveal information adverse to the insurer as the premium increases or, put another way, only the worst risks will seek insurance when the premium becomes high. Good risks may well opt for self-insurance.

Hazardous waste exposures, in the current insurance market, appear to have reached a point where the required premium is either unknown or so high as to be unattractive to all but the highest risks. Were the latter even to request insurance, it probably would be denied. The required premium could be so high that a meaningful transfer of risk would not take place.

Moral hazard. If an insured, as result of insurance coverage, becomes indifferent as to whether or not a loss occurs or as little incentive to avoid a loss, moral hazard is present. In extreme circumstances, the presence of insurance may even encourage the insured to bring about the loss; arson is a familiar example. The actuarial effect of moral hazard is to increase the underlying loss probabilities on which the original premium was based.

With most insurance, there remains a strong incentive to avoid the insured loss. At least some hazardous waste litigation suggests that the defendant may even have intentionally failed to take adequate precautions in the management of the waste. If that is true, the presence of insurance as a contributor to such behavior is a distinct possibility, though one that is difficult to measure. With unintentional hazardous waste losses, it is still conceivable that an insured's incentive for loss prevention may be altered. When that possibility becomes sufficiently distinct, insurance is not feasible.

Public policy and legal restrictions. The collective action of society or the community is frequently evidenced in in insurance coverages and markets. Insurance against gambling losses, for example, is deemed inappropriate even though odds are often calculable (at least those involving *a priori* probabilities). Insurance against entrepreneurial or business operating losses is unavailable and generally held as infeasible. On the other hand, where insurance is deemed an essential component of a social loss-sharing system, the law may see that it is made available. All states, for example, provide some mechanism for licensed drivers to obtain automobile liability insurance. A small number of states require health insurance pools, where insurers writing a given state agree to share those risks unable to obtain health insurance in the normal market.

As noted previously, the Resource Conservation and Recovery Act of 1976 required, through the operation of the EPA, financial responsibility of any organization involved with the management of hazardous waste. Insurance, of course, is the typical way of meeting financial responsibility requirements. Moreover, framers of hazardous waste laws apparently felt that insurers would be able to offer loss prevention advice and service, thereby benefitting both the insured and the public.

Cover limits. Restrictions on the amount of insurance underwritten or on the terms of coverage may make a risk insurable which otherwise would have been declined. Insureds obtaining automobile liability insurance through assigned risk plans or health insurance through state pools are typically offered less coverage and a higher price reflecting the elevated risk of the group. If a risk fails to meet other insurability criteria, however, combinations of price increases and coverage restrictions will not necessarily transform it into an insurable risk.

Liability insurance in general has, in recent years, been subject to increased premiums and reduced amounts of coverage through lower limits and higher deductibles. As described later, restrictions in hazardous waste coverage have generally not served to make the risk insurable.

CONTRACTS AND COURTS: THE INSURANCE CONFLICT

Insurance contracts stipulate the terms under which a risk is transferred from the insured to the insurer. Wording is often standardized across companies offering the coverage, with most of the commercial forms being developed by the Insurance Services Office (ISO) for use by its member companies. Non-members may also choose to use the same form and members may depart from it. Standardization can be of considerable benefit to both insurer and insured. Because contracts are complex, wide variation across companies could create a difficult position for buyers and competitors alike. Moreover, the sustained use of consistent language, together with consistent interpretation of that language by the courts, can stabilize the meaning of contractual terms.

It is well established that any ambiguity in the insurance contract is construed by the courts in favor of the insured. The contract is one of adhesion where the insured must conform to conditions typically prepared only by the insurer and not by the insured. It is not surprising the insurance contracts are litigated, especially where a dispute involves large sums of money. Such litigation has become increasingly common with commercial liability coverages and the effect has been to insert the court system as a major element in defining coverage. That courts should be available to settle coverage disputes is an inherent part of our legal system and, as noted, can stabilize the meaning of insurance contracts. On the other hand, inconsistent interpretations and perverting contractual language to the point of creating coverage where none was intended, creates pricing and loss-reserving difficulties not anticipated at the inception of the contract.

The insurance contract generally designed to protect firms from various liability losses is the commercial general liability contract. As with any liability coverage, it protects the insured for damages imposed under negligence law.

Insurance contracts. With respect to coverage for pollution and similar losses, the Commercial General Liability insurance contract has evolved over the past twenty years, with language changes prompted in part by court interpretation of coverage terms. The contract itself, first called the Comprehensive General Liability insurance contract, was developed around 1940. Prior to that, comprehensive coverage for liability was available only through more than one contract. Although the contract is complex, it is nonetheless important to focus upon two principal sections; the insuring clause and exclusions relative to pollution and similar losses. A convenient starting point is 1966, where coverage was broadened from and "accident" basis to and "occurrence" basis. Under the 1966 revision of the contract, coverage for pollution liability remained; it was not excluded. However, the 1973 revision excluded all pollution unless it was sudden or accidental. A pollution occurrence continued to be defined as an accident including continuous or repeated exposure to conditions which result in bodily injury or property damage, neither expected nor intended from the standpoint of the insured. However, the exclusion was written so as not to apply to events which were sudden and accidental releases. The intent was to exclude from coverage those losses which were gradual in nature, occurring over long periods of time and which were almost impossible to define or properly price.[12]

The exclusion just noted created a coverage gap for those losses which were gradual and ill-defined. Environmental Impairment Liability coverage, a separate contract from the Commercial General Liability contract, was developed by the London market in the early 1970's and was tailored to the characteristics of chemical hazards.[13] Environmental liability was its only focus. By 1981, a number of insurers had developed special pollution policies covering nonsudden occurrences and a new pool of insurers was formed to provide insurance for pollution liability coverage (the PLIA or Pollution Liability Insurance Association). The pool then reinsured all of its exposure. The expansion of the market for specialized pollution coverage at the turn of the decade came shortly after the passage of CERCLA or "superfund," with the coverage aimed directly at pollution and environmental risks. These focused contracts varied, but generally provided some form of coverage for the cost of cleaning property owned by third parties on which pollutants generated by the insured were found.[14]

Although the revisions in the Comprehensive General Liability contract were intended to limit pollution-type losses to those which were sudden and accidental, courts were interpreting the language of the contract to be much broader. The insuring clause promised to pay on behalf of the insured all sums which the insured was legally obligated to pay because of bodily injury and property damage caused by an occurrence. Some courts held that so long as the *results* were unintended, that even damages from intentional acts would be covered.[15] That view was not uniform among all courts, others holding it fundamental that an insurance contract should not be interpreted - in effect rewritten - so as to bind an insurer to a risk which it did not contemplate and for which it received no premium, unless the terms were ambiguous or obscure.[16]. The result was to create a legal environment in which underwriters were, at best, uncertain of the risk being undertaken.

The occurrence insuring clause provided coverage for damages from past events, even though the contract was no longer in force - so-called "long-tail" coverage. Courts have produced multiple decisions in defining the time interval of the coverage. Some have held that the contract in force at the time of exposure to the hazard provides coverage, others saying that manifestation of the disease is the coverage trigger, while another view hold that coverage exists from the time of exposure through the time of disease manifestation. More recently, it has been held that the trigger is the date that the victim's medical injuries actually occurred and not at the time of exposure or manifestation.[17]

To deal with the vagaries of the occurrence coverage, underwriters for other liability coverages (e.g., medical malpractice), developed the "claims actually made during the policy period. Prior losses would not be covered unless agreement could be reached on a retroactive date. Under that language, the insurer would be better able to estimate its losses in any given year. Only those with current contracts can collect on claims. The 1986 commercial general liability revision applied the claims-made wording to its coverage, although the specialized Pollution Liability and Environmental Impairment Liability coverages had always made use of the claims-made language. For any hazardous waste operator, the contract would need to be continued in force for as long as there was a possibility that claims might be forthcoming and would be required to pay any increased premiums in order for that coverage to remain in force. Under the claims-made language, there would be no insurance in force should the firm become bankrupt or cease to operate for some other reason, or if the insurance were canceled or not renewed.[18] Application of the claims-made language to commercial liability was controversial and the standard occurrence form is still in place, but with no requirement that underwriters employ it.

Language adopted in 1986 for the commercial general liability contract restricted coverage for pollution-type loses even further, with coverage essentially unavailable for all but the most innocuous of risks.

THE INSURANCE MARKET

Over the last twenty years, the market for environmental and pollution liability coverages has not been stable. Contracts have been revised, loss exposures have increased, coverage has been expanded by court interpretation and narrowed by contractual language. The number of insurers, once expanding to meet environmental coverage needs, diminished. Liability coverage for firms engaged in the business of hazardous waste management, at this writing, stops just short of being unavailable. Most insurers have considered the risk uninsurable.[19]

Some insurers are reported to provide liability coverage for sudden and accidental pollution, but only for exposures without significant risk. The London market is said to provide coverage for low to medium risk insureds under their liability policies. However, the contracts contain a provision which limits hazardous releases to a 72 hour period before coverage is applicable. Releases occurring over a longer time period would not be covered.

The pollution liability insurance association noted previously which provides reinsurance for its members has not become a force in the market and its presence does not appear to have impacted present insurance capacity to any meaningful degree.

The London marine insurance market provides pollution liability cover under its marine liability insurance contracts, but they are carefully underwritten and apparently limited to those insureds with significant marine operations. Presumably, a hazardous waste spill during marine transport would be covered under these contracts. On the other hand, if it is known that an insured contemplates the transport of hazardous waste instead of say, normal transport of oil, the contract might not be issued.

When the commercial market disappears, risk retention becomes a forced alternative. While self insurance is a commonly practiced form of risk retention, other risk-sharing devices have been encouraged by Federal legislation. The Risk retention, other risk-sharing devices have been encouraged by Federal legislation. The Risk Retention Act, passed in 1981 and amended in 1986, permits groups of companies to combine to form a risk retention group. That group, in turn, would be subject to the insurance laws of the state in which it was formed but would not require approval of policy forms, rates and the like. The concept evolved from the product liability insurance shortages earlier in the decade but is not limited to that exposure. The law also permits group purchasing of insurance as a way to achieve possible economies.

While some risk retention groups have been formed, they have not become a market force for pollution or hazardous waste exposures. Reinsurance for such exposures generally has not been forthcoming and that affects risk retention groups just as much as commercial insurers. In short, if a loss exposure is sufficiently high risk, changes in the way it is shared are not likely to make it any more acceptable.

Pure self insurance may not be acceptable to regulators who must enforce requirements of financial responsibility for hazardous waste operators. If regulators insist on liability insurance to meet such requirements, there is no choice but for the prospective insured to identify an insurer who will issue the contract and, given the uninsurable nature of the exposure, for and insurer to require guarantees of the insured that will reimburse the insurer for any claims paid. Such devices exist in the market but they are little more than a convenience for the insured with no risk being transferred and no risk being shared.

CONCLUDING OBSERVATIONS

The liability exposure of hazardous waste management (and other pollution exposures) has become uninsurable. It fails to meet sufficiently the requirements of insurability noted earlier. That such insurance became unavailable is not unexpected. Given current conditions, there is little prospect of meaningful increases in availability, although there is some evidence that Environmental Impairment Liability coverage is being carefully offered by a few companies.

Underwriting conditions have changed drastically since the exposure first appeared. Liability law has been broadened to make recovery easier, courts have interpreted contractual language to be broader than contemplated by underwriters and broader than contemplated in the rate structure. Liability for hazardous waste operators has become absolute and, under the joint and several application, falls upon any party associated with a hazardous waste site. Risk cannot be measured with any precision under these circumstances. The limitations of the insurance technique have been exceeded.

At least for the hazardous waste exposure, society generally has deemed it proper for all victims to be compensated, regardless of whether the traditional concepts of fault and negligence are present. Even the probability of injury has been a sufficient rationale for compensation. The catastrophic nature of the losses have compounded the insurance problem, enlarging it beyond the ability of the private insurance sector to handle. Hazardous waste sites apparently number in the tens of thousands and all are potential sources of catastrophic claims. If those exposed to loss are to be compensated, devices beyond the private insurance market are required.

EPILOGUE

Shortly after the work for this paper was completed, certain dimensions of the topic took on an even greater degree of newsworthiness. On March 22, 1988, the Attorneys General of Alabama, California, New York, Massachusetts, Minnesota, West Virginia and Wisconsin filed and antitrust suit against four major United States insurers in the direct market for liability insurance as well as 31 other organizations including reinsurers in the United States and London and the Insurance Services Office. The group is charged with a conspiracy to restrain competition and manipulate the cost of commercial liability insurance. Although the charges are complex, they appear to focus on allegations that the defendants conspired to force the change in liability coverage from the traditional occurrence coverage to the claims-made coverage. The suit also charges that the insurers forced coverage for pollution to be excluded from the comprehensive liability policy forms.

The suit and the charges are too complex and the time too short for such to be analyzed as a part of the paper. However, on an *a priori* basis, an international conspiracy of some 30 separate insurance organizations seems an unlikely explanation of insurance unavailability and terms of coverage problems. The market in commercial insurance is large, international and competitive. Increases in prices or restrictions in coverage can also be explained by the objective uninsurability of the hazardous waste liability and pollution liability exposures noted in the text of the paper. Uninsurability, in turn has been progressively impacted by damage awards and by court decisions altering the meaning of contractual language. The litigation is apt to continue for some time.

ENDNOTES

1. Carson, R.L. 1962. *Silent Spring.* Houghton-Mifflin, Boston, MA.
2. Huber, Peter 1987. "The Environmental Liability Dilemma," *CPCU Journal:* 34,pp. 207-216.

3. Hammond, J.D. and Shapiro, A.F. 1986. "AIDS and the Limits of Insurability," *The Milbank Quarterly,* Vol. 64, Supplement I, Cambridge University Press, New York, N.Y., pp 143-167.
4. Valovic, T.S. January, 1981. "Hazardous Waste Management: Risks, Regulations, Results," *Risk Management:* Vol. 28: p. 13.
5. Bergen, G.S. July, 1987. " A 1987 Superfund Primer," *Risk Management:* Vol. 34: pp. 21-25.
6. *Environmental Law Reporter,* Vol. 55: pp. 12-85.
7. Victims of industrial accidents need not prove fault to recover economic losses. The system of workers compensation replaced a restrictive tort law system nearly three generations ago and an injured worker need not sue to collect benefits. In states which have passed no-fault legislation for automobile accident victims, the same concept applies, though usually with considerable variation across such states but with all permitting the victim to retain some rights to sue the alleged wrong-doer.
8. Kleindorfer, P.R. and Kunreuther, H.C. 1987. *Insuring and Managing Hazardous Risks: From Severo to Bhopal and Beyond.* Laxenburg: International Institute for Applied Systems Analysis, p. 420.
9. *Ibid.*, p. 422.
10. Berliner, Baruch 1982. *Limits of Insurability.* Prentice Hall, Englewood Cliffs, N.J.
11. Dionne, G. 1983. "Adverse Selection and Repeated Insurance Contracts," *Geneva Papers on Risk and Insurance* 8:(6): 316.
12. Katzman, M.T. 1985. *Pollution Liability Insurance As An Instrument For Regulating Catastrophic Environmental Risks.* S.S. Huebner Foundation for Insurance Education, University of Pennsylvania, R.D. Irwin, Homewood, Illinois, p. 69.
13. These were written on a so-called claims-made basis rather than an occurrence basis, the effect of which is to cut off the loss exposure which remained for an insurer, even after a contract was cancelled or not renewed.
14. Jernberg, D.V. and Furse, M.C. July, 1987, Vol. 34. "Environmental Risk Insurance: Don't Count on It," *Risk Management:* p. 42.
15. Bradford, W.R. December 1985, Vol. 32. "Environmental Litigation and the Insurance Dilemma," *Risk Management:* p. 35.
16. *Ibid.*, p. 37.
17. "Toxic Waste Costs Could Outstrip Insurers' Funds," *Business Insurance:* Vol. 18, May 28, 1984, p. 76.
18. Kleindorfer and Kunreuther, *Insuring and Managing Hazardous Risks: From Severo to Bhopal and Beyond,* p.490.
19. There are instances where insurance contracts are issued and coverage appears to be in place. However, because of the presence of loss repayment promises from the insured, it is doubtful whether or not risk is actually being transferred.

Management of Hazardous Materials and Wastes: Treatment, Minimization and Environmental Impacts. Edited by S. K. Majumdar, E. W. Miller and R. F. Schmalz.

Chapter Thirty-Two

THE HUMAN CONSEQUENCES OF CHERNOBYL

PETER GOULD

Evan Pugh Professor of Geography
302 Walker Building
The Pennsylvania State University
University Park, PA 16802

Just imagine: none of us will be here, not even our great-great-great-grandchildren. All our cities will have gone . . . Even the pyramids of Egypt will be just a handful of dust, yet the sarcophagus around this reactor of yours will still be standing. The pyramids of the Pharaohs have been there for a mere five thousand years. But to contain the radiation your nuclear pyramid must remain for at least a hundred thousand years...That's some monument to leave our descendants, isn't it? . . . they'll see what *sort* of Raphaels, Michelangelos and Leonardos lived in the twentieth century.

Bessmertny, in Vladimir Gubaryev's *Sarcophagus*[1]

INTRODUCTION

At one hour, twenty-three minutes and forty-three seconds after midnight on April 16, 1986, Reactor 4 of the Chernobyl atomic power plant in the Ukraine went into a soaring and uncontrolled chain reaction. The steam and hydrogen explosion blew the thousand ton 'safety' cover off, and demolished two layers of steel-reinforced concrete casing 3.5 meters (11.4 feet) thick, releasing radioactivity equal to all the atomic bombs ever tested above ground, or one million times the emissions from Three Mile Island. An event of the physical world, caused by the human world, now rebounded back upon it. The details of reactor construction, the explosion, and the subsequent cleanup will not be considered here, nor the extraordinary courage displayed by those who fought the graphite fires under intense radiation for the next ten days. These early aspects of the accident have been well-documented,[2,3,4,5,6,7] not the least by commendably open reports by Soviet scientists and engineers under the new policy of *glasnost,* and by the International Atomic Energy Agency.[8] Less well-known are the human and biological consequences that followed in all their variety over the next two years, ranging across a broad spectrum of ecological, medical, political and economic effects. In the space available only a few of these can be touched upon in a rather compressed form, so the references are provided not simply as documentation, but to point towards further reading.[1]

THE CLOUD OVER EUROPE

The radioactive release from Chernobyl displayed two distinct peaks, a distribution recorded at hundreds of European monitoring stations. During the first 24 hours the damaged reactor released approximately 80 x 10^{16} becquerels[2], declining to 11 x 10^{16} bq by the sixth day. Unfortunately, by blanketing the pile with 5,000 tons of dolomite, clay, sand, boron carbide and lead, the heat could not escape, and discharges rose to 44 x 10^{16} bq on the tenth day, when hoses containing liquid nitrogen were finally forced into the reactor chamber to extinguish the graphite fires.

The initial explosion blew a mixed load of pulverized and vaporized radionuclides about two kilometers into the night sky, and released the entire inventory of noble gases with an activity equal to the entire discharge (1-2 x 10^{18} bq) over the next ten days. These gases are usually ignored in discussions about the accident because they combine with other elements only under special, and recently discovered, conditions. However, a large discharge of radioactive noble gases produces high levels of local gamma radiation until it is thoroughly dispersed by atmospheric turbulence.

Continuously fed for ten days, the radioactive cloud first moved northwest along a trough of low pressure across northeastern Poland and the Baltic Sea, at which point the upper levels diverged over Finland, while the lower levels, containing most of the fallout, moved west over central Sweden and Norway. (Figure 1). There they mixed with a wedge of cold air from the North Atlantic, and the resulting rainfall scavenged the fallout in large quantities. Swedish measurements indicate that rainfall scavenging increases the washout ratio up to five million,[9] and even 1 millimeter of light drizzle brings down particles that might have otherwise have passed through an area. The cloud also hit the mountain chain along the Swedish-Norwegian border, increasing the fallout in both countries.

Similar orographic effects occurred three days later over other countries of Europe as the original northward movement was blocked, and scavenging rain laid a great swath of fallout over the High Tatra and Carpathians in eastern Czechoslovakia and southern Poland, and along the alpine chain of Austria,

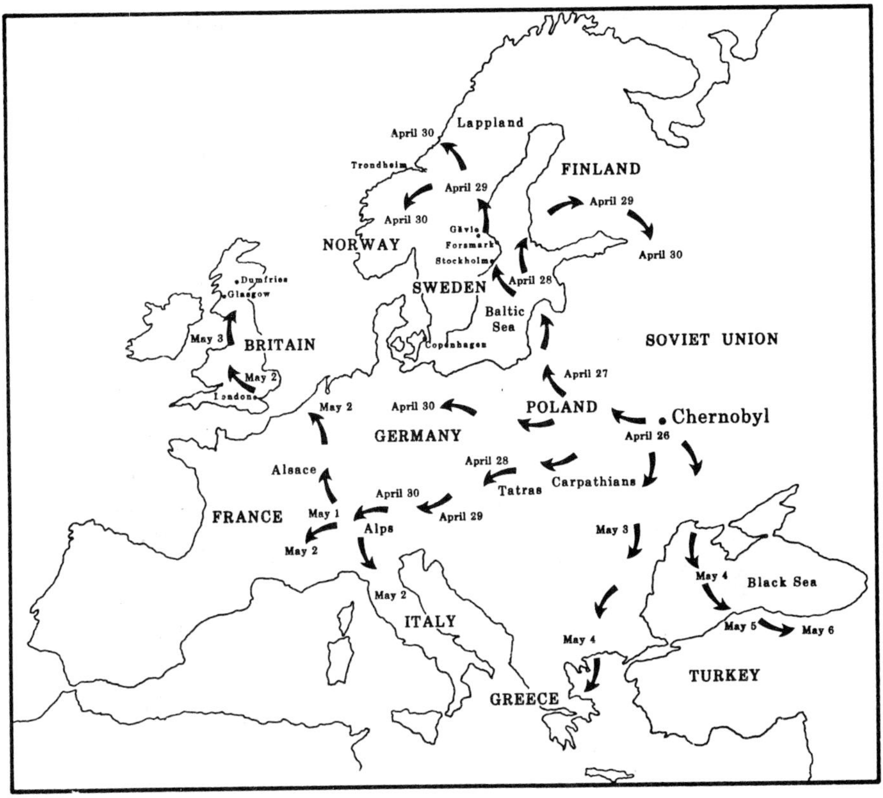

FIGURE 1. Generalized movement of Chernobyl "Plume" over Europe during the ten days of discharge.

Germany, Switzerland, Italy and France. The 'plume' then swirled north through Holland and Belgium and across England, scavenged once again over the mountains of northern Wales, Cumbria (the 'Lake District') and Scotland. The radioactive air mass then corkscrewed out over the Atlantic, only to hit these higher areas of the British Isles again. Meanwhile, the discharges from the second peak were being channeled by a complex and rapidly changing pressure surface northwards, to give Scandinavia a second dose, and south to the Balkans, Greece and Turkey.[10] Eventually the radioactive plume was widely dispersed over the northern hemisphere, with minute quantities of fallout being recorded as far away as the western United States and Japan. The entire discharge was monitored with extreme care by all European countries capable of making ground and air measurements, and it was the subject of intense atmospheric modeling.[11,12,13,14,15] Thanks to Rolf Sievert's network of 37 monitoring stations set up in Sweden in the sixties, and the extraordinary meteorological 'detective work' by the National Defense Research Institute, the Swedish scientific effort was outstanding. It provided the first concrete information about the accident, pinpointed its location, and gave Europe at least a day of warning in the absence of any Soviet announcement. This only occurred 80 hours after the accident, a terse three-sentence statement during the morning news of April 29, forced by strong inquiries by Sweden's ambassador to Moscow the previous evening.

THE CHERNOBYL 'TRACER'

The radioactive cloud that swirled over Europe also left behind turbulent eddies in the human and ecological condition below it. In modern medicine today, minute quantities of short-lived radionuclides are sometimes injected into human bodies to serve as tracers to disclose circulation problems, tumours and malignancies. In an analogous way, the radioactive fallout of Chernobyl entered equally complex ecological and social systems, disclosing problems of circulation and malignancies in bodies politic throughout Europe. The management, or, more accurately, the mismanagement, of the Chernobyl catastrophe by government bureaucracies shows neither democratic nor communist regimes at their best, and stands as a distressing comment on government in the post-modern world.

The information that there was a reactor accident at Chernobyl produced its own information crisis. All over Europe the switchboards jammed as people tried to get reliable information and advice. No one had made preparations for such a telephonic onslaught, perhaps because those charged with the responsibility of dealing with these events never really believed they could happen. Even countries with some sort of 'Radiation Protection Board' in place were unable to cope, for while a few had derisory plans to handle local evacuations, few had thought about any danger blowing across their borders. The first switchboards to be jammed were those of the World Health Organization's regional office in Copenhagen, as national agencies called to ask what they could do so that they could answer the same question in their own countries. As the cloud approach London, the telephone exchange at the National Radiation Protection Board (NRPB) locked solid when it opened at eight in the morning, and it remained jammed for days. Even official inquiries received a constant busy signal, and when a second 'hot line' was opened days later by the Ministry of the Environment it was also swamped immediately. As for the information itself, it was marked by incomprehensibility and contradiction as the spokesmen of various government departments issued a stream of conflicting advice. Reporters' suspicions were allayed or alarmed by picocuries, microröntgens, millisieverts, kilobecquerels and nanograys, none of which meant anything to the reporters, government spokesmen or even the scientists trained in a dozen different traditions. In England, newspapers described the situation as "little short of a shambles . . . with government public relations officials, with little or no knowledge of the subject, feeding incomprehensible statistics to equally baffled journalists in the name of diffusing public alarm." Meanwhile bland assurances were given by government spokesmen, assurances that environmental groups said "drifted between the patronizing, complacent, and downright incompetent".

In Italy, the Civil Protection Agency (CPA) said radiation levels were so low that 'there was no cause for alarm', a government phrase in many European languages that rapidly came to be interpreted as a signal that something was seriously wrong. This announcement was immediately contradicted by Italy's Ministry of Health, which banned the consumption of fresh milk by children and pregnant women. Some of the milk, laced with radioactive iodine 131 and the caesiums 134 and 137, turned up in yogurt and long-term milk date stamped before Chernobyl, and six months later independent scientists of the University of Milan found powdered milk in baby formulas to be "so significantly contaminated by caesium that it requires urgent intervention by health authorities". A stream of contradictory announcements created genuine panic in some parts of northern Italy, and pharmacies and supermarkets had their shelves stripped of pre-Chernobyl food and goods, including iodine in all forms, from which some people later died from self-administered potions. Having produced panic, dissimulation was then tried to calm things down. High readings from the Alps were spatially averaged by government 'scientists' with low readings from the south to give innocuous 'regional values', even as Swiss television was announcing fallout values 100 times higher across the border. Matters were not helped by Italy's Atomic Energy Commission refusing to release information about caesium in food, and some government scientists anonymously told reporters to contact university scientists "because they are not as restrained as we are". One spokesman of the CPA said "What do you want with those figures? By now the damage is done", and even public reports from WHO in Copenhagen were interdicted "for minor typographical corrections" when everyone else in Europe had access to them. Half the Italian population said they did not believe their own government.

In France, government spokesmen deliberately lied to the French people, the Minister of Agriculture saying that France had been completely spared, while the map for the evening weather news on television had STOP signs all along the eastern border, a Verduniam reminder that "They shall not pass". Meanwhile Alsace reported 35,000 bq/kilo in vegetables (while across the border in the Saar they had long since been taken off the market), and the French Electricity Board was recording up to 252,000 bq/m^2 in the French alps. Overturning the laws of physics, these figures has magically decayed to 42,000 by the time they were transmitted to the Radiation Protection Board. Three French scientists noted that "The public is kept in infantile ignorance", but as the director of France's Electricity Board noted so frankly "You don't tell the frogs when you're draining the marsh", a view shared by Finland's Minister of Home Affairs who said "There is no reason to inform, it will only create anxiety".

In Britain, trust in government statements was not enhanced when radioactive levels in food and milk approached, and in some cases exceeded, a set of newly recommended safety limits by the NRPB. These were still three and four times higher than those adopted by Germany and the United States, but as soon as measurements in Scotland approached the new limits, the government backpeddled vigorously and said they were not official, legal, and with statutory force. A high fallout area on an official map stopped just a pencil line short of Glasgow, although rainfall measured in the center of the city had the highest radioactivity anywhere in Britain. In Dumfries, iodine 131 entered the milk supply, and the NRPB noted babies would ingest the 'annual permissible limit', but then promptly cancelled plans to test children's thyroids in Glasgow "because parents would be alarmed".

Only West Germany, with a Federal structure giving strong local powers to the states, was able to deal with the crisis somewhat better than most. In Baden, the physics department of the University of Konstanz used most of its resources to measure and map radiation, and then conducted experiments that no one had thought of before—like washing vegetables to see if radioactivity was removed, a safety precaution recommended across Europe at the time. They discovered that vegetables, and grass used as animal fodder, retained 90 percent of the radioactivity, and with 194,000 bq/m^2 on the ground, and 125,000 bq/litre in the water, they advised local authorities to issue warnings and forbid sales. It was just as well that they did: a litre of milk would have taken a child's thyroid to the 'annual permissible limit', and a kilogram of leafy vegetables would have done the same for an adult![16] Unfortunately, just across the Swiss border in the Canton of Thurgau, no warnings were issued by either the local or Federal governments. All during this period, with readings of 315,000 bq/m^2 at Munich, officials in Stuttgart (the capital of Baden) were saying that there was no cause for alarm because their recorded values were 20 times lower, an indication of the extremely local and spotty nature of rain-scavenged fallout. Generally, however, states at the local level managed to provide accurate information and advice, particularly in those where the Green Party was represented in local government. Hess and Saar gave particularly reliable information, and the Greens even provided reports in a number of languages for visitors and foreign workers.

SUPPRESSING INFORMATION

Both during and after the Chernobyl crisis, enormous amounts of information were manipulated or suppressed by governments advised by atomic scientists and engineers. A number of people and organizations have commented

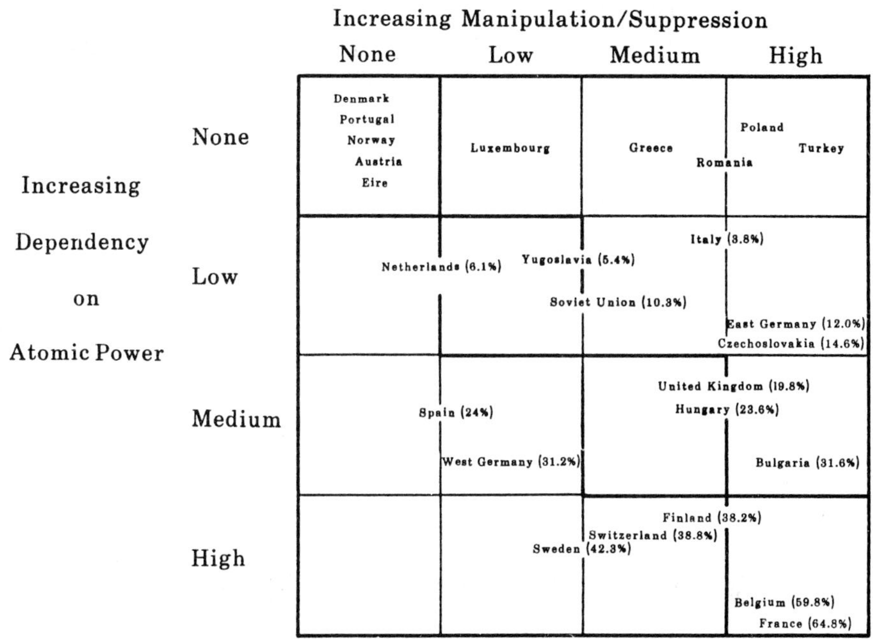

FIGURE 2. The general relation between information manipulation and suppression, and the national dependency on atomic power.

on the strong correlation between the degree to which a country had made itself dependent on atomic power, and the degree to which it employed the tactic of suppressing information. One large European consumer grouped noted "It is significant that those countries with the higher dependency on nuclear power tended to do the least (in the way of providing information), with the obvious exception of Germany".[17] "Suppression" may be a difficult thing to measure with scientific precision, but after working through 5-6,000 pages of official reports, documents, scientific treatises, eyewitness statements and media accounts, you recognize it when you see it, and you can at least order it from 'low' to 'high'. For example, during eighteen months of research and field work, I came across no attempt to suppress information in Norway, but a number of very strong pieces of evidence that suppression was distressingly high in Poland.

There is no question of a high correlation between information suppression and atomic dependency (Figure 2), although if it were perfect all the countries would lie in the diagonal boxes from Eire and her four non-atomic companions in the top left hand corner, to Belgium and France lower right. Some of the exceptions are easily accounted for: most of those in the top right hand corner are Eastern European countries whose governments already exert a high degree of control over the dissemination of information, and who have built, or who are building, Soviet-designed reactors. It is highly embarrassing to wake up one morning to find the main supplier of your reactors dumping heavy radioactive fallout on you. Greece and Turkey, without atomic power, also exerted a fairly heavy hand. In Greece, some of the durum winter wheat had to be destroyed, with radioactive levels far above the permissible limits, but the government provided a minimal amount of information for political reasons, and a group of doctors, lawyers and scientists formed an independent agency to test foods. In Turkey, the large tea crop was badly contaminated as the last stages of the cloud moved across the Black Sea to hit the high slopes of the tea growing areas. This information was strongly suppressed, because to replace the large quantities of tea consumed by the Turks each year would have cost a fortune in hard currency.

On the lower side of the diagonal—less suppression then we would expect from the overall rule or correlation—we find Spain with a quarter of her electricity provided by atomic power, but virtually untouched by the fallout. Finland finds itself in a delicate position, and is generally circumspect when it comes to broadcasting unfavorable information about the Soviet Union, especially when it is picked up by Estonian television across the Gulf. Nevertheless, pressure from the media and anti-nuclear groups produced somewhat less suppression. Switzerland is in the same box, but one acquires a rather strong impression of a government doing very little, rather than actively suppressing information. Sweden is a much more complex case, generally very open, but with some subtle and disturbing currents underneath. As noted already, West Germany, heavily dependent on atomic power, was remarkably open, thanks to strong local powers and pressures brought to bear by the Green Party. There is little point in suppressing information in Bavaria if the local papers are full of it next door; and if university scientists at Konstanz insist on publicizing their measurements, the bureaucrats at Stuttgart might as well grudgingly designate them as an official monitoring station.

FOODCHAINS: ECOLOGICAL AMPLIFICATION SYSTEMS

While the fallout may have moved metaphorically through bodies politic as a tracer to disclose bureaucratic blockages and malignancies, it also moved literally through ecological systems to disclose highly connected pathways between living things, including human beings. Much of the research on the ecological consequences of Chernobyl has been conducted in Sweden, Norway, and Britain, three countries hard hit by the radioactivity, mainly in the form of caesium 134 (halflife approximately 2.4 years), and 136 (halflife approximately 28 years).

A major difficulty, but a common one in science, is what we might call the 'signal-in-the-noise' problem, establishing a signal or effect in the presence of a great deal of natural fluctuation or background noise. For example, many predator birds (snowy owls, gyrfalcons, etc.) and animals (seals, otters, foxes, wolverine, etc.) are at the end of foodchains through which the radioactive caesium is amplified at each successive stage. But Swedish ornithologists, monitoring bird populations for eighteen years in undisturbed research tracts, have not been able to establish any immediate downward trend in the face of highly variable natural fluctuations. Otters and seals are also at the end of a strongly amplifying aquatic and fish food chain, but heavy environmental pollution, especially PCBs, has already devastated otter populations, not only throughout most of Sweden, but in all European countries. The organochlorides are also responsible for the steep decline in the Baltic seal population because of cancers of the liver, lips and other deformities.[18]

In the face of such an already disturbed environment, it is virtually impossible to marshall evidence for a fallout effect. Wild animal populations are notoriously difficult to monitor: for example, a wolverine in Norway will take roughly six wild reindeer each winter, and reindeer themselves are strong amplifiers of radioactivity because they feed on lichen. This 'airsponge' intercepts 86 percent of the fallout, and holds three-quarters of it in the upper half grazed by the reindeer. The 'in-the-plant-halflife' varies between 8-14 years, and the high fjäll of Norway had readings of nearly 200,000 bq/kilo. But how do you establish any Chernobyl effect in such a sparse wild animal population? As for the wild reindeer themselves, a highly experienced field zoologist specializing in these animals 'thought' that there were an unusual number of spontaneous abortions in the two years following the fallout, but even the most astute observations of a field zoologist do not constitute publishable scientific facts. How do you establish, with reasonable scientific certainty, the before and after spontaneous abortion rates in a wild animal population that result from the ever-amplifying effects through a foodchain?

Just how strongly these effects may be amplified can be judged by a series of studies in Sweden. Forsmark is the atomic power station on Sweden's Baltic coast where the monitoring scintillometers first picked up fallout radiation from the shoes of people arriving for work during that first morning. A broken ring of islands off the coast has been joined to form an artificial lagoon, and all the cooling water from two of the reactors, about 86,000 litres/sec, is pumped via a tunnel into the lagoon, raising the water temperature 8-10°C higher than the surrounding Baltic. Fine mesh covers all the gaps to prevent the fish from leaving, and the site forms one of the finest aquatic research stations in the world. Since it was built to monitor the almost undetectable radioactivity in the cooling discharges, Forsmark had valuable before and after data to make proper comparisons. Immediately after Chernobyl, the caesium 137 rose from barely measurable levels in May 1986, to nearly 4 bq/litre in June, an amount so small that it requires extraordinarily sensitive instruments even to get a reliable reading. By no stretch of the imagination could it be thought a hazard to human health.

The problem is that zoo- and phyto-plankton pick up the caesium before they become food for small shrimps, which then become food for small fish, which become food for larger fish, which . . . could appear under a delicate sauce bretonne flavored with just a hint of fresh dill. These larger fish ended up with over 3,000 bq/kilo by the following December, with the trend still rising linearly and sharply upwards. In other words, in six months the foodchain had amplified the radioactivity by three orders of magnitude (a thousand-fold), to ten times the limit allowed for human consumption.

Even these levels became almost derisory in fresh water fish, particularly the predators so highly prized by fishermen for their sport and taste. In the lakes around a fallout peak at Gävle, the radioactivity levels in bottom-feeding bream went from nothing to 2,000 bq/kilo in June, while the perch feeding on bream minnows hit 8,000 bq the next month, followed by the pike leading the upward trend at 10,000 bq by autumn.[19] In the north, peaks as high as 27,349 bq/kilo were recorded that first August, but it was in the fjäll of the mountain backbone between Sweden and Norway that levels in aquatic ecosystems soared out of sight. During that first summer, algae in high tarns and lakes contained 296,000 bq/kilo, while diatoms, forming the food for fresh water plankton, reached 300,000 bq, the highest levels of radioactivity ever recorded in biological material.

Similar effects were discovered in Norway, where research in the Høyshøn east of Trondheim focused on two 'water webs'. The first started with bottom feeders, either water plants, or animals feeding on them, while the second started with living things taking up the radioactivity directly from the water itself. Only ten weeks after Chernobyl, trout hit levels of 25,456 bq/kilo, hardly surprising when hatching mayflies at the outlet of the lake were over 19,000 bq.

SEEING WHAT YOU WANT TO SEE

The research on Chernobyl's fallout in Norway was fully supported by a virtual blank cheque from the government. In the first few weeks, the annual budget of the coordinating body, the Nature Conservancy Board, was completely blown, and as the director told me "I hadn't the faintest idea where the money was coming from. I just made agreements and signed the bills. But we called the Minister for the Environment in Oslo, and her response was immediate—Do whatever you think is necessary and we'll find the money". Later the Norwegian Parliament passed a special appropriation, and put 'Chernobyl research' on a firm financial footing until 1990.

In a rather stark contrast, scientists in Sweden (about 45 percent dependent on atomic power), experienced some difficulties unless their research had been approved by a lengthy review process that was used to guide Sweden's 'normal science'. These review procedures had produced excellent scientific research under ordinary conditions, but sometimes new events from Nature are not willing to wait. For example, biological sampling in the high fjäll was not supported by Stockholm, and had to be done by limnologists sympathetically assigned to their new 'duties' by local government officials who happened to be environmentally concerned. The actual radiation measures were undertaken by friends in the armed forces in what were euphemistically called 'off duty hours', because the Swedish army has some of the most modern, sophisticated and sensitive instruments in the world. Even one of Sweden's most distinguished ecologists was unable to get research funds to study the effects of acid rain on fallout, thinking that it might produce a lipid, a water soluble form much more easily taken up by living material, including the hard carapaces of insects. As we shall see, his early, and very astute educated guess was probably right on the mark.

Insects are not usually thought about much as links in human food chains, but they enter aquatic systems (trout and mayfly), and they are vitally important for breaking down vegetable material, particularly cellulose. But one insect does enter the human food chain immediately, for beekeeping is an important local industry, not only crucial for orchard pollination, but also producing significant amounts of honey. At Gävle, which had readings as high as 750,000 bq/m^2 in the first few weeks, the bees were collecting the pollen from the new spring flowers whose petals had formed small collecting bowls for the iodine 131, the caesiums 134 and 137, and other radionuclides. Bee larvae were feeding on pollen at 20-25,000 bq/kilo before going through the critical stage of metamorphism during the 9th to 14th days. A large sample of 800 larvae was taken from 140 hives around Gävle, but research funds to analyze the possible effects were then denied by Stockholm. Two years later the larvae still lie frozen in liquid nitrogen, and the scientists are so pressed with other educational duties, long-term genetic experiments, and the threat of a new disease on the island of Gotland, that they cannot undertake the detailed and careful analysis without modest funds and some additional help.

As for the fallout itself, two Swedish physicists, who had worked with the atomic power industry, examined air filters for radioactive particles from the fuel pulverized during the initial explosion or expelled in the heat plume. Using standard volumes of air, they detected only one or two particles, but a highly respected radiophysicist from Finland, working under the same conditions, detected 10,000 particles using special film and an exposure time of three days. Some of the particles were pure uranium and ruthenium capable of killing cells in lung and other tissue completely within 0.1 millimeters, and damaging others with their alpha particles up to 7 mm away. "You can see the clear regions of dead cells around the particles with the naked eye in Petri dishes", he said. The danger arises not so much from the dead cells, because the body can repair such small and chronic lesions. The main danger is to the cells in the small spherical 'killing ground' around the particles where the cells are only wounded. It is these damaged cells that can turn malignant and it was in these that genetic changes in RNA were observed.

What we might charitably call the 'conservative' approach to Chernobyl research in Sweden has directed scientific attention towards some things and away from others. After the first few months, one scientist wrote a review of what was known about the effects of Chernobyl for an impeccable scientific journal, and noted that even small doses of radioactivity could disrupt rather sensitive aquatic systems. This simple, well-documented, and apparently innocuous statement was considered much too bold by two of the reviewers, and the editor demanded the removal of both the offending sentence and the documentation. A year later, as we shall see, national and international commissions lowered 'safe' levels in human 'systems' three to five fold.

Documenting Chernobyl's effect becomes increasingly difficult because no systematic record is being kept for the thousands of measurements made by medical and biological scientists, often at local levels. This is true not simply in Sweden, but all over Europe, despite pleas by prominent scientists and commissions in the early weeks to set up a centralized data bank, what a geographer would call a GIS, or Geographical Information System, recording every measure in its spatial and temporal coordinates. When asked why such a scientific resource had not been established at a quite trivial cost, a prominent member of Sweden's Institute for Radiation Protection looked at me suspiciously and said "But what would you want such a data bank for?".

The fact that a majority of the members of Sweden's Academy of Sciences are physicists and chemists, and 80 percent are pronuclear, may be one further reason for minimizing research on Chernobyl and hoping that in time the problem will go away. But even scientists are not always logically consistent. One member of Sweden's Academy, a physicist who was strongly pronuclear, and who had held a high office, provided one of the finest examples of the NIMBY-NOOS complex. NIMBY-NOOS is a grave inconsisto-contradicto-malady that strikes human beings with great swiftness, otherwise known as 'not-in-my-back-yard' or 'not-on-our-street'. Owning a sommerstuga on Sweden's south coast, and hearing that an atomic power station might be built nearby, he declared he would sell out at once if the wretched thing were approved.

FRESH FACTS FOR THE MODELERS

Whenever a catastrophic event appears in Nature, it alters some of the conditions of the ecological systems that may be in a rather delicate balance. We can often learn a great deal from such changes if we are prepared to act quickly and boldly, and somehow prevent bureaucratic structures responsible for funding science from grinding on in their unimaginative, 'business as usual' ways. And these days, when computer modeling frequently deteriorates into the scientific version of a trivial computer game, some fresh facts from Nature can have an astringent effect upon the sort of mechanically-trapped thinking that finds it increasingly difficult to dintinguish between a simulated model and a concrete natural system. There is nothing like going back 'to the things themselves', as the philosopher-mathematician Edmund Husserl noted at the beginning of this century. We can learn a lot when 2-3 x 10^{18} becquerels are injected into Nature, and the fallout moves like a sharp radioactive pulse through complex ecological systems.

In Alaska, for example, marine zoologists have discovered that the radioactivity from atomic bombs, especially the hugh Soviet tests on Novaya Zemlya in the sixties, is taken up annually by the bowhead whale, and then laid down as strata in the layers of baleen.[20] From these we now realize that grievous errors have been made in estimating the age and sexually maturity of this endangered species. Chernobyl's fallout has a special 'signature' different from atomic bombs, so if we want to date surviving creatures over the next half century we should look for the mark of Chernobyl in tooth and bone.

In the eastern Mediterranean, models estimating sedimentation rates were also upset by the Chernobyl 'tracer'. Oceanographers were surprised how the fallout deposited on the surface drifted down to the bottom sediments in a few days, rather than the years predicted by their models. It turned out that the radionuclides were quickly taken up by the zooplankton feeding on the surface, and then incorporated into their fecal pellets making up 70 percent of the sediments.[21] Not only were the sediment modelers forced to return to their drawing boards, but it meant that the radioactivity was taken out of the foodchain much more rapidly than expected.

However, it is on land that some major discoveries have been made, a number of them still puzzling. In Sweden, for example, the caesium quickly penetrated to the root zones of the fireweed and heather, two favorite flowers for the bees gathering nectar. It then rose systematically through the plants, to raise levels in honey to over 3,000 bq/kilo, or 10 times the level allowed for human consumption. And with a halflife of 28 years, the caesium 137 it is not going away. The following year the same dramatic rise in August and September appeared. The caesium was not being held in the lattice structures of the clay particles as soil scientists had hoped, but was freely available to the plants.

It was the Norwegians who first pointed to some of the problems ahead, showing the now 'obvious' and well-accepted fact that soil composition and structure were major determinants in locking up or releasing caesium to plants. It was not so obvious two years ago when models calibrated by the British on the clay soil of southeastern England gave totally false predictions for the high humus and high acid soils of Wales, Cumbria and Scotland. Just by chance, the Norwegians had two research stations at Atna, where the soils were poor in nutrients, and at Knutsjø, where they were relatively rich. The stations had been set up to study the effects of acid rain, and they provided those valuable before and after data that are essential to tracing and documenting change.

As a result of the totally inaccurate predictions made by the original models, the British have put a great deal of research into caesium movement in soils. Sheep farming in the highland areas has been stymied for nearly two years because the caesium is so readily available to the foraging animals due to the low clay content of soils formed from the old Cambrian and pre-Cambrian base. Recently, an even more unpleasant and totally unexpected surprise came to light. In the course of being incorporated into the soil, the fallout has become much more easily absorbed by animals, and the transfer rate across the gut is now 4-10 times what it was two years ago. Tragically for many farmers, even though the caesium is slowly decaying, the amount being retained in the muscles of sheep is still unacceptably high. No biological model had predicted, or even suggested, this, because the possibility had never 'come to mind'. Like any internally and logically consistent scheme—a system of mathematical equations, or a piece of computer software—what can appear in a model is already contained in what has been set down. Despite the recent, and possibly important interest in chaos theory, there are no real surprises in logic. Only Nature tempers our hubris with the previously unthinkable.

FALLOUT AND HUMAN HEALTH

Quite apart from the somewhat detached scientific interest, the most immediate concern after Chernobyl was the effect of the fallout on people. And it is here that we enter an area fraught with controversy and strongly conflicting interpretations. We still tend to think of science as a cool and objective enterprise, an intellectual endeavor in which people of reason agree once the facts are known. But even cursory reading in the field of low level radiation should dispel such notions, for here we see science as a socially embedded institution, where interpretations are strongly shaped by extrascientific preconceptions that raise questions of morality and ethical conduct not informed by the logic of scientific thinking. If you are pronuclear (power or bombs), you may accuse me of uninformed hyperbole at best, and outrageous dissimulation and distortion at worst. If you are antinuclear, you will point to my distressing lack of moral fibre, my abandonment of responsibility to future generations, and my failure of nerve to make a righteous commitment to truth.

Which is why Professors (of the truth as they understand it, what else?) have tenure—and tough hides. In this most controversial of areas, I shall take that responsibility to truth seriously and do the best I can.

What is involved? What are we concerned about? Certainly not exposure to high levels of radiation. There is no argument here. Controversy arises when we try to evaluate the cumulative effects of exposure to low level radiation, and it is here that assumptions and speculations are only being slowly displaced by long term studies whose interpretations may well be informed and shaped by those moral and social preconceptions that are constitutive of being human.

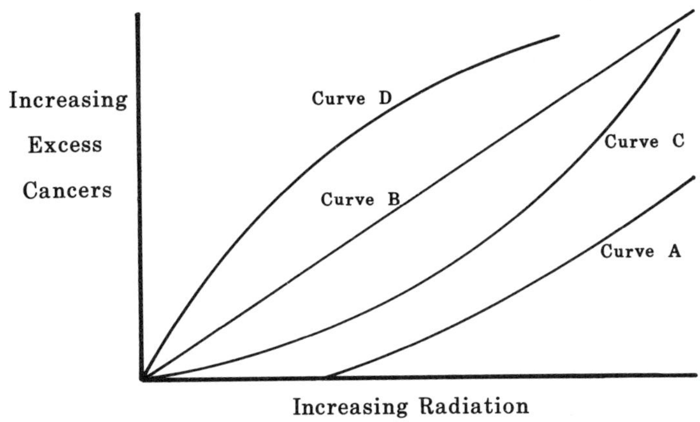

FIGURE 3. Possible relations between excess cancers and low level of radiation.

What are the possibilities? Imagine a graph of cumulative exposure to radiation (X axis), plotted against the probability of excess cancers over and above those normally expected according to current rates (Y axis). We are not concerned here with the top right hand corner: we know what happens there, and whether there is a 63 percent chance or 81 percent chance of becoming cancerous is really not the point. Any civilized person agrees that these levels are simply too dangerous to fool around with. But down in that lower left hand corner, near the origin of the graph (Figure 3), the controversy becomes sharp.

We might ask whether really minute doses of radiation have any effect at all. This would mean that we were on Curve A, which only starts to rise once a certain threshold value has been reached. Leaving aside literally quantum level effects, I do not think any responsible scientist holds to this idea today, especially in the face of the Oxford Survey of Childhood Cancers, a classic of epidemiology that established radiation damage to fetal brain cells from a single x-ray photo taken during critical weeks of embryo development. Perhaps the lowest energy level is that which can break a molecular bond, a level far below the energy released by a single fissioning atom (roughly 15 MeV, or mega-electron volts). Any curve must start rising from a point essentially indistinguishable from zero.

So what about Curve B, the linear hypothesis that says that the probability of cancer rises in direct proportion to the dose? Many in the atomic power industry strongly supported this, even though they may have preferred Curve C. They magnanimously granted that they would accept the conservative linear hypothesis for the sake of argument, implying that the risks were so small anyway that it did not make much difference. Perhaps you can feel the way 'worlds' and their preconceptions inform questions and interpretations?

But slowly and patiently the careful and detailed evidence is being gathered that Curve D may well fit the known facts best of all,[22,23] based on research of epidemiologists associated with the original Oxford Survey, using data collected over 33 years from the Hanford plant in Washington making plutonium 239. This evidence was very carefully examined and accepted by the Sizewell Inquiry (UK) by a distinguished member of Britain's NRPB who said: "I believe the evidence she produced is sound . . . and if her data had not been scrutinized they might have dismissed." Seven years earlier, at the Sellafield Inquiry, they actually had been dismissed as unpalatable to the then accepted doctrine. This means that the risks of low level radiation have been underestimated 5-10 times for adults, and perhaps 10-15 times for children.

Other evidence comes from a recent and very thorough reassessment of the Hiroshima and Nagasaki data by the International Commission on Radiological Protection (ICRP), a scientifically conservative and highly respected group that sets most international standards. We now know that the two bombs dropped were very similar in their radiation effects, that moist air produced greater scattering, and that buildings afforded more shielding from gamma radiation than originally thought. Previous estimates of exposure have been halved, and we can also pool the two populations for statistical purposes, and make judgments against the meticulously kept tumour registries. People who were ten years old at the time now have lung, liver, cervical, urinary and prostate cancers at 8 times the rate of those who were over 35,[24] and new blood scanning techniques allow us to see protein damage to red blood cells in victims over 40 years later.

So convincing was this evidence that Britain's NRPB did not wait for the ICRP to work out their recommendations. "We can't go on waiting," said the director, "We should put people on notice. It would be wrong to wait two or three years." Annual doses for people in atomic industries were cut to one third of previous 'acceptable' levels (50-15 mSv), and to one half for ordinary people (1.0-0.5 mSv).

Safety levels are usually measured in units of absorbed dose equivalents called millisieverts (mSv), units to which the pioneer radiobiologist and physicist Roft Sievert in Sweden gave his name. These take into account the different forms of energy from a fissioning atom—alpha and beta particles, and gamma radiation—and weight the massive 'heavy calibre' alpha particles 20 times for their effects on living cells. Although the mathematics is essentially simple bookkeeping, the specific effects of the various radionuclides in fallout are complex, and the ICRP publishes everything that is known in great detail in its regularly updated handbooks.[25] The issue is complex because the amount of radiation absorbed by the body depends on at least five different things that are quite specific to a particular radionuclide. First, the radionuclide itself, and how many becquerels it emits. A gram of plutonium 239, for example, emits about 2 billion becquerels, and it not a terribly healthy thing to have around. Second, the physical halflife of a radionuclide, about 24,000 years for plutonium 239, 28 years for caesium 137, and 8 days for iodine 131. Third, the propensity of a radionuclide to be absorbed by the body, about 100 percent absorption of caesium, especially in the increasingly soluble forms, and roughly 10 percent for strontium. Fourth, the propensity for different radionuclides to migrate and lodge in various parts of the body: caesium to the muscles, iodine to the thyroid, strontium to the bones, silver to the liver, and so on. Fifth, the biological halflife, the time the body takes to get rid of half the radionuclide it has absorbed: about 70 days for caesium, to 50 years for strontium. All these 'ins and outs' have to be added up over time, so the specific complexity is usually reduced by health physicists to a convenient 'rule of thumb' that converts becquerels ingested daily to absorbed dose equivalents in sieverts. Eating 300 becquerels per day for a year (considered by many to be rather low safety level), would mean absorbing a dose equivalent of 1.424 mSv, or three times the new upper limits recommended by Britain's NRPB. A reindeer herder in Lappland, ingesting foods daily that might well contain 4,000 bq, would be 40 times over the recently revised limits. Even after one year, some young herders had body counts as high as 102,000 bq, and the curve was still rising exponentially upwards.

So what are, and more importantly what will be, the overall effects of Chernobyl's fallout on people? The answer is that no one knows, and no one may ever know for two rather basic geographical reasons. First, we lack the geographical, place-specific knowledge about fallout and people, the steady accumulation of data so necessary for large-scale epidemiological work, place specific data that is simply not being collected systematically. As in most epidemiological analysis of this sort, estimates have to be made statistically on large populations, and it is doubtful that we could ever point to a particular person and say "Chernobyl caused your cancer," unless muscles and kidneys were so contaminated with caesium that we could not put commonsense aside. The second, thoroughly geographical reason is what a geographer calls 'spatial averaging.' If you want the problem of heavy fallout to disappear, you take a reindeer herder in Lappland at 14.2 mSv, add him to nine farmers in Skåne in southern Sweden where these was negligible fallout, divide by 10 and declare "there is no cause for alarm." As one thermodynamicist said to me: "It's like putting your head in the freezing compartment of the refrigerator, your feet on the stove, and telling me that you feel pleasantly warm."

Yet dozens of such estimates and predictions were made all over Europe after Chernobyl, and you could tell immediately where the person doing the arithmetic was coming from, what the assumptions were, and what preconceptions informed them. Pronuclear proponents grabbed one set of assumptions that went into their arithmetic at one end and produced no cause for alarm at the other. Antinuclear proponents took another set of assumptions and found their arithmetic generated Armageddon. One pronuclear physicist said that Three Mile Island would only produce one cancer death, so by his own assumption Chernobyl, releasing one million times more radioactivity, should produce a million cancers. But he would be the first to heap scorn on just such an estimate by a man who has spent his entire life in medical physics, and who tried to use the best available geographic and scientific evidence to arrive at this higher figure.[26] Most estimates of deaths and malignancies lie between one thousand and one million, a thousand-fold range within which the most rabid pro- and anti-nuclear proponents should find what they are looking for. The Soviet Union is conducting a huge epidemiological survey on the 135,000 people evacuated, and has promised to cooperate fully with international medical authorities. The health-identity cards of the people evacuated have been marked with a special symbol so that they can always be identified as a member of the Chernobyl study. But it will be 20-30 years before really reliable estimates can be made.

THE DEMOCRATIC CONSEQUENCES OF CHERNOBYL

Perhaps the greatest casualty from Chernobyl was the damage to public trust in government, a matter that strikes at the taproot of democracy. So many aspects of atomic power have become a matter for secrecy, particularly in Europe where most civil servants are not hampered by a Freedom of Information Act, and have, in fact, vigorously opposed the passage of anything faintly resembling it. The result is a skilled and ready willingness to dissemble and lie, using all the apparatus of government to avoid telling people the truth. The French and the British are particularly skilled at invoking such privilege forms of dissimulation, and the *privus lex* (private law) makes the government-industry combination a law unto itself. In France, for example, it has been called 'a state within a state,' responsible to no one but itself. Such cynicism simply breeds a sense of numb futility on the part of many, a feeling devastating to democratic participation in government at all levels.

The litany of deceit is very long, and includes falsification of medical data, the application of security classifications to hide accidents, the denial of scientific conclusions that proved embarrasing, the computation of pseudoprobabilities, and the corruption of officials in the atomic energy industries. An appalling scandal broke recently in West Germany and Belgium, involving officials at nuclear plants and facilities for reprocessing nuclear fuel. The German manufacturer admitted to $12 million in bribes, and the Belgian killed himself after police accused him of spending another $9 million more. I recall at one international meeting on the effects of Chernobyl the hurt voice of physicist saying "I just don't understand it. We know what we're doing, so why don't people trust what we say?" "Because," said a forthright geneticist "Because you're the ones who have been telling people it couldn't happen, and now it has. Why are you surprised they don't trust you anymore?."

MOVING FORWARD TO ARISTOTLE

Pulling together many threads from the physical and human worlds in a particular place, and at a particular time, is an old geographic tradition much neglected of late, particularly in an academic world increasingly fragmented into its nineteenth century administrative boxes. But synthesis is an honorable intellectual tradition, particularly valuable when academic partitions and specializations break connections in a world that the Comte Buffon, in the eighteenth century, insisted was all connected up.[27] Chernobyl was an event that intimately linked the physical with the human, and demanded that the specialized knowledge of scores of fields, from atomic engineering to zoology, be connected up again.

At first sight, it might seem strange that eighteen months of research across many specialized scientific fields should lead someone to reading books III-VI of Aristotle's *Nicomachean Ethics*[28] again, but a concern for human consequences invariably leads to a concern for human knowledge and politics, and to all the 'ought' questions of how these should be conducted in a moral and ethical way. These are not 'old' philosophical questions answered once and for all, but deeply human questions constantly renewed and reexamined in light of an always changing human world. Chernobyl, in its human ground, was about knowledge and responsibility, and raised questions of technical skill, prudence, intelligence, wisdom and judgement. Aristotle called them *techne, phronesis, nous, sophia,* and *gnome.*

Chernobyl brought us face to face with a dilemma, and if any good may be said to have come from the catastrophe, it is the possibility that it has opened up the path of reflective thinking once again. The dilemma comes from the fact that it is not easy to think prudently, for the choices are agonizing. We have built a civilization on energy. Our movement, our warmth, our food, our communications, our material comfort . . . all require prodigious amounts of energy, and we have fallen into profligate habits. Even the stability of nations, and therefore the world system made up by them, become questionable when the energy sources are suddenly placed in jeopardy. If civilization is to continue to unfold in more humane ways, the energy to fulfill the hopes and necessities of civilized life must be there.

And so the last swirling effects of that Chernobyl cloud are in human thought itself. Here we stand at the center, and it is from this center that thinking, in its most deliberative and reflective sense, flows to consider choices and alternatives. We have inherently safe and commercially feasible forms of atomic power,[29,30] and have had them for over 20 years. It was only 30 years ago that we stood at a fork in the road, and for reasons of *thrasutes* (rashness), *philotimia* (ambition) and *asotia* (prodigality) chose the most dangerous form of atomic power, the light water reactor—what one atomic engineer has described as "the most complex and unforgiving" system. Do we now, in greater prudence and wisdom, aware of our responsibilities to future generations, consider other choices that might meet even Aristotle's approval? The choice is ours. It is up to us.

NOTES

1. References are to English sources only, and are confined to those readily available through academic and scientific libraries. References in French, German and Swedish may be found in (2).

2. In conformance with common scientific and international practice today, radiation is measured in becquerels. A becquerel is the spontaneous decay of one radionuclide atom per second, releasing approximately 15 MeV (mega electron volts)—8 MeV from the alpha particles, 4 MeV from the beta, and 3 MeV from the gamma radiation. In fundamental energy terms, 1 becquerel is equal to 24×10^{-6} ergs. Simple linear scalars convert these basic physical units to a variety of other measures devised as various specialized branches of science and medicine developed over the years—grays, rads, curies, röntgens, etc.—and when the alpha, beta, and gamma energies are weighted to take into account their effect on living tissue, sieverts and rems result. An older unit, the curie, is equivalent to 37×10^{9} becquerels, the number of spontaneous emissions per second from a gram of radium. Such an amount, held in the hand for one second, would give approximately 16 times the lethal dose to a human being. Since there is a deep concern today for the effects of radiation on living tissue, and a growing realization that small, but accumulating effects have serious consequences, the 'finer' unit of the becquerel is now considered the most appropriate. The standard adoption of the unit is important if radiation measures are to be meaningful across many scientific and technical fields, and to a general public informed by mass media. The plethora of units in use as recently as two years ago (what a French report called *une belle salade radioactive*), caused enormous confusion in the immediate days after Chernobyl, not simply among journalists and the public, but between scientists from different fields and traditions.

REFERENCES

1. Gubaryev, V. 1987. *Sarcophagus.* Penguin Books, Harmondsworth, pp. 81.
2. Gould, P. 1989. *Fire in the Rain: The Democratic Consequences of Chernobyl.* (forthcoming).
3. Edwards, M. 1987. Chernobyl One Year After. *The National Geographic Magazine.* 171:632-653.
4. Flavin, C. 1987. *Reassessing Nuclear Power: The Fallout from Chernobyl.* Worldwatch Institute, Washington, D.C., pp. 91.
5. Hawkes, N. *et al.* 1987. *Chernobyl: The End of the Nuclear Dream.* Vintage Books, New York, pp. 246.
6. Marples, D. 1986. *Chernobyl and Nuclear Power in the USSR.* St. Martin's Press, New York, pp. 228.

7. Wilson, R. 1987. A Visit to Chernobyl. *Science.* 236:1636-1640.
8. International Atomic Energy Agency (IAEA). 1986. *The International Nuclear Safety Advisory Group Summary Report on the Post-Accident Review Meeting on the Chernobyl Accident.* IAEA, Vienna, pp. 106.
9. Persson, C., H. Rodhe, and L. De Geer. 1987. The Chernobyl Accident: A Meteorological Analysis of How Radionuclides Reached and Were Deposited in Sweden. *Ambio.* 16:20-31.
10. Alexandropoulos, N. 1986. Chernobyl Fallout on Ionnina, Greece. *Nature.* 322:779.
11. ApSimon, H. and J. Wilson. 1986. Tracking the Cloud from Chernobyl. *New Scientist.* 1517:42-44.
12. Barneby, F. 1986. Chernobyl: The Consequences in Europe. *Ambio.* 15:332-334.
13. Horrill, M. 1987. *Chernobyl Fallout in Great Britain.* National Environmental Research Council, London, pp. 176.
14. Jaworowski, Z. 1986. *Chernobyl Accident: Emergency Monitoring and Protection Measures in Poland.* Central Laboratory for Radiological Protection, Warsaw, pp. 11.
15. Smith, F. and M. Clark. 1986. Radionuclide Deposition from the Chernobyl Cloud. *Nature.* 322:690.
16. Hohenemser, C. *et al.* 1986. Chernobyl: An Early Report. *Environment.* 28:6-13, 30-43.
17. Bureau European des Unions de Consommateurs. 1986. *Chernobyl: The Aftermath.* BEUC, Brussels, pp. 94.
18. Bergman, A. and M. Olsson. 1985. Pathology of Baltic Grey Seal. *Finnish Game Research.* 44:47-62.
19. Petersen, R., L. Lander, and H. Blanck. 1986. Assessment of the Impact of the Chernobyl Reactor Accident on the Biota of Swedish Streams and Lakes. *Ambio.* 15:327-331.
20. Pain, S. 1987. Age Miscalculations Threaten Bowhead Whales. *New Scientist.* 113:21.
21. *Anon.* 1987. Chernobyl Fallout Aids Seabed Studies. *New Scientist.* 115:30.
22. Kneale, G., T. Mancuso, and A. Stewart. 1984. Job Related Mortality Risks of Hanford Workers and Their Relation to Cancer Effects of Measured Doses of External Radiation. *British Journal of Industrial Medicine.* 41:6-14.
23. Stewart, A. 1985. Detection of Late Effects of Ionizing Radiation: Why Deaths of A-Bomb Survivors are so Misleading. *International Journal of Epidemiology.* 14:52-56.
24. Rotblat, J. 1988. A Tale of Two Cities. *New Scientist.* 1594:46-50.
25. International Commission for Radiological Protection. 1978. *Limits for Intakes of Radionuclides by Workers.* Pergamon Press, Oxford, pp. 512.
26. Gofman, J. 1986. *Assessing Chernobyl's Cancer Consequences: Applica-*

tion of Four "Laws" of Radiation Carcinogenesis, Symposium on Low-Level Radiation, Anaheim, California, pp. 57.
27. Foucault, M. 1970. *The Order of Things: An Archaeology of the Human Sciences.* Vintage Books, New York, pp. 387.
28. Aristotle. 1953. *The Nicomachean Ethics.* Penguin Books, Harmondsworth, pp. 383.
29. Lidsky, L. 1987. Safe Nuclear Power. *New Republic.* 197:20-23.
30. Lidsky, L. 1988. Nuclear Power: Levels of Safety. *Radiation Research.* 113:217-226.

Management of Hazardous Materials and Wastes: Treatment, Minimization and Environmental Impacts. Edited by S. K. Majumdar, E. W. Miller and R. F. Schmalz. © 1989, The Pennsylvania Academy of Science.

Chapter Thirty-Three

THE BHOPAL ACCIDENT: ITS AFTER-EFFECTS

GEETA TALUKDER and ARCHANA SHARMA

Human Genetics Unit
Centre for Advanced Study in Cell & Chromosome Research
Department of Botany
University of Calcutta
Calcutta 700019, India

1. *Background*

Sometime between late night of December 2 and early morning of December 3, 1984, water entered a tank of the Union Carbide pesticide plant in Bhopal, India. The tank contained the chemical compound, methyl isocyanate (MIC). The ensuing intense reaction led the lethal contents, turning from liquid to gas, to flow over the town of Bhopal and its sleeping inhabitants, causing the greatest industrial disaster known in history. The official figures place the number of deaths at 2,850 (it was 2,344 in October, 1986) and of the seriously afflicted between 30,000 to 40,000, the final figures of death and damage. The pathological aftereffects of the survivors and residual disabilities are still conjectural. Despite official and unofficial enquiries, the scientific data published are even now inadequate.[1,2,3]

The township of Bhopal, in 1978, sprawled untidily over an area of about 6,878.5 square miles at an altitude of 500 to 600 meters above sea level. The average temperature ranges from 10°C in winter to 40°C in summer, with occasional lower and higher figures. The last Census in 1981 gives the population to be 672,329 and that at the time of the accident to be approximately 100,000. Nearly seventy percent were immigrants from the countryside in search of jobs, adding to the problems of overcrowding, malnourishment and inadequate facilities for living and medicare. An estimated number of 351,214 persons lived in the 30 municipal wards of the city adjacent to the Union Carbide Factory.

The factory was set up in 1969 on a five-acre plot, where alpha-naphthol was manufactured and mixed with methyl isocyanate (MIC) imported to make the carbaryl insecticide, Sevin. MIC is produced by the interaction of phosgene with monomethyl amine phosgene. Phosgene, in turn, is produced by the reaction between carbon monoxide and chlorine.

In 1975, the Union Carbide Company obtained permission to manufacture

MIC from the Government of India and the MIC Unit became operational in 1980. By 1981, the factory complex occupied 70 acres. Though initially the area where the factory was located was regarded as an industrial belt, yet by 1984, the township had engulfed it. The surrounding hutments were congested, badly organized with inadequate sanitation and lack of basic amenities. The Railway Station was close to this area.

2. *The agent*

Methyl isocyanate (MIC, $CH_3 \cdot NCO$) is one of the most toxic chemicals known in the gaseous form. Under normal conditions it is a volatile colorless liquid, with a b.p. of 39°C and a specific gravity of 0.96, being heavier than air. It dissolves in water, the reaction being instantaneous and highly exothermic, with the formation of monomethylamine (MMA) and CO_2. Some of the MMA reacts with the remaining isocyanates to produce alkylureas, with the release of heat. An unstable intermediate product is methylcarbonic acid. The water-soluble MMA can further react with nitrites and nitrates to form the known delayed carcinogens—nitrosamines. MIC is also highly reactive, producing trimers in the condition of stagnancy. It reacts with itself, liberating 54 KCal of heat per molecule of trimer formed.[4] Thus, both MIC and its degradation products have a high potential for damaging biological systems on both short- and long-term basis through reacting with multiple functional groups and producing heat energy by exothermic hydrolysis. According to the World Health Organization, the final product of MIC—Sevin—was classified as moderately hazardous. However Temik (Aldicarb)— another end-product in which MIC is an intermediary—is regarded as extremely hazardous. MIC has a threshold limit value of 0.02 ppm or 0.05 mg/M^3 on skin, that is the maximum concentration to which a worker may be exposed for 8 hours per day during 5 days per week. Brief exposure above 20 ppm is likely to cause death in human subjects. It is highly irritating to the skin, eyes and mucous membranes. The irritation in human subjects is felt from 2 ppm upwards, becoming unbearable at 21 ppm. In sensitive individuals, even very low concentrations may induce an asthmatic response.

The isocyanate group (-NCO) of MIC, as of other related isocyanates, is highly reactive. It combines with the -OH and $-NH_2$ groups in biological macromolecules, including proteins and enzymes, blocking or inactivating their functions. For example, interaction of MIC with the -OH group of the enzyme acetylcholinesterase leads to an inactivation of the enzyme. The consequent accumulation of the protein, acetylcholine, in the body results in increased transmission of effects of nerve stimulation. It produces symptoms of lachrymation, irritation of the eyes, excessive mucous formation in the windpipe and increased bladder and intestine movements. The ultimate effects are congestion and asthma of the lung and uncontrolled urination, defecation and ulceration of the mucous surface. Theoretically, these effects are reversible.

3. *The accident:*

Between midnight of December 2 and early morning of December 3, about 42 tons of a cream-colored gas escaped. From the numerous statements, official and non-official, claims and counter-claims, certain facts emerge. MIC was stored in three steel containers of 60 tons capacity each. The containers were provided with valves to prevent the contact of the highly toxic chemical with rust and water, both of which are believed to produce violent exothermic reactions. On the night in question, the temperature started rising in the tank No. 610 and the pressure increased rapidly far above the danger point. The intense heat generated caused the concrete above the tank to crack. The gas poured out at an initial rate of 180 pounds per minute and raced to a maximum pressure of 720 pounds per minute. The safety measures, like alkali vent gas scrubber, flare tower and the water spraying system, were faulty, inoperative and/or inadequate and the gas could not be neutralized or contained.

4. *The Immediate aftermath*

4.1 *Inadequate information*

In dealing with the after-effects, there was an initial uncertainty about the nature of the gas—whether it was MIC alone or a mixture of several gases. A suggestion was that a mixture of hydrogen cyanide and carbon monoxide might have been formed through dissociation of MIC at the high temperature (250°C) recorded before leakage. An earlier record of a leak of phosgene in the same factory led to further speculations. Since the gas had dispersed by the morning, the exact nature could not be confirmed. The soil or the biota did not give any clue.

Data on the toxicological effects of MIC were also lacking because the chemical had not been assessed due to its use as a captive intermediate. Some tests had been carried out on its mutagenicity after 1980 but little information was available on effects on human or animal systems after exposure.[4,5]

4.2 *Immediate effects on human populations*

Initial figures released, following the night of suffocation and stampede, were 1754 deaths, 173 hospitalized and 14,000 diagnosed as seriously affected (October, 15, 1985). Later, the official death toll rose and kept on rising to the latest figure of 2850 as more cases died in their sleep. These deaths included mostly women and children. The official machinery was geared up effectively only on the third day after the accident. By that time a large number of bodies had been disposed of.

The major cause of death was by immediate asphyxiation, accompanied by severe irritation of the eyes and throat. During the first two days, the death rate in some of the worst affected areas was as high as 20 per thousand persons. During December 1984, the frequency of deaths was 24 per thousand as against a normal frequency of one per thousand for the same period in other years. Most affected were victims below the age of 5 years with mortality figures of 33 per thousand. Postmortem showed death to be due to asphyxia, with thick froth at the nose and mouth, accompanied by congestion, edema and consolidation of the lungs. A cherry red coloration of the blood was noted. Spectroscopic studies of the blood showed absorption bands of carboxy hemoglobin.

4.3 *Effects on animals and plants*

A survey by Indian Veterinary Research Institute gave 1047 deaths and 7000 cases of hospitalization of large animals 8 to 9 days after the accident. All animals were affected up to 6 km from the site. Lethality was recorded up to 4 km in domestic animals like cows, buffaloes, goats, pigs and horses. Chickens, parrots and vultures were also killed.[6]

The effects on plants were more selective, the rate of destruction being 50 to 100% in areas close to the factory (up to 3-4 km) and 25 to 50% in regions farther away. The effects were mainly thermal. Certain plants like *Musa, Saraca* and *Mangifera* were less damaged than *Azadirachta* and *Ficus religiosa*, but these observations are not conclusive.[7,8]

Water samples were negative for MIC, cyanide, dimethylurea and phosgene, although destruction of fish and phytoplanktons had been reported. Chemical oxygen demand (COD) had increased 3 to 4-fold on the first day and returned to normal level by the fifth-day.

5. *Subsequent stages*

5.1 *Mechanism of toxicity*

The major problem faced after the accident was the difficulty in medical treatment of the affected survivors for toxicological effects. The main reasons were:

(i) Until several weeks after the accident, the nature of the gas leaked could not be specified, whether it was MIC, MIC+CO or phosgene.

(ii) No data were available to the local doctors on the toxicology of MIC or methods for dealing with exposure to MIC.

(iii) Since information on phosgene was available, following human exposure during World War II and later, cases were treated along similar lines.

(iv) No data were available on the possibility of the formation of CO and CN by dissociation of MIC. Following a suggestion that such dissociation may lead to the production of hydrogen cyanide, carbon monoxide and cyanogen chloride, sodium thiosulphate was used for therapy only after February 4, 1985.

5.2 *Sodium thiosulphate treatment*

Following the administration of sodium thiosulphate, marked symptomatic relief was recorded, accompanied by a 6-to 8-fold increase in urinary thiosulphate in double blind trials carried out by the Indian Council of Medical Research.[9] The baseline levels were significantly higher in exposed populations as compared to unexposed ones and remained appreciably high in spite of progressive decrease with time.

The content of free amino ($-NH_2$) groups in the lysates of hemoglobin from exposed persons was reduced by 20 to 40%, as shown by trinitrobenzene sulphonic acid method. Such reduction was also recorded in cord blood of children of exposed mothers.

These findings were taken to indicate that MIC produced effects similar to anoxia caused by carbon monoxide.

5.3 *Study of specific effects*

1. *Respiratory system*—Autopsy showed that the entire tract was damaged. The major ailments were— bronchitis, pneumonia, hemorrhage, edema with necrotising bronchiolitis and damage to lung parenchyma. After three years, residual tissue anoxia and impairment of pulmonary function persisted, as supported by radiological studies.[10]

2. *Ocular damage* was severe immediately after exposure with severe conjunctival inflammation, with residual corneal opacity and dryness of eyes. In different mammals, similar effects of MIC were recorded as well.[11,12,13,14]

3. In the *hematological* system, hemoglobin levels were decreased. The immunological profile showed enhanced IgM and IgA and reduced IgE. The absolute T-cell and TH-cell counts were altered and T-cell response to mitogens was depressed. Subsequently MIC-positive antibodies were observed. Laboratory experiments on mice exposed to MIC induced suppression of mixed leucocyte response[15] and myelotoxicity.[16] The effects were eliminated after variable time periods.

4. *Teratogenic and genotoxic effects* — Epidemiological studies carried out at different periods following exposure recorded a greater fetal loss, as indicated by abortion, stillbirth and neonatal death in exposed women as compared to unexposed controls.[17,18,19,20]

Chromosomal alterations were reported in exposed individuals even three years after the accident.[21] Such alterations were also recorded on laboratory animals exposed to MIC.[22,23,24]

6. *Suggestions for evaluation, prevention and remedy*

6.1 *Management of chemical accidents* requires specific guidelines.[25] In general, there are four levels of responsibility:

(i) operator;

(ii) local/community;
(iii) regional/national and
(iv) international

The *precautions* for prevention need:

(i) identification of possible accidents both from prospective and retrospective studies;

(ii) continuing planning for accidents, which include designation of responsibility; preparation of a suitable action plan; liaison with external authorities; identification of resources in emergency; preparation of a suitable information system; health services.

Management in disasters needs emergency response; choice and implementation of required protective and remedial measures; determination of the extent and dissemination of the information to the public, authorities and health services.

The next step is *rehabilitation.* The final stage is *follow-up,* which includes monitoring of effects with time and researches on epidemiology.[26,27]

6.2 *The situation in Bhopal* is a case where most of these components were either absent or inadequate. Out of the mass of conjectures and collected information, some findings have been confirmed; such as,

(i) The causative compound was a gaseous substance, which was initially predominantly unaltered MIC. The earlier involvement of other gases, like phosgene and chlorine, in chemical accidents suggests that gaseous products are possibly the most dangerous of industrial pollutants.

(ii) The initial absence of basic information on the nature of gas leak and its possible degradation products delayed appropriate detoxification measures, thus raising the levels of morbidity and mortality. These effects could have been avoided if the relevant information was available.[28]

(iii) The location and subsequent expansion of the Union Carbide factory in a densely populated region magnified immensely the dimensions of exposure to the escaping gas. This problem is liable to recur frequently in view of the rapid unplanned industrialization and flocking of rural job-seekers to towns without adequate accommodation or civic facilities.

(iv) Facilities for neutralization and containment of the effluents and for analyzing the pollutants were ineffective, if not non-existent.

6.3 *Some recommendations* can be made to ensure that such disasters do not recur and in case of an accident, the effects be neutralized immediately. Most of these recommendations are routine ones but are not followed, through lack of awareness, carelessness, callousness or lack of adequate training.

(i) Industries with gaseous products or effluents or intermediates should have fully organized systems of neutralizing the chemicals round the clock, manned by trained personnel. The latter should carry out regular exercises, checking each stage of the procedure.

(ii) The levels of the pollutants in the atmosphere and soil and the rate of excretion in the exposed workers should be measured routinely.

(iii) The capacity of storage tanks for volatile substances should be inversely proportional to the toxicity of the chemical. In France, MIC is stored in small stainless steel drums of a maximum capacity of 213 kg. In one location, not more than 280 drums can be kept.[26] Each container should have safety values and cut off mechanisms, in addition to arrangements for optimal temperature and pressure.

(iv) In case of shutdown, intermediate products should be neutralized or destroyed immediately.

(v) All products—primary, intermediate and final — should be tested for mammalian toxicity, teratogenicity and mutagenicity. In order to minimize the effects of minor accidental exposure, factory workers should be routinely given long periods of recovery in exposure-free duty shifts following limited period of work under conditions of possible exposure.

(vi) To minimize the effects of accidental spills in transit, detoxifying measures should be documented in detail and be kept available for the general public at short notice.

(vii) The need for hazardous chemicals should be fully assessed before permission is given for manufacture or storage. In India, carbaryls are being replaced by cheaper pesticides. The high residual levels of organochlorines — like DDT and BHC—render them unsuitable for further use. A conscious long-term attempt is needed, firstly to replace hazardous pesticides by relatively non-hazardous ones, like synthetic pyrethroids; secondly, to train the consumers in the correct doses and methods of use and finally, to replace progressively, chemical methods by integrated biological methods of pest control.

(viii) The factory environments should be surrounded by a green belt, preferably of plants with known resistance to the particular product and the substances involved in its production.

(ix) There is a dangerous gap in the available knowledge regarding the effects of long-term exposure to subtoxic doses of chemicals on human populations—exposed both through occupation and residence in the industrial zones. In the absence of *in vivo* human data, such information can be obtained through studies *in vitro* from human tissue and leucocyte cultures and *in vivo* from mammalian test systems. Toxic, subtoxic and genotoxic effects can be studied and threshold levels worked out. Possible containment through nutritional supplements can be worked out.

In conclusion, all measures for containment will be of no use unless the will to prevent such accidents is present—the will of the industry to put safety before profit; the will of the government to look at longterm plans for human welfare and the will of the people to enforce the safety measures.

REFERENCES CITED

1. *Report on the factors related to Bhopal toxic gas leakage.* Government of India, December 1985.
2. *The Trade Union Report on Bhopal.* The International Confederation of Free Trade Unions, Geneva, July 1985.
3. Hazarika, S. *Bhopal: the lessons of a tragedy.* Penguin Books India Ltd., New Delhi, 1987.
4. Kimmerle, G. and A. Eben. Zur Toxicit[8]at von methylisocyanat und dessen quantitativer Bestimmung in der Luft. *Arch. Toxicol.* 20:235-241, 1964.
5. NTP/NIHES Report: The toxicity of methylisocyanate. *Environ. Health Perspectives,* 72:1987.
6. *Report of Scientific Commission for continuing studies on effects of Bhopal gas leakage on life systems:* Vegetation around Bhopal, January, 1986.
7. *Report of Scientific Commission for continuing studies on effects of Bhopal gas leakage on life systems:* Animals and aquatic ecosystems, January, 1986.
8. Mukherjee, A.K. Effect of MIC on plant systems in Bhopal. *J. Econ. Tax. Bot.,* 7(3):568-72, 1985.
9. ICMR Report on *Health effects of exposure to toxic gas at Bhopal,* December, 1985.
10. Kamat, S.R., A.A. Mahashur, A.K. Tiwari, P.V. Potdar, M. Gaur, V.P. Kolhatkar, P. Vaidya, D. Parmar, R. Rupwate, T.S. Chatterjee, M.D. Kelkar, and S.G. Kinare. Early observations on pulmonary change and clinical morbidity due to isocyanate gas leak at Bhopal. *J. Postgrad. Med.,* 31(2): 63-72, 1985.
11. Bucher, J.R. The toxicity of methyl isocyanate: where do we stand? *Environ. Health Perspect.,* 72:197-8, 1987.
12. Gupta, B.N., S.A. Stefanski, J.R. Bucher, and L.B. Hall. Effect of MIC on eyes of Fischer 344 rats. *Environ. Health Perspect.,* 72:105-8, 1987.
13. Harding, J.J. and K.C. Rixon. Lens opacities induced by methyl isocyanate. *Lancet* 8431, March 30, 1985.
14. Salmon, A.G., M. Kermuir, and N. Andersson. Acute toxicity of MIC—a preliminary study of the dose response for eye and other effects. *Brit. J. Indust. Med.,* 42:795-8, 1985.
15. Tucker, A.N., J.R. Bucher, D.R. Germolec, M.T. Silver, S.J. Vore, and M.I. Luster. Immunological studies on mice exposed subacutely to methyl isocyanate. *Environ. Health Perspect.* 72:139-142, 1987.
16. Hong, H.L., J.R. Bucher, J. Canipe, and G.A. Boorman. Myelotoxicity induced in female B6C3F1 mice by inhalation of MIC. *Ibid.* 143-148, 1987.
17. Varma, D.R. Anatomy of methyl isocyanate spill in Bhopal. In: *Hazard assessment of chemicals,* Vol. 5, J. Saxena Ed. Hemisphere Publishing, New York, pp. 233-289, 1987.
18. Bang, R. Effects of the Bhopal disaster on women's health on the December

1984 *Union Carbide Incident,* V. Pinto and S.D. O'Leary Eds. Washington Research Institute, San Francisco, CA pp. 35-37, 1985.

19. Shilotri, N., M.Y. Raval, and I.N. Hinduja. Report on gynecological examination in Medical Survey on Bhopal Gas victims between 104 and 109 days after exposure to MIC Gas. *Nagrik Rahat aur punarvas Committee,* Bhopal, pp. 16-17, 1985.
20. Varma, D.R. Epidemiological and experimental studies on the effects of MIC on the course of pregnancy. *Environ. Health Perspect.* 72:153-157, 1987.
21. Talukder, G. Unpublished data on *Three year survey of cytogenetic effects in exposed populations in Bhopal,* ICMR project.
22. Kligerman, A.D., J.A. Campbell, G.L. Erecson, J.W. Allen, and M.D. Shelby. SCE in lung and peripheral blood lymphocytes of mice exposed to MIC by inhalation. *Environ. Mut.,* 9:29-36, 1987.
23. Shelby, M.D., J.W. Allen, W.J. Caspary, S. Haworth, J. Ivett, A. Kligerman, C.A. Luke, J.M. Mason, H. Myhr, R.R. Tice, R. Valencia, and E. Zieger. Results of *in vivo* and *in vitro* genetic toxicity tests on methyl isocyanate. *Environ. Health Perspect.,* 72:183-88, 1987.
24. Tice, R.R., C.A. Luke, and M.D. Shelby. Methyl isocyanate: an evaluation of *in vivo* cytogenetic activity. *Environ. Mut.,* 9:37-58, 1987.
25. Vainio, H. Current trends in biological monitoring of exposure to carcinogens. *Scand. J. Work Environ. Health,* 11:1-6, 1985.
26. Anon. *the Bhopal disaster Mazingira,* 8(4):2, 1985.
27. Sorsa, M., K. Hemminki, and H. Vainio. Biologic monitoring of exposure to chemical mutagens in occupational environment. *Terat. Carcinogen. Mutagen.,* 2:137-156, 1982.
28. *Bhopal gas tragedy:* clinical and experimental studies. *Indian Jour. Exper. Biol.* 26(3):149-242, 1988.

Management of Hazardous Materials and Wastes: Treatment, Minimization and Environmental Impacts. Edited by S. K. Majumdar, E. W. Miller and R. F. Schmalz. © 1989, The Pennsylvania Academy of Science.

Chapter Thirty-Four

ASHLAND OIL SPILL: A CASE STUDY IN EMERGENCY MANAGEMENT*

Introduction by
LIEUTENANT GOVERNOR MARK S. SINGEL
Commonwealth of Pennsylvania
200 Main Capitol Building
Harrisburg, PA 17120

JOSEPH L. LaFLEUR
Director, Pennsylvania Emergency Management Agency
P.O. Box 3321
Harrisburg, PA 17105

INTRODUCTION

Mark S. Singel
Lt. Governor of Pennsylvania

Let me say first that we should all be grateful to the several hundred dedicated workers at the municipal, county, state and federal level and in the private sector for their efforts to respond to and recover from this spill. Their quick response limited the impact this terrible spill had on the residents of Allegheny, Washington, Beaver and Butler Counties. These people worked long, hard hours—in several cases more than forty hours without rest—under extreme weather conditions to restore normal water supplies and to identify and secure alternate water sources. Their dedication to this task and their commitment to the residents of southwestern Pennsylvania communities may never be fully documented and recognized. The time that I spent in the incident area and in the state Emergency Operations Center in Harrisburg gave me an appreciation of the hard work which took place behind the scenes.

The tank failed shortly after 5 p.m. on Saturday, January 2nd at the Ashland Oil bulk storage facility in southern Allegheny County, along the Monongahela (Mon) River. It was a Saturday night on a major holiday weekend, it was dark and there were nearly 4 million gallons of #2 fuel oil on the ground restricting access to the site. The initial response at the state level was by the regional staff

---

*Any opinions, findings, conclusions or recommendations expressed in this article are those of the authors and do not reflect the views of the editors, or the Pennsylania Academy of Science.

of the Department of Environmental Resources (DER) in Pittsburgh.

For the first several hours we were faced with a massive release of fuel oil which could have had a significant environmental impact, but the threat to the public was limited. Two events changed the scenario dramatically later that night. The fuel oil flowed into the storm sewers and had access into the river. Also it was believed that a nearby storage tank containing gasoline was leaking from a valve. The Allegheny County Haz/Mat team began efforts to stop the gasoline leak. As a precaution, officials evacuated nearly 2,000 residents.

The Commonwealth through the Pennsylvania Emergency Management Agency (PEMA) was prepared to support the counties with equipment and personnel. This was not necessary until Monday when an unexpected development on the river forced the closing of all water intakes along the Mon and Ohio rivers. However, when I was informed at about midnight Saturday that people were out of their homes, I directed PEMA to activate the state's Emergency Operations Center in Harrisburg. This was done to monitor and support the county and DER activities if necessary and to establish better communications between the DER team at the site and the state's operations center in Harrisburg.

Throughout the evening I briefed Governor Casey on conditions. At Governor Casey's direction, I flew to the Ashland site at 5 a.m. Sunday morning. DER was working with the Coast Guard and Ashland officials to ensure that the clean-up contractors who were already on-site were acting in a timely and proper manner. Booms were flown in and were in place on the river in several places by late Sunday morning. We were optimistic that Sunday morning in Pittsburgh that the booms would collect the bulk of the oil and the spill would have no significant impact on residents down stream. Preliminary reports from DER and the water authorities indicated that water intakes were low enough to avoid the oil or that it could be filtered out of the water. On Sunday they found this was not the case.

The oil mixed with the water as it went over the dams along the Mon and formed an emulsion, somewhat like salad dressing after oil and vinegar are shaken. This emulsion did not stay on the surface, but contaminated the entire river from surface to the river's bottom. This forced the closing of water intakes from the Ashland site to the Point in Pittsburgh and all along the Ohio River in Beaver County.

By Monday morning, the pieces were in place for a major disaster in western Pennsylvania. We were faced with the potential that several hundred thousand homes and businesses might be without water. A number of large hospitals and nursing homes in suburban Pittsburgh were affected, as was the supply of water needed to fight fires in dozens of communities. The state Emergency Operations Center was fully activated with all appropriate agencies present. The Governor signed a proclamation declaring a Disaster Emergency to exist and Monday morning Joe LaFleur, PEMA's director was dispatched to the scene. Collocating with Allegheny County, Mr. LaFleur brought together the principal players in the county EOC. A fast survey by DER found water in storage to be in short supply. Before the day was out, the Robinson system was dry and more than 13,000 people were without water service to their homes for 5 days. We were in danger of losing three much larger systems as well. A public appeal for conservation was made and we began identifying other ways to bring water into those communities without water or with very low pressure. Allegheny County pressed its limited supply of water trailers, known as buffalos, into service. The Governor directed that all state resources be made available, and more than 100 PA National Guard water trailers of varying sizes were brought from across the state to the Greater Pittsburgh Airport where they were dispatched into affected communities. More than 100 Guardsmen worked around the clock at the airport relocating the trailers to communities and filling the buffalos on a regular basis.

Throughout Monday night into the early hours of Tuesday morning, county and state personnel made contacts across the state to locate larger trailers which could haul water into these affected communities. In addition, other options were explored including railroad cars, canned water, and emergency pipelines. By first light Tuesday morning the first complement of trucks was in place dispensing water. The effectiveness of these larger trucks, anywhere from 1,000 gallon to 9,000 gallon capacity, was limited because they were not designed to dispense water to individuals. The County Maintenance Division solved the problem by designing a system to install faucets on plastic pipe which ran down both sides of each large truck.

In addition, the City of Pittsburgh installed several temporary interconnects with the West Penn system using fire hose and fire trucks to pump the water. A planned large interconnect was expedited and, working around the clock, the work was completed before the end of the week. Through various means the City of Pittsburgh provided at least 12,000,000 gallons of water daily to the West Penn system.

Tankers were also positioned at area hospitals in the affected area to feed the boilers and maintain heat if water service was lost. St. Clair and other hospitals were notified to reduce admissions to essential cases only in case evacuation became necessary.

Heat was a concern as well. Many homes are heated by hot water and boilers, and there was the real potential for homes without water service to lose heat or that electrical service would be lost. The county emergency management agency and the Red Cross set up a mass care center for those displaced due to heat loss. Fortunately, no one found it necessary to go to the center. Remember that this was one of the coldest weeks of the year. Temperatures hit zero at night and the wind produced a chill factor of more than 20 below zero.

Firefighting was a special concern. A major fire would have drawn down the little water left in the systems and there wouldn't be sufficient supplies for effective suppression. State Fire Commissioner Chet Henry worked with local fire companies and the county to establish a separate water tanker fleet to supply water to fight fires during this disaster.

The state teams worked with the county emergency management staff to develop fliers to let residents know where they could go to get water from trailers and a special hot-line was established by the county for residents who experienced problems. Special efforts were made to address the needs of the elderly through the county social service units.

DER and EPA technical staff devised a way to filter the oil out of the river intake when the parts per million contamination reached a certain level. This experimental use of special charcoal filters was first tested with the West Penn system and it worked. There was constant testing to insure safe, fresh water was discharged from the plant. Governor Casey was very specific, in his Wednesday visit, to Director LaFleur that we would not compromise on having a predictable, safe water source. The intakes were not to be prematurely opened. This same filtration application was used in the Midland system in Beaver County just hours before that system would have been completely dry. Again with success. We passed on this knowledge, as well as some equipment, to our neighbors down river in Ohio and West Virginia.

The news media played a key role during this disaster. Allegheny and Beaver county radio, tv and newspaper reports, editors and management distinguished themselves by their commitment to the residents. Throughout the ten day emergency period, news reports were thorough, accurate and balanced. Errors were quickly and cooperatively corrected. Beyond this, special programming and reports provided the public with essential information on the location of water trailers. I believe the news media contributed greatly to the public's voluntary conservation of water. The state team, through Joe LaFleur, and the county conducted regular press briefings to keep the working press, and thus the public, informed about the status of water systems and developments made to correct this difficult problem.

Because of the severity of the potential, as well as the problem at hand, the Pennsylvania Emergency Management Council instituted mandatory restrictions on water usage in the three county area. There was some limited impact on business, but as the largest water users, their reduced consumption was necessary to keep the systems operating. Those mandatory restrictions remained in effect until Monday, January 11th, when all systems were back to a safe level of normal and independent operation.

As the emergency phase passed with the restoration of water, we began the necessary documentation of extraordinary expenses incurred by municipal, county and state agencies which responded to this disaster. PEMA is the lead agency in coordinating this assessment and in recovery of costs.

Director LaFleur met several times with EPA officials to urge them not to leave the site without a long-term order or agreement with Ashland as to future cleanup. EPA agreed and PEMA has passed the technical aspects of the agreement development to DER. In addition, we were very concerned about the integrity of the remaining tanks and pipelines. We immediately retained a private contractor to perform a detailed inspection to assure future safety and avert more problems. The Governor created an investigative team charged with looking for causes of this disaster and what we can do to prevent it from happening again.

This summary covers only the major issues and cannot begin to address the range of activities which involved state, county, municipal and federal staffs. In the interest of space, let me say that this emergency reached potential disaster proportions, but a crisis was averted. This was due in large measure to the reasons I have just mentioned and to the dedication of hundreds of government workers, private sector efforts, and citizen cooperation at all levels.

ASHLAND OIL SPILL: A CASE STUDY IN EMERGENCY MANAGEMENT*

JOSEPH L. LaFLEUR
Director, Pennsylvania Emergency Management Agency
P.O. Box 3321
Harrisburg, PA 17105

FAILURE INVESTIGATION OF ASHLAND OIL TANK NO. 1338 BY BATTELLE MEMORIAL INSTITUTE AND ASHLAND OIL

Summary of Findings Issued May 16, 1988

Ashland Oil Tank No. 1338 was relocated from an Ashland facility near Cleveland, Ohio, and reassembled at Floreffe, PA. The tank was placed into operation at Floreffe in August, 1987. For the next several months the level of product fluctuated with the peak levels increasing. On January 2, 1988, a level of 45 feet, 10¾ inches was reached for the first time at the new site. Around 5:00 p.m., and approximately 3 minutes after reaching this level, the tank wall failed. Battelle was asked by Ashland Petroleum Company to conduct an independent investigation to determine the probable cause of the failure.

A site visit was made in the morning of January 6, 1988. At that time major portions of the tank, bottom plating and exposed area of the foundation were accessible for close viewing. Initial examinations of the failed tank section indicated that the tank was failed by brittle crack propagation emanating from a small flaw below a weld joining the lowest and second level of steel plates which form the wall of the tank. However, it was not known at that time whether the observed fracture was the primary failure or a secondary failure produced by some other occurrence, e.g., explosion, foundation subsidence, or overpressure.

Battelle's overall approach to the failure investigation was to consider all possible failure mechanisms, eliminate those for which a sound scientific rationale could not be established, and arrive at a reconstruction of the events leading to the failure. This reconstruction of the failure event was supported by engineering and physical principles consistent with all physical evidence, information and data. This was accomplished through site/debris examination, review of design/construction/operating information, determination of material properties, and fracture, stress, and foundation analyses.

---

*Any opinions, findings, conclusions or recommendations expressed in this article are those of the authors and do not reflect the views of the editors, or the Pennsylania Academy of Science.

Battelle has concluded that the tank failed by brittle fracture of the tank wall originating at an existing flaw in the lower level plate. The conditions necessary for brittle fracture were present in the tank and were sufficient to produce the failure, as is evident from the following determinations:

—*Initial point*—A flaw was found in the steel plates of the tank at the origin of the fracture of the tank wall. This flaw, approximately the size of a dime, was located close to a weld in material embrittled by the stress and heat of welding. The orientation of the flaw in the wall made it susceptible to failure when under stress due to filling of the tank.

—*Temperature and material state*—Temperature of the steel at the time of failure was calculated to be in the temperature at which a brittle fracture would occur. Thus, once the initial flaw began to grow, it would propagate through the thickness and height of the tank without arresting.

—*Stress*—Stresses due to filling of the tank, coupled with weld residual stresses, were sufficient to initiate a crack from the existing flaw. Once the crack initiated, stresses due to liquid in the tank were sufficient to propogate the crack.

Brittle fracture produced by these conditions was the primary and sole mechanism which produced the observed failure. No other failure causes or contributing mechanisms were determined to be significant.

All physical evidence suggests that a crack initiated at an existing flaw in the tank wall. Based on chemical and metallurgical analysis of the flaw surface, the flaw was determined to have been present in the plate prior to the initial assembly of the tank and to have been produced by an oxyacetylene torch. This process initially created an embrittled region around the flaw. This embrittlement was furthered by additional welding in the vicinity of the flaw during reconstruction of the tank at Floreffe.

With embrittlement, this existing flaw was sufficient to initiate a crack under the static tank wall stresses. The stress, due to liquid in the tank, increased as the level of liquid in the tank rose. Consequently, on January 2, 1988, a new maximum stress was achieved. Combined with the residual stresses produced in the steel by welding, a critical condition occurred in the embrittled region around the flaw. The embrittled region failed, a crack was initiated, and the crack propagated. The fracture propagated downward into the lower level plate and upward through other plate levels, causing the tank wall to part.

The driving force for propagation of the fracture was solely produced by the stress of pressure due to filling of the tank. The fracture did not arrest because the temperature of the tank wall, calculated at approximately 38F, placed the steel in its brittle range. Consequently, once initiated, the fracture did not arrest.

The initial fracture, originating from the existing flaw, triggered a dynamic chain of events:

—The tank wall fractured in less than a second and the walls began to separate rapidly.

—The bottom-to-wall joint failed.

—The tank wall moved from its original position due to the force of the existing oil, taking the tank roof with it.
—Oil rushed out of the failed tank in a relatively low height stream, eroded the tank foundation, and a portion shot over the dikes like a skier off a ski jump.
—The stream of oil splashed against other tanks hitting one with such force that it buckled.
—The tank wall settled some 120 feet from its original location with the roof still partially attached. That portion of the oil, which shot over the dikes, entered storm sewers which carried it to the river.

MULTI-AGENCY CRITIQUE SUMMARY

Six weeks after the crisis ended, Pennsylvania Emergency Management Director Joseph LaFleur called together representatives of the myriad organizations from the public and private sectors, which were involved in those dramatic ten days, for a critique in Pittsburgh. The participants were the key decision-makers who had been in charge of various portions of their agency response throughout the period. The purpose was not to find fault or fix blame for things that might have gone wrong, but to examine the response effort with a view to improving the state-of-the-art for use in future such incidents.

In his opening remarks, Director LaFleur outlined five guidelines for the critique. "First, we want to review the process and the capabilities that were involved. Make an assessment of our capabilities. How did we go about things; why did we do what we did when we did it.

"Second, it is not our intent to Monday-morning-quarterback decisions in various locations at various times. Organizations were making decisions with limited information, with very short time limits to make those decisions.

"Third, I propose to conduct an academic exercise. I think if we go about this today, in a collegial environment, we'll be able to gain a great deal from the event and be able to supply those people who talk to us in the future from national, international, local, community, or whatever level it might be, with some good solid information.

"Fourth, we are not here to point fingers at individuals. We don't want to do that. There's not one person who was responsible for anything that was great or anything that was really terrible in what happened. It was a very intricate, complex series of connected events which took place. There's no one person who is responsible for anything that may have gone wrong or could have been done better. We were all in the learning mode in this situation.

"The fifth goal is a finalization of the response phase. A critique is like going through a grieving process. Fortunately there was no loss of life. But, there have been some people pretty stressed-out by this event. They worked long hours and worked very hard. So we want to bring the response phase to a finalization, and say, okay, there's a recovery effort that's gong to go on and there's an environmental management effort, but we are bringing community emergency response to finalization.

"I think we need to talk about it because of the three Rs that we have: we have responsibility, we have readiness, and we have response. And those are the three Rs that we go to school on if you're in the emergency management business, and I guess we all are at one time or another."

Decision-Making Processes

The deciphering of control and coordination responsibilities and identifying for political entities, the media and the public "Who Is In Charge?" is important. And, in a multi-agency, inter-governmental response (which this was) is an issue which experts reviewing and critiquing from the outside should also understand. Incident management is transitory in nature, depending upon the stage of the incident and the problems at hand. Most large scale on-going technological incidents are transitory as witnessed in events such as the Hyatt Regency skywalk collapse in Kansas City and the Air Florida crash in Washington.

Determining the officials responsible for initial response coordination was not difficult and seldom is. First on-scene personnel are responding with the essential mission of immediate life safety concerns which not only include public health and safety but also mitigation at the incident scene, if possible. Each agency at the federal, state or local level not only responds in its public health and safety venue, but also from its regulatory perspective. Until the essence of the emergency is identified and the point of real or potential human suffering is reduced from critical, the primary action focus is on the first local responding agencies' leadership with the appropriate equipment.

The municipal officials and local volunteer agencies in the municipal jurisdiction responded and took what actions they could under the circumstances (low temperature, night-time conditions, minimal information to the extent of the flow, products at the tank farm, etc.). As they began to decipher the magnitude of the incident, outside resources from the county were requested.

Within the first 12 hours, the incident essentially remained under control of the county and its hazardous materials response team who worked cooperatively with Ashland officials to gain additional information as to the magnitude of the oil spill and the problems caused by the gasoline leak. Technical assistance was also being provided during this time-frame by the State Department of Environmental Resources (DER) for minimizing the impact on the environment as well as a review of their regulatory responsibilities under the circumstances. Federal action centered on cleanup assistance of navigable waterways provided by the United States Coast Guard (USCG) and their agreement with the Environmental Protection Agency (EPA) to provide direction and advice to the spiller (Ashland) and its contractor cleanup teams. By agreement, the Coast Guard was the lead federal clean-up agency until the EPA on-scene coordinator, Jerry Sasseen, could arrive. State government (the Governor's Office) in the initial 36 hours was requesting "federalization" of the spill by the EPA. During this period, EPA was assessing the capabilities of Ashland to pay for the spill cleanup and mobilize the necessary contractor resources to implement the cleanup. EPA's determination during this 36-hour period was that Ashland was capable of handling the spill site and its cleanup. However, it seems likely, that even if EPA had brought the event under federal control, the state emergency management capabilities and responsibilities would have been brought into play for the "people problems" that developed, based on the loss of water and potential loss of power and other resources.

Government Response

Mark McClellan, Deputy Secretary, DER: "There was a lack of study ahead of time by all of us in terms of knowing clear procedures and protocol, and understanding and delineation of the roles and responsibilities of all different levels of government."

Cmdr. Gene Miklaucic, USCG: "Concurrent jurisdiction" of local, county, state and federal agencies can cause some difficulties in coordination. Acknowledging that the question of "Who is in charge" posed problems in the Ashland spill, he said, "a response organization evolved ultimately."

In discussing the issues of declaration, LaFleur stated," "It became clear with the facts at hand about closing water intakes in the river by 8 a.m. on Monday, January 4, and with the lack of movement for a federal declaration of some type, that the state needed to do two things: first, assert itself to handle its responsibilities (those charged to the elected and appointed officials through statute for life, health and safety protection of the public) and, second, invoke the highest level of emergency authority available, which would be through a state declaration, in order to bring organization to an event of unknown length and magnitude."

In the 9 a.m. January 4th conference call conducted between LaFleur, Lieutenant Governor Mark Singel, who was also acting in his capacity as Chairman of the Pennsylvania Emergency Management Council, and Walter Giesey, representing Governor Casey, it was clear that the situation was going to get worse before it improved. A formal on-site, high level state presence of a continuing nature was required. It was determined during the 9 a.m. call that LaFleur now needed to go to Allegheny County. When the Lt. Governor left the incident scene Sunday, it was under belief that the spill would not become as serious as it eventually became. LaFleur requested that an emergency declaration be invoked by the Governor to provide coordination and control on a multi-county basis and allow LaFleur to take extraordinary actions as well as call into effect the capabilities of other state agencies in a coordinated fashion. The Lt. Governor strongly supported this position. Giesey concurred and requested that the Declaration of a Disaster Emergency be drafted and sent over within the hour. Giesey would insure the Governor's signature and appropriate dissemination. By 10 a.m., the declaration was signed and LaFleur was preparing to depart for Pittsburgh with press officer John Comey and State Fire Commissioner Charles Henry. Upon arrival at Pittsburgh, LaFleur was to team up with DER Secretary Davis who was providing information on recommended water conservation measures. LaFleur took Commissioner Henry with him knowing that without water, fire protection could be limited and the need to interact with the fire community for other potential manpower requirements and equipment would be necessary. Comey was assigned to coordinate state and county media information.

Cmdr. Miklaucic, USCG: "The initial response on the local scene was by the local firemen and local county personnel; they set up their command post at the Floreffe fire hall. That's very typical in an emergency response. When I arrived on site I saw what they were doing. They were doing everything that local county response personnel do. They were addressing public safety, health concerns, that type of thing. After viewing that for about twenty minutes to a half hour and seeing that it was as under control as it could be in the emergency situation, I felt we had to address the spill.

"At that time the communications at that site were lacking and we had, in my opinion, to move over to the Ashland site, the Ashland offices. Their power had been restored about midnight and we needed better communication. Recognizing a split of command posts was not the best thing to do, but the only resource, the only option we had at that point. I knew it was a split, but we had to do it. The locals were taking care of the spill at the site itself and site safety. We wanted to address the spill itself which was traveling down river the whole time. And so we moved over into the Ashland site.

"Basically, we developed three different issues to address. And you might even consider that possibly even a fourth command post developed; the recovery sites down in South Side Park. It would have been ideal to have one central command post, and I had looked and contemplated this several times throughout that night, but I did not see any one single spot as being the best place to do it.

"You need to have facilities, communications, and quite a variety of other essentials in order to make, say, a command post such as this room. It would have been nice to have something like this in a central location, but because of the spreadout nature of the spill, and the water quality issues which spread it out on twenty-five to thirty miles of river, I think a centralized location would have been very, very difficult to handle."

Mr. Fabian, Pittsburgh Regional Director, DER; "One of the concerns I had initially was trying to get twenty-four hour staff coverage at both Ashland and

Penn Avenue and then subsequently up at Beaver Falls. We wound up with a rather odd emergency compared to other natural disasters that we had where we have a focus at a given location.

Here we had a response necessary at the scene to guide cleanup effort and that was paramount in terms of protecting the Monongahela from any further discharge of oil. I think it was very appropriate to have an on-scene headquarters for the Coast Guard, EPA and DER relative to those cleanup activities.

The actual water supply impact and the efforts that had to be carried out downstream, basically in the suburban Pittsburgh area, needed a different location. We did not want to be down at Floreffe. And it made a lot of sense to have that EOC in operation at Penn Avenue."

Mr. Johnson, Director of Response & Recovery, PEMA: "First, a little bit about the state EOC. In this case we were an information, support and command center. We're not always a command center. We're totally automated with over twenty microcomputers which have digitized maps of Pennsylvania and can zoom into street level, in Beaver or any county. We have a volume of data in the automated system that relates to hospitals, schools, old folks homes, jails and all resources, in state and out of state.

"So, with that I will give you our two primary missions in the state EOC. Number one, it's our mission to provide the Governor and Lt. Governor with real time updated information in disasters, so they can direct state government, and that's a difficult thing to do.

"Number two, which is equally important, is to provide state agency assistance to the county at all costs. This can be proactive assistance before the disaster strikes; in some cases if it's a flood oncoming or during a disaster, we provide assistance and recovery after the disaster. So those are our two key missions in the state EOC.

"Now, to Ashland. We were notified at 2030. That's 8:30 p.m. The reporting from there to midnight was sporadic on how much oil, because we get a lot of oil spills state-wide every night, every day. And we got from—I think what was portrayed here was from 100,000 gallons to three million. However, because of the late notification, (our adrenaline is normally high), we came down quickly because we were three hours late being notified by DER and the county.

"So we—and some of these are criticisms of ourselves—we evaluated that as a cleanup. We had an oil spill. DER was on-scene, the spiller and Coast Guard were on-scene, and they were going to clean it up. From there until midnight, we continued that kind of an evolution.

"At fifteen minutes past midnight, we were notified by Allegheny County that there was a gas leak in a hundred thousand gallon tank and that some twelve hundred people were going to be evacuated. Now we saw a different thing. This is life and property. This is gasoline. So at midnight the Governor ordered the secretaries of Health and DER to come to the area and the Lt. Governor ordered the EOC to go to full operation. When we go to full operation we can have over twenty state agencies assisting us. From that time until 6:30 in the morning, we saw the evacuation and the gas leak as a threat. That's how we portrayed this.

"The Lt. Governor went on-scene, did his tour and came back. Around 7:35 a.m., the Coast Guard was ordering in more booms from the Atlantic (region) and we saw it still as a cleanup, that they would handle it. When the Lt. Governor came home at noon he came into the EOC and we decided at that time to release the state agency response teams and to stand down in the EOC (return to duty officer coverage), because the evacuation had been lifted and the gas leak had been stopped. Now you can see we didn't analyze that quite right because that evening, Pittsburgh reported that West Penn could be out of water.

"We saw some water inter-connections taking place. We thought that Pittsburgh could provide West Penn with water, so we didn't see a threat on water at that time. However, at 7:30 p.m. we received a report that West View Township Water Authority was going to shut down and had a forty-eight hour supply. Robinson was going to shut down and had a twenty-four hour supply.

"The connect between Pittsburgh and West Penn was having trouble and we reactivated the EOC. Now we knew we had a water problem. So when we reactivated, we notified Ohio, West Virginia, state agencies and federal agencies in Washington of our problem areas. It's difficult to assess when there's no threat of life. Very difficult.

"Reports came from various sources, different numbers of EOC's and different splinter groups reported and had their own information centers. The county often was too busy to give us real time information. Some agencies were going off on their own without coordinating with the state EOC, or the Governor's Office.

"Assessment of a water problem later on was very difficult. We got reports that we had water at five gallons per minute, so many feet in the tank, so many thousand gallons per day, very difficult to relate that to the Governor or Lt. Governor.

"Recommendations—improve communication from the County EOC's, discourage splinter groups from going out on their own, provide information to PEMA. PEMA will provide counties with a liaison officer on the scene and in the counties to assist in information flow. Once federal authorities are on-scene in any disaster, we must have a liaison officer from that agency in the state EOC."

Water Response Develops

One of the first things LaFleur realized, probably by 10:00 a.m. on Monday morning as the event was beginning to unfold, was that water was going to be in shortage. It was going to be a problem for an extended period of time.

LaFleur discussed what he learned during his initial briefings. "We were already at the point where Robinson Township, Moon Township, and the North Fayette area may have pressure problems and some loss of water. The discussion was picking up strongly about water rationing, about water reserves, water in the tanks at this time, and water tank capacity.

"So we started looking for some alternatives and began to discuss them Monday afternoon. I arrived and began to get oriented to the situation. The information started to become clear between county Health and DER and after some initial information from EPA late Monday afternoon that we were going to have to shut down the systems, that the spill was of such magnitude that we were going to have a loss of water for an extended and undetermined amount of time.

"It became clear that if we lost the West View system we were going to have several hundred thousand people without water for an extended period. It was, (at least in my position and my role in the scene, and everybody had a role and everybody was an actor as this played out) clear that if we went from that threshold of fifteen to twenty thousand people to two hundred fifty thousand people, we had a really serious event on our hands.

"So we started looking at alternatives and one of those alternatives turned out to be pipeline. We have about eleven miles of eight-inch aluminum pipe. We had a mile of it brought to the scene, to one of the county engineering shop locations, from the PEMA state storage center in Harrisburg. This could help us with interconnects, fire safety and distribution."

Allegheny County Deputy Emergency Management Coordinator, John Kaus, organized a meeting of all affected fire companies that evening. Over 150 people attended. LaFleur opened the meeting and described their role of backup fire service and their need to help maintain public confidence that the citizens could be protected from fires even if water was scarce. Fire Commissioner Henry assisted in development of mutual aid and equipment sharing programs to protect areas with no water or low pressure. John Comey and the media team spread the word on fire safety vigilance during this critical period.

Joe LaFleur, PEMA: "I asked the state Operations Center to start looking for tankers, particularly railroad tank cars. Obviously they had to be clean. I first wanted railroad tankers because of the extensive track and spur system in the area. We could put large quantities of water in good distribution places for people. If we got the tankers, we needed nearby parking and multiple spigots for dispensing. I didn't want traffic jams to further irritate people who might have to do this for weeks. The railroads today are hauling things like corn syrup, sugar products, and dry goods. But the railroads we talked to, CSX, Conrail and others, said they don't get into the water hauling business anymore in rail cars. So they just don't have the number around that they did at one time. They sold most of them off and they had a hard time figuring out where they were if they did have some. Next, we told the National Guard to mass their water buffalos (500 gallon field water trailers) and tank trucks from armories across the state to a central location in Allegheny County. They did this at the airport.

"We looked at the water reserves in place. We decided that if the water reserve went down after people went home that night and they didn't conserve as we requested, we were going to have to start to take action. We saw that the reserves were down after 7:00 that night. West View had less than a twenty-four hour water reserve and if things didn't turn around, we were going to have to be ready. We didn't want the public to panic. We wanted to have the ability to show that we were, in fact, ready, that officials in the government were dealing responsibly with this issue. Disappointingly, usage didn't slow—they didn't do as well as we hoped.

"So we worked all night. Some of those people were up thirty-six hours straight. That's bad emergency management. We were still behind the power curve. We found about thirty or forty tankers in the private sector in addition to those from the National Guard.

"This was a yeoman's effort by the county. I give the county great credit because they got on the phone and called. Jack Messing was their team leader the first 24 hours. They called fire departments and milk haulers—anybody who had a stainless steel tank that could be sanitized. We had thirty or forty of those line up, both for fire protection (which was an issue) as well as public drinking water.

"And we had to get those trucks in by 4 a.m. the next morning. The county developed a multi-spigot system, which was able to work on distribution so we didn't have traffic jams and we had a way to distribute the water relatively conveniently. John Kaus arranged for the drivers to be met and registered at the EOC. There was a street full of trucks. The city provided a heated garage for the county plumbers to put on the pipes and faucets. The county also set up the distribution locations. There was a police escort to get the trucks out to their distribution site.

"Public perception of our ability to manage and their feeling of personal safety was what I was dealing with. Clean water, safe water, was the issue in this process."

Joe Moses, Allegheny County Superintendent of Maintenance: "As far as the water distribution was concerned, Allegheny County had twenty-eight of its own water buffalos. Eighteen which are categorized in large quantity-size from twelve hundred to thirteen hundred fifty gallons. We had ten smaller ones that had the capacity of four hundred fifty gallons.

"Through PEMA and the County's Emergency Management Center, a number of tankers were ordered into the area. Throughout the night, nineteen of those forty tankers were retrofitted with multi-faucets, so to speak, in order to keep the people who were in need of water from waiting in long lines and to expedite the water distribution process. Those faucets that were mounted on the trucks ran from probably twelve faucets to twenty-one faucets on each individual tanker.

"From the private sector, certainly we commend those people also. Coca Cola,

Pepsi Cola, Miller Brewery, Stroh's, Iron City, Anheuser Busch, and Coor's sent in truckloads of canned and bottled water.

"Through the help of the Health Department and Adult Services, water was distributed to the elderly by way of the Meals on Wheels Program. Several hundred volunteers came forth and were able to distribute water to those people who were not able to go out and get it on their own.

"One other situation arose in the course of the operation. I think it deserves a comment. Some of the institutions in the area such as hospitals, nursing homes, and elderly citizens' complexes were affected. We had a team, a composite crew of plumbers and steamfitters, go out and inspect their heating systems to determine what types of heating systems they had and, in the event they were ordered shut down, what could we do. The last thing we wanted to do was to evacuate these people.

"They made a cursory inspection and they came back and indicated that, should the order go out, that these hot water heating systems would have to be turned off, but we could retrofit the system using tanker trucks to keep their supply of heat going for them because, as you recall, the temperatures were well into the zero area.

Russ Chiodo, Coordinator, Beaver County Emergency Management Agency: "We got called Saturday on the 2nd of January that there was a little spill from a hundred thousand gallons to three million. It really put me on the alert. But we had the luxury in Beaver County of knowing what was coming down the river and we had a lot of time; Allegheny County didn't.

"On Monday we attended the meeting in Allegheny County conducted by PEMA. We returned to Beaver County and informed the commissioners and agencies of the situation. (In addition to the fire meeting, LaFleur had called a meeting of emergency management and water companies on Monday.)

"We had two affected areas of people that had to shut down their water supply, which was Beaver and Midland. Midland has their direct intake right out of the river. They were fifteen feet below level and they had to shut down completely. I had a problem because they supply water to our geriatric center which houses seven hundred fifty elderly people and our main hospital. I think there were five hundred fifty patients in our hospital. Those were the areas that I was really concerned about. How was I going to get water to them. When water buffalos are used to provide emergency water they must be spotted in indoor locations to avoid freezing."

Joseph Chnupa, Regional Community Environmental Control, DER: "I think one of the key decisions that was made was to get as many of the water buffalos and emergency tankers lined up and staged out here in Pittsburgh as soon as possible."

Joe LaFleur, PEMA: "On Monday afternoon, when evidence began to appear that the region's water supply might be in jeopardy, PEMA, with help of Allegheny County, began looking for alternative water sources. Moving that quickly was an important aid the next day when parts of the system ran dry."

Lt. Col. David Smith, DMA: "This incident is a reminder that when water tankers and trailers are used, water specialists must perform chemical tests for potability before people draw water from them. It is more expeditious for the military to respond quickly to requests relayed through the Emergency Management channels than other channels. We're better able to serve you that way."

Steve Jarvela, EPA: Described masses of data on river flow, oil content and other information obtained in continuous river monitoring, but saw a related problem: "The area where we were the weakest in this response dealt with dissemination of that data. Some information was reported, we had a format, but I think that it was not as clear as it could have been."

Al Brunwaser, Director, Allegheny County Health Department: "I see the need for strengthening and expanding our water plant contingency planning efforts, which should include the industry, government and the academic community." He also urged expansion of a program to interconnect water systems, including storage capacity; and to assess the vulnerability to water shortages on institutions like hospitals and nursing homes.

Joe LaFleur, PEMA: "We were also talking with the Department of Defense about U.S. Army resources in the event that we might have seven hundred thousand people, for a prolonged period of time, without water capability. And that was a real concern.

"Obviously, you can ask milk haulers and everybody else to come in and deliver water, but then we're going to take away their livelihood. Some of these people are coming from thirty, sixty, ninety miles away with these trucks. If we're going to tie them up for a week or two weeks, we're cutting down their ability to earn a living. We're spreading out the economic impact of the disaster. So we were talking about Army Reserves, Air Force Reserves, possible active Army resources for water hauling. I was trying to look at scenarios to see how the situation might play out."

INVESTING IN EMERGENCY MANAGEMENT: Planning—Exercising—Training Payoff

Investing wisely for our personal future, that of our families and our communities is something we hear about often from insurance advertisers in the daily information blitz by multimedia marketers. Investing in a comprehensive emergency management program also pays off. As case studies highlight, planning, researching, training and exercising save lives and protect property because they routinize and therefore hasten actions during periods of emergency. During critical emergency stress periods decision makers and on-scene responders must think clearly and respond with all senses functioning. Coupled with the foregoing must be hard earned knowledge gained through hours of study.

The field of emergency management today, more than ever before in its history, has a body of knowledge which has become a basis for quickly restoring lifestyles and economic vitality of our citizens and communities. This body of knowledge has begun to develop a profile and critical mass at the Emergency Management Institute (EMI) of the Federal Emergency Management Agency (FEMA) Learning Resources Center (LRC) both located on the grounds of the National Emergency Training Center (NETC) in Emmitsburg, MD. From 1984-86 the number of case studies completed, those on which development was begun, and those used by lecturers in EMI, expanded dramatically.

One of the reasons the response to the spill went as well as it did under those most difficult situational characteristics as earlier outlined was because of the ongoing vigilance, pure persistence, and deep personal commitment of many people. All current, future and past public officials should take an oath on budgetary and organizational commitment and doing their emergency management homework in advance. They may sleep better knowing that if a known or unknown emergency event unfolds, that whatever happens it is going better than it might have with no preparation.

Mr. Joseph Dinkel, Assistant Executive Director and Plant Manager, West View Water Authority: "Our notification occurred, as Mr. Brunwasser said, somewhere between 8:00 and 10:00 on Saturday night. I have to tell you that it was met with a little bit of skepticism.

"We're normally informed when there's a one or two-gallon spill at X milepoint on whatever river. It was no big thing at the moment. But approximately fifteen minutes later, when it was reaffirmed, it became a full blown catastrophe, and a lot of pandemonium set in. All of our emergency operating procedures at that point came into focus, and we were damn glad that we took the time to develop a highly documented emergency operating procedure.

"And I'm happy to inform you that it did work and it worked well. One thing we did find. We have tried to design that emergency operating procedure based on worst case scenarios, but we never thought a worst case scenario would be the volume of oil that was coming down that river. We now add an extra twenty percent on for future planning.

"Much 'tado' has been made on the opinion of some that the Environmental Protection Agency was ill-prepared to react to a catastrophe of this dimension. Others have proclaimed that the U.S. Coast Guard was too slow to react to contain the spill. Both of these groups, in my opinion, and especially the EPA, performed magnificently against insurmountable odds.

"What one must consider in evaluating the situation is that it is without equal precedence. The Allegheny County Health Department, DER, EPA, West View, West Penn, and we were meeting in the wee hours of the morning. I can't tell you that they were rosy meetings, that everything was business as usual. There were some heated discussions, but it was all with the consumer in mind, and everyone's interests were well protected.

"Someone alluded a while ago about the Coast Guard scenario. Interestingly enough, about two years ago, the Coast Guard conducted an exercise based upon the fact that a chlorine barge sunk in the Monongahela River with one million gallons of chlorine. Interestingly enough, the ramifications of that spill closely paralleled exactly what took place with the oil spill. Therefore, everyone that participated in that scenario was the benefactor of their planning.

"Pragmatic thinking on the part of our Board and planners in providing contingency connections with the City of Pittsburgh and redundant sources of supplies in our groundwater wells, combined with the efforts of our consumers and mandatory conservation efforts put in place by the Governor, enabled us to get by this crisis with no loss of water to any consumers."

Marsh Johnson, PEMA: "Since there were no emergency plans designed specifically for the January 2 incident, the state used other disaster plans, like nuclear facility plans, to prepare to handle a potential mass evacuation."

McClellan, DER: "Communications need to be improved because we had different communication networks going on. How we communicate among agencies, federal and state, in these kinds of circumstances and with whom, at what levels, is critical."

Joe McCary, FEMA: "Various agencies (in the state EOC) whether they were state, federal, local or county, had their own system of communicating. They also had their own system for providing a situation report, and their own definition of technical terms, and nomenclature. Case in point, if we're talking about capacity of a water treatment plant, if we say they're fifty percent capacity, what does that fifty percent mean? Does that mean fifty percent for January 4th, when maybe usage would be that high; is that fifty percent of the time for an alternate supply of water, or is that fifty percent of the capacity of the plant totally? I just bring that point out to try to emphasize the type of problems that occur inside EOC and the type of problems that you would have in a coordinating function."

Johnson, PEMA: "Allegheny County was unable to feed information directly to PEMA's automated system in Harrisburg. A PEMA secretary with a portable computer terminal was sent to Pittsburgh to provide this capability."

Bob Yeager, Allegheny County: "Trying to keep up with logs while answering phones in EOC is not possible in real time. Two or three days later, yes, it's easy to do."

Cmdr. Miklaucic, USCG: "The single greatest advantage we had was our working relationship with the other involved agencies which has developed in our annual drills."

Tom Voltagio, EPA: "...I'd like to pick up on what Commander Miklaucic mentioned in his response, and that was the need for more participation at response drills and simulation. If we can get people higher up in the organizations to understand what's happening here, we might be able to alleviate the misunderstandings or the misconceptions that might occur if an actual release occurs. I am going to be asking the Regional Response Team to do more outreach to the higher levels of state and local government so they can understand that, as well. That's something I want to take up in this critique."

Joe LaFleur, PEMA: "Then the difficulty of the intergovernmental multi-agency impact in the overlay of the political impact comes into play. And as those of you know, who try to get the attention of your boss periodically about emergency preparedness and emergency plans, the two of you ought to sit down and decide which scheme we're going to operate under. Well—that can be difficult.

"And when you have an event of this magnitude, obviously the Governor becomes concerned and it is a very high political profile; it is a public safety issue. If you look at all the statutes, I think you will find that, not the President, not the Chairman of the County Commissioners or Mayor, but the person most singly charged in law after law is the Governor.

"The other difficulties that we had, they are simply of government, if you will. We needed to deal with the political profile, the public safety profile, and the coordination issue.

"So we were educating as well at the state level. In our Governor's office, in our Governor's staff, in our Lt. Governor's staff, in trying to get the attention of the elected-appointed officials to understand that these relationships have to be developed in advance. You have got to have the money in the budget for communications equipment, whatever it might be. That's got to be there every year. You can't suddenly buy a whole bunch of it now and think you're going to be prepared. That event is over."

Site Control—Spill Cleanup

The spill cleanup knowledge of the EPA, USCG, the contractors used by Ashland, DER, and Ashland itself speeded the spill control and cleanup effort. Over 2 million gallons of oil were recovered at the site and through river operations. DER estimated that 11,000 fish and 2,000 waterfowl died on the Monongahela and the Ohio Rivers. Walleye, muskellunge, muskrats, ducks and other river birds were affected. Over 100,000 gallons of oil were recovered from the river. Ice, current, and temperature conditions made work miserable on the river. Hydraulic equipment used in the skimming and suction operations broke down due to the extreme conditions of use. Every drop of oil that was stopped from going into the river or recovered from the river reduced the impact on Pennsylvanians, the citizens of Ohio, Kentucky, Indiana, all of the respective response forces, and the river wildlife. In the words of the old cliche, "every little bit helped".

The national and regional experience of federal officials was certainly beneficial. This is where the federal role is proper and most necessary. They knew what approaches to use. EPA assigned Jerry Sasseen of Wheeling, WVA, one of their experienced, on-scene coordinators, to coordinate spill cleanup and interaction with Ashland. LaFleur agreed to this role of EPA as long as DER was able to monitor all events, have personnel at any locations desired, and have access to all information. LaFleur kept in close contact with DER Regional Director Fabian to monitor the working relationship.

Sasseen also requested a PEMA presence at the site to coordinate state interaction and speak for LaFleur. With all the aspects of site security, tank collapse investigators, threats from unknown callers directed at site workers, need for state police liaison, etc. Sasseen felt PEMA should coordinate state activities. LaFleur chose Josef Jordan, deputy director of Plans and Preparedness from PEMA, Harrisburg, for this task. Jordan had been working the night shift in the state EOC. Jordan was instructed to work with Sasseen and his staff to tie up loose ends with DER and orchestrate necessary actions. Computer to computer contact via phone lines and modems was established and message logs were communicated daily. Key DER staff on-site included Sam Harper, from DER's regional office in Pittsburgh. Within 48 hours, cellular telephones and added hardwire phones were in use at the site by multiple agencies.

LaFleur was concerned with the discussion about construction negligence and had requested state police security on the tank site until a permanent tank collapse review task force chairman was named. As each shift changed, Jordan met with the new officers and briefed them. On January 9th, Keith Welks, DER Chief Legal Counsel, was named to chair the task force. LaFleur briefed Welks via phone that evening on status at the site. On Saturday the 10th, LaFleur and Welks met in the state EOC to further discuss key contacts and issues. LaFleur provided Welks a copy of the video tape the state Commonwealth Media Services made of the site. This included close ups and perspective shots. LaFleur then arranged, at Welks' request, for "crime scene" shots to be taken on the site and from the air by the state police. Jordan was released from his role by LaFleur at the end of the week, after Welks agreed with Ashland officials that tank wreckage could be moved.

Terry Fabian, DER: "We did not realize the potential for the emulsification of the oil until we started getting reports that there was, in fact, oil down in the water column at lower depth in the river and not just on the surface."

Terry Fabian, DER, and Tom Voltaggio, EPA speaking at the critique agreed that, Ashland Oil's assistance on site by the second and third day in getting up the spilled oil in the diked areas and removing it sped up efforts to reopen the nearby water systems.

Mark McClellan, Deputy Secretary, DER: "I think what we realized very quickly, under our responsibilities of the spill, was the lack of planning ahead of time by all of us in terms of establishing clear procedures and protocol and understanding of the delineation of the roles and responsibilities of all different levels of government, not just federal and state which we're discussing here.

"The Department really has multiple responsibilities in a spill like this case, dealing first with emergency response ourselves to the spill: our regulatory and investigative responsibilities under the Act, which may have been violated; our concern as well for regulating water suppliers and the need to notify them. Our responsibilities for overseeing any cleanup and assessing the environmental impact, both immediate and long term. Our continuing cleanup and monitoring of the problem to insure that it's not a recurring or continuing problem as a result of this spill.

"It was very clear to me from the beginning when I arrived on the site early that morning the lack of preunderstanding by all of us of the intergovernmental roles. We had somewhat of a unique circumstance here compared to other counties in that not only did we have EPA, Coast Guard at the federal level, and DER at the state level, but because we delegated our regulatory responsibility to the county for water supplies and other programs, we had another, if you will, regulatory agency with some authority in the county as well, not to mention the local emergency office and its independent authority, as well as PEMA itself.

"I recognized clearly that although within the state we clearly understood our roles and responsibilities and at the local level they clearly understood theirs and the federal relationship between the Coast Guard and the EPA was well understood, the over-arching agreement among the local, state and federal was simply not there. This resulted at the beginning in the possibility of delay in response, some confusion, some failing to understand right from the beginning who was responsible for what; who had the authority to do what; who could act independently; who had to be consulted before we could act; who was responsible and who had the authority to order the water supply to close or to open up, or to direct them to do anything.

"I think one of our principal concerns which we had expressed both in Congress before a House Committee and to the federal government, as well, is the limitations we feel that the 311K Fund and the National Contingency Plan placed on the Coast Guard and EPA which, in our judgment, prevented what we feel would be a quicker and more responsive action by the federal government.

"Their limitations, we believe, and we think they're either procedural or in the Act itself, limit the ability of the federal government to respond quickly without meeting and discussion and protocol or discussions with other agencies. While oil can flow down a river, we felt the federal government on-site coordinators and the Coast Guard were somewhat limited in their ability to act. And I think they acted probably as quickly as they could under their understanding of that protocol.

"Protocol, I think, needs to be set up for those kinds of things now rather than after the fact as we tried to do it in this spill. Overall, I think the intergovernmental relationship in this emergency worked well and after the first twenty-four hours, I think everything worked well, splendidly together. And as a result, a lot of problems did not occur that could have. Nevertheless, the potential for that remains unless we resolve these now."

PUBLIC INFORMATION AND MEDIA COOPERATION

The public was able to help lessen the impact of this disaster upon itself through water conservation. The potential for the public to take action to get out of harm's way in technological events and in slower developing natural disasters has always been evident to emergency managers. Motivating the public to take preparatory action until they perceive the actual threat is very difficult. Marketers say the public is best motivated by "free" enticements and discounts. It is hard to develop programs with government funds to use these two criteria for motivating the public to prepare itself. Prior to most events there isn't enough preparedness time to act in a controlled fashion and not be trampled by the evacuating hordes.

Emergency managers, those newly emerging professionals looking at the wide range of hazards we are susceptible to and the multiplicity of functions to be performed, along with multi-agency performance and cooperation required for response, know that the public reaction/response to warning and direction is usually unmanageable and often unpredictable. If an emergency management system is in place and the personnel who perform the functions can do their job, be flexible and can function without ego turf battles, then the operation can be a success.

LaFleur assigned Comey to head up emergency public information based on the Governor's decision to put PEMA in charge of the state response. Immediately, Comey instituted the Commonwealth's Emergency Public Information System which is an annex of the state Emergency Operations Plan approved by the Governor. Under this annex, affected agencies' press personnel and others unaffected, but required to help meet necessary manpower requirements due to magnitude are assembled either at the state EOC, the disaster site or in both locations. In this incident, Comey went to Pittsburgh where the national and international media were assembled and D'Arcy Charney, Comey's assistant, stayed in Harrisburg to monitor events. Assigned to assist Comey were six state and county press secretaries. This group developed press releases, organized interviews and press conferences, developed copy for flyers on water distribution locations, and, importantly, also acted as a news/information gathering entity for LaFleur and other officials. The ability of this group to establish good working relationships with the media and to aid in the back and forth flow of information was critical to the success of the event in that a major water shortage was averted through conservation.

Two press briefings each day were conducted by state-county officials at the county operations center. EPA and USCG maintained their press briefings at the site. Sometimes LaFleur and Fabian attended the federal press briefings to supplement information. The distance between the two operational locations made it difficult for the principles such as Bob Kroner, Allegheny County Emergency Management Coordinator, Health Department officials, other county department heads, LaFleur, Fabian and state personnel to carry out their duties and be at three official press briefings daily besides do special TV events. LaFleur used PEMA Western Area Director Manclark, his deputy for operations, and Comey to conduct other interviews and briefings.

The participants in the critique were enthusiastic about the very supportive role of the media during the emergency. For their part, the media said this support was possible because of the quantity and quality of information supplied by government spokesmen. A suggestion was made to centralize the press center in the future to improve information flow, make coverage easier and avoid possible contradictions from government sources.

John Comey, Press Secretary, PEMA: "I have never seen the kind of commitment by the members of the working press in your community to the residents of this area. It is extraordinary for the print, the radio and especially the TV people to provide the continued public service and public information material over a two-week period. It's most unusual."

Steve Joyce, KDKA-TV: "The primary purpose or one of the primary purposes that I'm here for is to let you know that we are your friends, we're on your side in this. Many times the media is perceived as the enemy and many times you have been ambushed by a maze of reporters when you first arrive on a scene before some of the information is available to you, but I do thank you for those kind words earlier.

"The reason I believe that we are invited here is that you have seen the effectiveness of how we've handled this. We want you to understand that we perceive ourselves in a situation like this as being a direct extension of you, of the government emergency management agencies. And, I want to make it clear that during the initial phase of an emergency, whether it be a tanker spill or a derailment or spill as we have in this incident, that our primary objective during that initial phase is to disseminate the information as rapidly as possible and as accurately as humanly possible when the situation requires the cooperation of our viewers and of your citizens.

"We're here to serve the same people as you are, and as I said earlier, we're your friends. We're not the bad guys. We're not out to ambush anyone. Tomorrow, we'll go after the people that are responsible and we'll do our head hunting, so to speak, after our viewers and your citizens are safe and their health is secure.

"When you think back to those early days and that long night of January 2nd, you know that some nightmares had to go along with those memories, and we too have nightmares. Your nightmares may be the ambush that you faced by those reporters, but I want to give you an idea of some of the mobilization that we do and some of what's required on our behalf to effectively get that message across to the people in the county.

"We also have to wake up reporters and photographers and editors and technicians and writers and production people and anchors, and those people all have to get moving. We too have to mobilize equipment, such as microwave trucks, satellite trucks and the portable equipment that goes along with producing newscasts or a special cut-in from a remote area such as Floreffe.

"I was born and raised in Allegheny County and that was a new one, just where Floreffe was. We also have the responsibility of completely changing, on a moment's notice, a very complex program schedule blowing out commercial breaks that cost the station lots and lots of money and putting together half-hour specials at 6:30 in the morning and 11:30 at night without much notice.

"So there's much change and there's much preparation that we also have to put our people through. The reason I mention this is so that when you are ambushed on those occasions, you realize that we're stretched and we're stressed also, but the bottom line is we're trying to do the best job we can to take care of the health and safety of your citizens.

"When you're confronted by those hungry newsies on those occasions, and I want to make it clear that we do feed our people (we run food out from Denny's), we don't primarily serve them diets of parts of bodies of former emergency management people. When you do get those calls, and when you're confronted by that, keep in mind that every occasion when you're confronted by a reporter or if 'Betsy's on the phone' and that phone doesn't stop ringing, keep in mind that every confrontation you have, we are hammered on those phones by hundreds if not thousands of people that are understandably concerned about their safety and their health.

"Before I get into some of the recommendations or criticisms, as you may call them, I also want to mention some accommodations and some of the things that made doing our job very easy and very effective.

"I want to thank the people from West Penn Water that hired a special media consultant. Whatever time of day or night we made a request, we had someone from West Penn Water that was available. Whether it was the 6:30 a.m. breaks that we had and the half-hour shows or the shows at midnight, they were very gracious in providing us someone that could bring us the facts as they were available at the moment.

"We want to thank all of the state and county agencies for the regular scheduled news briefings so that we didn't have to probe people excessively, that people and officials were made available to us.

"This included specifically during the last weekend of the problem, the fact that water supplies at West Penn Water had reached what they termed 'safe levels,' actually higher than their reserve prior to the incident, and yet the state insisted that the Governor's mandate was still in effect. We had a high-placed county official who said that was not the case and no one should lose any more work or be inconvenienced any further because of the cooperation they had given us thus far. That puts us in a pinch, folks. All we had to do then is to say one government agency has said this and another says this, and really no one is put in a desirable position, in a desirable limelight. And, that's not an attractive position for us to be in so we would ask that you cooperate as fully possible as you can in the future.

"While we're on confusion, one minor matter, we must have had a thousand calls about 'what is conservation? Can I do my dishes, can I do my laundry, how many gallons am I allowed to spend?' And, if you're able to be specific with us, we can likewise be specific with people that are very good-hearted in their intentions and really wish to cooperate to help us out.

"We had a problem with the news centers as well. If we had incidents at 11:00 a.m., we had a news conference in Floreffe at 11:00 a.m. We had a 10 a.m. news conference at county emergency management and at 10:30 we had a news conference in Mount Lebanon which was West Penn Water. All very important information, but very difficult to reach all at the same time."

Maddy Ross, Managing Editor, Pittsburgh Press: "As I sat here all day today, I was struck by the fact how often your problems were identical to our problems in this disaster. This was an unusual disaster for us as well. The disasters that we're really good at are the ones that have short moments of chaos followed by long periods of cleanup.

"In this disaster, we had the chaos and we had the cleanup but then the chaos continued floating down the river and striking community after community. Each time it hit a new community, we started all over again. Like in your situation, that strained our resources as well. We had people like you working literally around the clock and, in fact, the reporter who was doing our main story was working the story 37 hours straight by the time he finished his story, which I think accounts for the way that story read. It was cram-packed with information but it won't win the Nobel Prize for Literature.

"But the point is, we were there with you working at the same hectic pace. Just as you had hordes of reporters descending on you for information, we also had hordes of the public descending on us.

"In the beginning stages of this disaster, we could not give enough information to the public, they wanted to know every last detail including 'which is my water department?' And, just as you were dealing with the multiple agencies, the PEMA's, FEMA's, DER, EPA all the other letters that you can think of, just as you were trying to figure out who was doing what, we were trying to figure out who was doing what. We began to make mental charts putting each of you in this federal category, this state category, this is the local category, and then when we needed reaction, we'd take one from column A and one from column B and one from column C. But, it was very difficult for us to know to which of you to turn for specific pieces of information. It was confusing.

"But, the other thing that was unusual about this disaster was that in preparing for this meeting, I could not get our staff to complain about you. That's notable because our staff usually complains about you, but in this case, they were delighted with the quantity of information that flowed during this disaster.

"And if it's true, if we deserve the kind words earlier that the media did a better-than-usual job on this disaster, I'm going to throw it right back at you and say the reason for that is the quantity and the quality of the information that you freely supplied this time. Rarely did we call anybody that we got the reaction, 'you'll have to talk to the PR man, I'm not allowed to talk to the media, or you have to talk to my boss.' The kind of usual runaround that the media gets on certain situations was not there this time.

"We preach the sermon all the time to the public sector saying that the more open you are, the better the information is that you will read. I know that ordinarily sounds self-serving, but I think this is a case in point that proves the more information you give us, the more accurate the report is going to be. And, it's something to keep in mind for the future.

"In most stories, you know, we can't get somebody to talk to us. In this story, we couldn't keep you quiet. We had more information than we could handle in most cases. You would call us with information. Sometimes, we didn't even have to call you. We had more press conferences than we could get to, as was mentioned, three simultaneously, for example. A lot of you went way beyond the call of duty to help us out.

"The Coast Guard: when we couldn't get through on the telephone lines, because there was so much action on the phone lines, we had taken to calling the Coast Guard office downtown and asking them to radio people out at the Floreffe and ask the questions for us. And, they were kind enough to do that.

"And, when we wanted to put Reporter Mary Pat Flaherty on a barge to follow the slick down the river, you accommodated us and she went out on the barge in sub-freezing temperatures. As she says, she hasn't had that much fun since we sent her to Haiti, and she got caught in the revolution.

"There are three things really that were problems that could be fixed. One were the phone lines, as you already know. Don't underestimate the interest of the news media. If you estimate that there will be X number of reporters wanting those phones you can easily double that and you'll still be short. If you have to have multiple command centers because that's the way the disaster is handled best, we can understand that, but don't have your press conferences simultaneously. That's a problem.

"And, the last problem—well, the third problem is that the quality of the information on Sunday, the day after the disaster, was a problem at the site of Ashland. When everyone descended upon that site, they went behind closed gates and locked the media outside in sub-freezing temperatures and not much information was flowing at that moment.

"Keep in mind that when you know the least is when we need you most. So even though there wasn't a lot of substantive information at that point, we needed to be in communication with you to even tell us what, at least, was being pursued.

"And, the last problem was our own, the Pittsburgh Press'. We need to develop a better disaster plan for oil spills because we found ourselves duplicating our efforts and bothering you twice as often because two reporters were asking you

Dick Winters, Allegheny County EMA: "EOC swamped by hundreds of calls per hour from public. Not enough phones or people to answer them."

LaFleur, PEMA: "You have got to have the money in the budget for communications equipment. That's got to be there every year."

James Manclark, PEMA: "The agency solved the problem of communicating with its key people in the field by renting several cellular phones. Other agencies later followed suit, with similar good results."

Bob Yeager, Allegheny County: "The phone company has already begun a survey of the EOC to design a better phone system, which we realize is going to sit there stagnant 90% of the time."

James Manclark, PEMA: "Joe asked me to say a few words regarding state and county communication on a day-to-day basis and how it worked into this affair. Insofar as PEMA goes, liaison with the county is part of our daily routine, and it's our mandate to work with all the counties. In my particular case, it's 24 western area counties. We work with them with respect to mitigation programs, preparedness, response and recovery programs, and we do this through providing technical assistance in terms of planning. We have training programs that were mentioned, quarterly training programs. We implement an exercise program, which is essentially an exercise in the Emergency Operations Center in response to a scenario which throws a lot of disasters their way. On a day-to-day basis, we're trying to establish a good working relationship with all the counties so we can get the daily business done. Also when we get into a situation, like this disaster, it makes it a little easier on everyone to have had that good working relationship and come in and get the job done.

"In an emergency, the PEMA role doesn't necessarily change, but then we take on the added role of state coordinating officer. Joe, in this particular case, was the state coordinating officer, and I was his deputy. And, what we're doing there is trying to pull together all of the state agencies and make effective and efficient use of all the resources available. And, often time, it's not easy for other state agencies to come under this umbrella because, let's face it, you know, they have their own chain of command from day to day, and they normally report through a whole different channel so it takes a lot of adjustment on everybody's part to adapt to the EOC concept. I would hope that the response that we had in this particular drill reflects the working relationship that we at PEMA have had with Allegheny County all along, and we do appreciate all of those efforts."

EMERGENCY DECLARATION

During the first forty-eight hours the issue of some type of disaster declaration or a federal designation of emergency was discussed between Pennsylvania officials and various levels of EPA, FEMA and Ashland personnel. As already noted in the comments of Mark McClellan of DER, Pennsylvania officials believed that by "federalizing" the spill, questions about the financial viability and commitment of Ashland to pay for the spill would not have been necessary. Ashland officials, up to and including the Chairmen of the Board, provided assurances to EPA and demonstrated to the satisfaction of EPA that they could and were going to manage the spill. Governor Casey conducted telephone calls with Governors of surrounding states and EPA Administrator Lee Thomas discussing the multi-state aspects and environmental damage of the spill and the need for federalization in case Ashland failed to maintain its commitment. EPA remained firm in their position of not to "federalize" the spill unless Ashland changed its posture. The difference of opinion between Pennsylvania and EPA officials remained.

By January 6th LaFleur was becoming concerned that without the EPA designation federal attention to the long term cleanup responsibilities of Ashland would wane. Second, Commonwealth authorities would have to deal with Ashland for months or years to come on cost recovery for local, county, state and volunteer agencies. A declaration of some type would maintain some clout. Further, LaFleur was concerned that special technologists and equipment could be needed from the federal government at some future time. Without the weight of some kind of federal action, beyond the emergency powers of the Governor's declaration, there might not be enough collective force to keep Ashland's or the federal government's attention. The Governor's declaration would eventually expire, but the need for long term action in Pittsburgh could not be allowed to sink to the bureaucratic red tape level in the event future problems, such as contaminated wells or auxiliary chemical dumping, became an issue.

Joe LaFleur, PEMA: "In this case, I think the Governor was, at first, very concerned about the issue of EPA and Coast Guard: who was going to be in charge; who was going to be 'the' federal agency; what did this mean about federalization of the spill; what was the cost recovery process going to be.

"There is a level of comfortability which comes in dealing on a government to government basis during an incident of this magnitude. Certainly there was contact right from the very highest levels of the Ashland Corporation and that kind of assurance I know was appreciated by Governor Casey. But nonetheless, when you're dealing with unfamiliar territory like this, there is a great desire to have some kind of an official ongoing process.

"We then looked at the issue of a designation of—for an emergency through the Federal Emergency Management Agency. Understand that at no time did the state ask for cash, or give us the money to manage this situation. That was not the idea. We wanted technical assistance. We wanted the ability to get specialized equipment."

LaFleur had the state EOC staff discuss the issue of a declaration with FEMA's Joe McCary in the state EOC under the provisions of the Federal Disaster Assistance Act of 1974, as amended. LaFleur's request was not for cash but specialized equipment and personnel. McCary responded after seeking counsel with FEMA Washington staff. Pennsylvania staff clearly felt it fell within the provisions of the Act. The FEMA Washington response indicated that they did not support the PEMA position. They also responded that all PEMA had to do was make a request for federal technical expertise and they (PEMA) would get assistance. PEMA EOC staff, at LaFleur's direction, requested engineering help to inspect the remaining tanks at the site. FEMA sought assistance from EPA for the FEMA request. EPA's response was that PEMA would have to pay for the help. This response, plus EPA's arrival at the state EOC five days after the initial request, did not enhance the overall Pennsylvania attitude toward FEMA and EPA decisions during the incident.

Following the discussion with FEMA, LaFleur began discussing the need for some type of a permanent consent order against Ashland by EPA so as to define the future of the government-Ashland relationship. LaFleur asked Steve Wassersug of EPA Region III to review this option. He and Tom Voltaggio of Region III discussed this issue with LaFleur and agreed that this vehicle should be used. LaFleur met with Joe Jordan, PEMA site coordinator and Voltagio on January 12th to confirm the process. LaFleur stated that DER would be the Pennsylvania team member in reviewing and negotiating the agreement.

COST RECOVERY

Cost recovery following technological emergencies, in the United States, has developed several patterns. First, there is the route through the courts. One party sues to establish liability by proving negligence and thereby obtain a court ruling directly assessing damages or setting up a fund with an administrator to review the claims and approve them for payment. An example of this is the Three Mile Island nuclear power plant accident and subsequent rulings for payment of "damages" to local governments. Another possible outcome of the court process is that the charged party does not admit negligence but agrees to set up special funds for future improvements, also an outcome of the TMI incident. A third process is for a court to establish the amount of an award from an existing fund, such as the federally enacted and utility funded Price Anderson fund for nuclear power industry accidents.

In the chemical side of technology emergencies, this pattern of incident payment also exists in federal and state laws in that the "spiller pays" philosophy is written into the past and developing body of the law. Environmental or spill insurance is becoming common place in the chemical industry. Ashland had a $400 million dollar policy which was administered by Crawford and Co. following the event.

In the three months preceding the incident, the state emergecny management agency had begun a major effort to motivate, train and increase the size of county damage assessment teams. Not only was this training for natural disaster incidents but an emphasis was placed on the ability to use these teams and systems for cost recovery outside of the traditional physical damage assessment historical process.

After arriving on the afternoon of January 4th in Pittsburgh, PEMA Director LaFleur requested that the Agency disaster assistance staff be deployed to get an on-scene early look at where the "cost centers" were developing, even though physical damage was minimal. LaFleur's concern was relayed to team leader Karen Critchfield as, "a mission to identify business costs and economic dislocation, identify and record state, county, local government and volunteer agency costs, and to ensure that a process for business and individual claimants to register claims was instituted with Ashland/Crawford".

LaFleur held a meeting with Ashland representatives, county emergency management directors, and staff at 7:00 p.m. on the 11th of Janaury in the state trailer at the spill site. An outline of the process and identification of the principles was agreed upon.

LaFleur, PEMA: "the concern that drove me in the months preceding, when we were beginning our training drive, and the issue at hand here was the potential budget impact on county and local government. This was the first week of the new fiscal year. Some of the responding entities had expended 20% of their annual budget. Their ability to raise taxes was limited. By October they would have to begin limiting services. Obviously that was unacceptable. Second, I knew that if they all filed suit against Ashland the process could take years to resolve—not a good way to solve an impending financial problem. We needed an expeditious positive environment for moving forward quickly. Fortunately Ashland was agreeable.

"The time period for emergency response expense recording was during the emergency phase only. The time covered would be the length of the Governor's emergency declaration. It would not cover long term environmental costs. Those costs would be determined by DER. I wanted the team to start right away so they wouldn't have to reconstruct data and reporting and recording would be uniform. We intended to follow some parts of the federal and state recovery portion of the federal disaster assistance program.

"The cost recovery organizational and administration process was labor intensive. Potential business and government claimants received telephone calls from PEMA the first week inquiring about potential impacts, cost data, and apprising them of potential avenues of action as well as emphasis on the need to conserve water. A follow-up mailing was provided to business claimants on January 20th and to government and volunteer groups on January 21st. The mailing contained key information, assistance telephone numbers, a description of the claims process and a diagram depicting the organizational flow. Key state agencies called in by PEMA to assist claimants included: Department of Insurance, Department of Labor and Industry and the Department of Commerce.

Karen Critchfield, PEMA: "It was decided that teams consisting of the county representatives, a PEMA damage assessment staff person, and one insurance adjuster would visit each county and local government reporting spill-related costs. The process allowed the public sector claimants to present the costs and the supporting documentation directly to Ashland through its insurance adjuster, Crawford and Company, for possible reimbursement. The process also enabled county and state to gather public sector financial impact data. By January 20th, only eighteen days after the accident, five teams were in the field visiting the affected communities. This prompt response was due to the coordination and cooperation by all levels of government.

"The county emergency management coordinators, the Allegheny County Department of Development, and the Council of Governments were instrumental in identification of claimants. I might add the Allegheny County team members were from the Department of Development.

"Public sector claimants from six counties—Allegheny, Beaver, Butler, Fayette, Washington, Westmoreland—were identified and visited, resulting in over seventy claims totalling nearly one million dollars. By close of business February 24th, our on-scene mission, for the most part, had been accomplished. Nearly all claimants had been identified and given the opportunity to file a claim for legitimate spill-related costs through an organized and coordinated cost reporting process.

"The next step was Ashland's offer of reimbursement. Although PEMA monitored the claims process, each public sector claimant was responsible for the final claim review and the signing of any partial or full releases at the time a reimbursement offer was made. The claimant often worked directly with an adjuster. Each claimant was asked to provide PEMA with any additional claim costs as well as actual reimbursement figures. The information will be incorporated into the overall public sector financial impact report.

"In summary, the overall process has helped communities identify spill-related costs, provided an organized method of claiming those documented costs, and has set a precedent for future incidents. The success of the process was measured once payments began to flow into the affected counties, municipalities, and other public sectors for the claimants. As of December 15, 1988, there were 104 state, county, local government and volunteer claimants who had received $1,533,139 of $2,515,710 requested. Of the requested amount, the twelve state agencies had requested reimbursement of $761,730 and received $415,841. The initial emphasis on repayment was to Pittsburgh area agencies."

Richard Thomas, Vice President, Ashland Petroleum: "We have never been involved in anything like this before, of the size that this encompassed, all the way down the river. Some claimants may be involved in the class action suit and, therefore, are not coming forward with claims. I think it's worth noting that of the communities up here that were affected, I have to thank them for not flooding the company with frivolous claims, claims that are nothing more than an effort to see if there's some money in the office downtown. The community here has been wonderful.

"Some people that were evacuated were asked if there was anything we could do, and a lot of the evacuees, who lost possibly a night's sleep because they had to get up in the middle of the night, said, no, that there really wasn't. They did not consider it a problem that would require money to be paid to them, and I think that's something to be said for the community and the people of the Pittsburgh area.

"We asked some claimants for additional information. We have gotten most of that additional information. Crawford is doing a good job helping all of us through this period. Now the question is, not how much are you going to have to pay, that's fine, but when are you going to pay? The process of evaluating these claims is long, they have to be audited to a certain extent, not only by our carrier but by ourselves internally. And, hopefully, Crawford will support us in saying that we will be able to get back to some of the claimants shortly with our opinion on some of those claims. I think it's fair to say that some claims do appear to be a little bit difficult for us to pay but we are considering all claims, anywhere up and down the river. Any questions that any of you have, feel free to call the Pittsburgh office."

Dennis Neary, Allegheny County EMA: "As Karen, and many others have said today, it was an absolutely unprecedented incident. And, a system needs to be set up for cost assessment and recovery, which it was. The key to that whole system was its simplicity.

"Number one, it was the cooperation of Ashland Oil, who opened the door intially, and secondly, it was the organizational system that was set up between PEMA, the county emergency management, and Department of Development and Crawford and Company in order to set up a procedure to make sure that all public agencies, government agencies, fire departments, ambulance services, all agencies who would have expended monies directly associated with the spill, that their costs were reported in a simple way. They were. Teams were set up and visited sites, actually visited locations rather than the individuals coming in. One of the key things that fire companies and so forth commented on, over and over again, was the fact there weren't three, four, five, six pages of forms to fill out. It was a very simple procedure that was set up. "What's your claims, these are in the parameters of the claims and what are your justifications?"

CRITIQUE SUMMARY

Joe LaFleur: "It was very obvious that we averted a disaster, but we had an emergency. If the typical gun is loaded with six bullets, I think we dodged five of them.

"Number one, when the incident took place, and in succeeding days, there were no other tanks that opened up. Other tanks were dented. There was gasoline in those other tanks but none of them opened up. I credit the Ashland people for letting the agencies come on site, the reports that we did, the outside consulting engineers PEMA contracted who did find that there were some welding problems, potentially some fatigue problems in some of the other nearby tanks. If they had been full or they stayed full we could have had a second collapse. I compliment Ashland because they drew down very quickly on those additional tanks. Another night it could have let loose again. We dodged bullet number one.

"The second bullet was that there was a gasoline leak and it didn't ignite. If we'd had fuel oil and gas ignition, we'd have had a heck of a fire. That would have been a real difficult one for the emergency response community. That was bullet number two.

"The third bullet was that we didn't lose the West View Water System. We stayed with fifteen, twenty, twenty-five thousand people without water, or with low pressure, but we didn't go to that next threshold of two hundred thousand plus people without water. In one case, we were within two hours of losing some of those systems. We dodged it again.

"Fourth, we didn't lose the electric power systems after Penn Electric Power said they had to shut down because their intakes were fouling with oil coming in. We knew that the Beaver Valley nuclear power plant only had one reactor up and it was running 100 percent. The Perry nuclear plant in Ohio in the western part of the grid system was totally down. The coal plant up in Shippingport had to go down for maintenance. We didn't lose power because the grid reacted, but there were about four tense hours about which I don't think the public is readily aware, between 11:00 p.m. and 3:00 a.m. until we knew the grid could handle it. Otherwise, we would have had to manage one of the most massive evacuations, certainly since the spontaneous departures at Three Mile Island, and perhaps one of the largest and most difficult ever, because of the temperatures that we had at that time. If you wanted emergency management problems, we would have been into it in a serious, serious way.

"The fifth is that the main water systems held up, and we didn't face a major fire. Under these conditions a break or a major draw-down would have enlarged our area of vulnerability. We had many new interconnects going. The system was at tensile strength, no doubt about it.

"I think the conservation order was a good decision. I know that there were people that had difficulty with it, but I thought one of the water people said it well; that there was a need to maintain the conservation order for a period of time until we knew that that system was going to work. Lt. Governor Singel displayed great courage by keeping the order on the additional 48 hours so the entire system could stabilize. That's a tough job to stand up in the middle of that public controversy, when everybody wants their water turned on and kept on. We had to wait until we were fully in the clear system-wide.

"So, there were a number of bullets, some very major ones that we dodged.

This could have been a much more difficult situation. Usually Murphy's Law is at work, but not this time.

"In doing this report we have tried not to 'Monday-morning quarterback.' There weren't any individuals, I think, who were unfairly credited with doing something, anything, that was certainly not in their control, because we were all so dependent on other people during this situation.

"I do believe that this is a benchmark report. It is developed in the tradition of the Missasaugua, Canada train derailment report. That's why we want it word for word in places so we know where we should go in the future, what we should continue to do, and who we should talk to.

"Emergency management agencies are not finished with this event. Emergency management agencies must be a conscience in this process. Hopefully, not to nag at people but to remind them about the need to put that money for communications in your budget next year. To exercise, to plan and mitigate. We can't let the bureaucracy forget. Don't let the emergency management priority slip too low. There needs to be a conscience to this process. There are many human frailties that contribute to emergencies. One of the most serious of those will be that we don't have time, its not high on our agenda, we forgot what it was like during that emergency period of time.

"We do need to give this event introspective and continuing thought. It's very clear. The complexity of a technological event and the information transfer difficulty is overwhelming as opposed to other types of disasters and emergencies. If there is anything that could be akin to serious wartime events and a threat to this country, at least in terms of the information transfer process, this was that kind of event.

"I think that there's no doubt that the Ashland Company, by the way they conducted themselves in everything heard since the event, shows that they have learned well in the corporate sense, from what took place at Three Mile Island, Chernobyl, Tylenol I, Tylenol II, the shuttle disaster, and Bophal. There's an accumulation of lessons in that process. In each of these disasters, and now the Ashland case, human error created the incident.

"Don't forget those three R's that we talk about in emergency management. The responsibility that we all have, the public depends on us. Readiness and response when the incident does happen. Vigilance is needed. You are all very important people in this process. The most important thing you can do now besides the readiness effort is to give of yourselves. Go out and talk to other people about emergency preparedness. Be part of the information transfer process and go and teach them this event so that we all do better in the future.

"You won't know where you saved a life, you won't know where you protected some property, but there's no doubt that it can happen. It's worth your effort. You can't always figure out when or where your efforts had an impact, but rest assured you've contributed to the welfare of your fellow man."

Management of Hazardous Materials and Wastes: Treatment, Minimization and Environmental Impacts. Edited by S. K. Majumdar, E. W. Miller and R. F. Schmalz. © 1989, The Pennsylvania Academy of Science.

Appendix

CLEANING UP WITH BIOTECHNOLOGY*

JOHN DOUGLAS
EPRI Science Writer

The emerging science of genetic ecology is being pursued for in-situ treatment of toxic wastes. Using established principles of molecular biology for breaking down organic substances, the new technique avoids many of the technical and policy hurdles of genetic engineering.

The destructive capability of microorganisms can be startling. Virtually any organic material, from wood and coal to pesticides and PCBs, can be decomposed under the right conditions by a combination of bacteria and fungi. In any given soil sample, at least a few microorganisms probably have the particular genes necessary to control the decomposition of some specific compound, and nature has provided a variety of ways to multiply these genes and enhance their activity when the opportunity arises.

Some preliminary attempts have already been made to harness these natural processes as a means of destroying toxic substances in the ground. Contaminated soil may be plowed and sprayed with nutrients, for example, to stimulate the growth of bacteria that destroy a certain toxin. The results of such cleanup efforts have been rather unpredictable, however, since both the genetic variables and the environmental conditions that influence them are not well understood.

Now efforts are under way to study in more detail how environmental processes are controlled at the genetic level in microorganisms, especially bacteria. These pioneering efforts represent the first stage in the creation of a new scientific approach, called genetic ecology. Although still in its infancy, genetic ecology appears to hold great potential for leading to better ways of destroying

*Reprinted with permission from EPRI Journal, Volume 13, No. 6, pp. 15-21, September, 1988.

toxic wastes where they lie, in both land and water. Other potential applications range from enhancing the biological gasification of coal while it is still in the ground to preventing microbially induced corosion of condenser tubing in power plants.

"We're breaking new ground by taking existing concepts of molecular genetics and ecology and applying them to some pressing environmental problems," says Robert Goldstein, program manager in the Environment Division. "If we can enhance bacterial action to destroy toxic organic waste materials, then we can significantly lower cleanup costs. It's not going to be a panacea, but we may be able to lower costs by as much as a factor of 10. Within five years we should be able to provide utilities with practical tools they could use in bioremediation of some compounds at waste sites."

COLLABORATING WITH NATURE

Natural processes that break down complex organic molecules into harmless subunits usually involve many separate reactions, each catalyzed by a specific enzyme. In turn, production of each enzyme in a microorganism is controlled by the activity of a particular gene. The group of genes that controls a specific sequence of enzymes is called an operon, which is turned on or off as a unit and may also be duplicated or transferred from one microorganism to another as a unit.

Learning how to enhance the natural duplication and transfer of operons that control the degradation of specific toxic materials is the ultimate aim of EPRI's present work on genetic ecology. Unlike genetic engineering, which involves the artificial manipulation of operons in a laboratory to create genetically novel microorganisms, genetic ecology seeks to understand and eventually control the behavior of microbial operons in a natural setting in response to environmental changes.

"The point is that we're going to use organisms already present at a site," says Betty Olson, a professor at the University of California, Irvine, who is the principal researcher in EPRI's genetic ecology work. "These organisms have a better chance of surviving than those you might create in the lab and then introduce to a site. In addition, I believe we'll encounter fewer regulatory problems with this type of experiment than if we tried to engineer new organisms."

Olson's general approach begins with determining the genetic potential of a particular site—that is, finding out how common the specific operons needed to break down a particular toxin are among microorganisms already present in that environment. Several microbial species may be involved; but out of a billion bacteria, Olson says, perhaps only a few hundred or thousand may have the desired operons. Further, not all of the operons present may be turned on.

The next step is to find ways of causing the bacteria to make more copies of the required operons and to turn more of them on. Samples of soil from the site are taken to a laboratory and subjected to a variety of experimental conditions, such as the addition of different nutrients, changes in temperature, and oxygen enrichment—a process called microcosm studies. After varying periods of time, the effects of these changes on gene amplification and activity and on toxin breakdown are assessed.

From the information gathered during these microcosm studies, a model is developed to predict the effect of producing similar environmental changes at the site. Actual field trials using these environmental modification techniques at the site will be conducted during the third year of EPRI's planned research. Three projects are getting under way this year. Two focus on degradation of specific hydrocarbons at town gas sites, and the other involves control of organic mercury compounds in lakes and streams.

TOWN GAS SITES

During the Gaslight Era, towns all over the United States manufactured their own gas for local consumption. In the East this gas was usually made by heating coal with steam; on the West Coast the thermal cracking of petroleum was favored. Both processes left behind a thick residue containing many potentially toxic materials. Most of these materials were simply buried—following the accepted practice of the time—and as electricity and natural gas became more popular during the early decades of this century, many of the sites where so-called town gas had been produced became the property of a new generation of utilities.

As more has been learned about the environmental and public health hazards that may be associated with such wastes, present-day utilities have begun to consider various options for cleaning up the old town gas sites they inherited. Conventional techniques—which usually involve digging up the wastes, followed by incineration or treatment with chemicals—would most likely be very expensive. Current estimates of the costs range from $50 million to $300 million per site, depending on the levels and types of contaminants.

Because techniques developed in genetic ecology may be able to reduce these costs substantially, one EPRI project is focusing on a group of toxic substances causing particular concern at town gas waste dumps—polycyclic aromatic hydrocarbons (PAHs). Although they are well known for their potential toxicity and carcinogenicity, PAHs can be rendered harmless by bacterial action.

The first compound selected for study was naphthalene, a two-ring PAH whose degradation by bacteria is well understood. Two operons—one with six genes, the other with seven—control microbial production of the enzymes that drive the 13 reactions required to transform a molecule of naphthalene into pyruvate, a common, harmless substance naturally found in plant and animal bodies.

Information gathered from genetic experiments related to naphthalene will be used to build a model for stimulating microbial destruction of PAH molecules with as many as five rings. David Gibson, a professor at the University of Iowa, is working on an EPRI project to identify and sequence the genes in operons capable of controlling destruction of heavier, multiple-ring PAHs, which can be the most toxic.

Two town gas sites, one in the East and one in the West, will be chosen for

Fine Distinctions in Biological Cleanup

Bioremediation is the use of microorganisms in the soil to render potentially toxic materials harmless. A very broad approach is usually taken, in which soil is plowed to increase aeration, then sprayed with water containing nutrients. Such treatment encourages microbial growth in general, and little attempt is made to select particular bacteria or genes for stimulation.

Genetic ecology attempts to refine bioremediation by identifying specific natural toxin-destroying genes and stimulating their activity. Soil samples are tested to see what genes are present in native microorganisms. Then environmental factors are modified—first in the laboratory and later in the field—to determine the optimum conditions for gene reproduction and expression.

Genetic engineering creates new microorganisms in the laboratory through the manipulation of genetic material; that is, genes are introduced into microorganisms that do not naturally contain them. This approach appears less promising than genetic ecology for bioremediation because of the likelihood of greater public resistance and the difficulty of keeping laboratory-created microorganisms alive under field conditions.

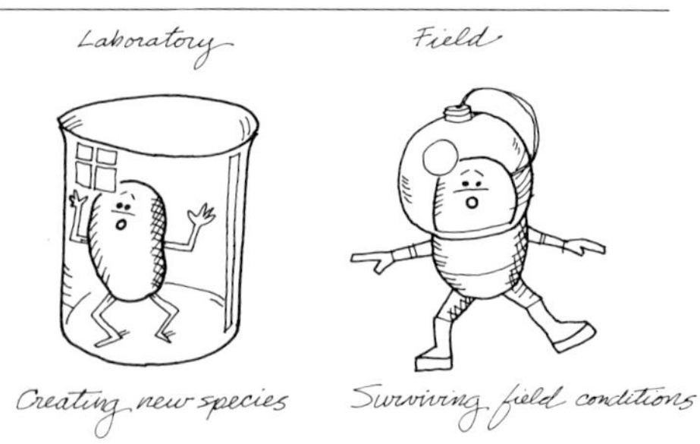

the sampling of naphthalene-destroying genes and the testing of environmental variables that influence their activity. The critical variables at the two sites may well differ, since the growth of soil bacteria is often limited in the East by cold temperature and in the West by lack of water. Other factors to be considered include the lack of specific nutrients—such as nitrogen, phosphorus, or potassium—and the effect of adding more oxygen to soil, perhaps in the form of hydrogen peroxide. The timing of environmental modifications may also be important, Olson says, since bacteria may be more likely to exchange genetic material during periods of freezing and thawing.

MERCURY POLLUTION

The second genetic ecology project is concerned with altering the transport of mercury in aquatic food chains. In some cases of high mercury levels in lakes and streams, the original cause is again quite old. Mercury was mined extensively in California during the Gold Rush, for example, for use in extracting gold from ore. More recently, however, concern has been expressed that mercury now being found in much lower concentrations in remote lakes may be the result of utility activity. Specifically, some scientists argue that some of this mercury comes directly from air pollution caused by burning coal or that its biological availability increased at low pH, possibly as the result of acid rain.

Once in the water, mercury forms a variety of compounds, including methylmercury, which is easily incorporated into the bodies of aquatic animals. Not only is methylmercury highly toxic, it is also difficult for fish to excrete; they tend to accumulate it from the environment. The possibility that people might be poisoned by eating fish caught in waters polluted with methylmercury has become a growing public health issue.

The transformation of metallic mercury to methylmercury and back again is at least partially mediated by bacterial action. Again the genetic control system is understood well enough for researchers to assess the potential for reducing the toxic substance (methylmercury) by changing the environmental conditions that affect bacterial growth and genetic activity. In the case of aquatic systems, pH appears to be a key variable, as does the lack of oxygen that occurs at the bottom of many lakes during summer months. The presence of certain ions, such as calcium and sulfate, may also be important.

This project is being coordinated with other EPRI work that explores the transport and tansformation of mercury in aquatic ecosystems. Project manager for that research is Donald Porcella of the Environment Division, who says, "Whatever Betty can find out will be useful to us as soon as it becomes available. We're not sure which organisms are doing what, or which processes are the most important. Quite aside from the bioremediation aspect of genetic ecology, the information being generated can help us better understand a very complex process."

Sites include a reservoir near an old mercury mine in California, a pond and stream near Oak Ridge National Laboratory (ORNL) in Tennessee, and several remote lakes in Wisconsin. In the first two cases, bottom sediments appear to be a significant repository of mercury, with 1-13 ppm mercury being found in sediments at the California reservoir and up to 600 ppm at the Tennessee pond. (Mercury was used extensively at ORNL during World War II to enrich plutonium.) What relationship this sediment-bound mercury has to the level of methylmercury in the food chain will be a particularly important question for both sets of EPRI researchers.

Primer on Genetics

The hereditary information of every cell in all organisms is carried by molecules of deoxyribonucleic acid (DNA). These molecules are composed of two long chains containing four kinds of building blocks, called nucleotides. The arrangement of these nucleotides represents a code, which tells the cell how to construct proteins. Two key steps in the building of a protein are the transcription of the genetic code from DNA to RNA, or ribonucleic acid, and then its translation into a sequence of polypeptides that make up the protein molecule.

Enzymes are large protein molecules that act as catalysts for most of the chemical reactions that take place in living organisms. For example, some bacteria can break down naphthalene to salicylate through a process of six reactions, each catalyzed by a specific enzyme. Similarly, salicylate can be broken down to pyruvate by another seven enzymatic reactions.

The sequence of enzymes involved in the degradation of naphthalene—and the corresponding sequence of genes that control their formation—is well known. The genes, labeled nah A, nah B, and so forth, occur in two discrete sets. The first set is called the nah 1 operon; the second is called the nah 2 operon. For the breakdown of many other organic materials by bacteria, such detailed gene sequences have not yet been identified.

In order to control the breakdown of naphthalene in contaminated soil, genetic ecologists must first determine the genetic potential of bacteria in the soil; that is, they must measure how frequently nah 1 and nah 2 operons occur in the natural bacterial population. This procedure involves matching DNA from the bacteria to radioactively labeled probes containing the desired operons. Next, an RNA probe is used to measure the rate of genetic expression—that is, to determine to what extent the bacteria are actually using their nah 1 and nah 2 operons to make enzymes. Once these two factors, the genetic potential and the rate of genetic expression, have been determined for a degradation process at a particular site, genetic ecologists can explore ways to enhance the process by modifying environmental factors. □

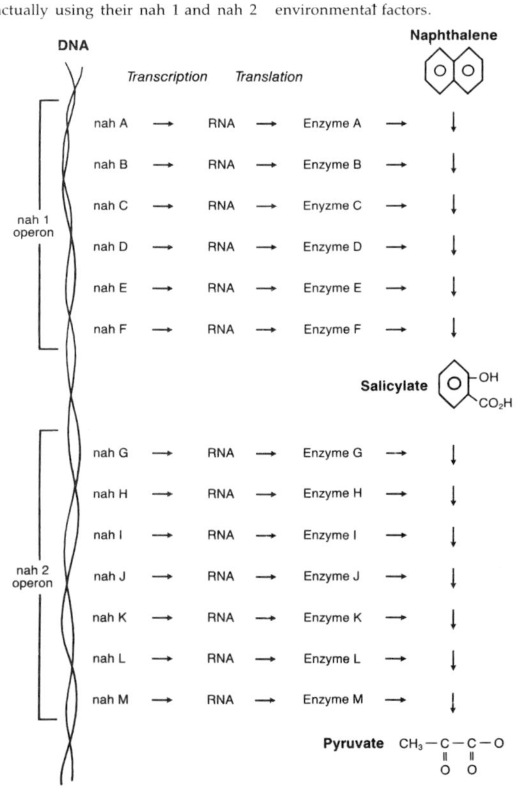

The Wisconsin lakes present a different problem. Although mercury levels in sediments there are low, levels in fish are high enough to prompt warnings against their consumption. EPRI researchers hope to identify the source of this mercury.

DIVERSE REACTIONS

Even while they speak enthusiastically about the potential of genetic ecology

---

Genetic Ecology in Action

Genetic ecologists are exploring the use of environmental modification to enhance the degradation of toxic substances in soil by bacteria. This work includes experiments on small samples of soil, called microcosms, taken from the field site under study to a laboratory. Through such experiments, researchers seek to learn how to enhance the natural duplication and transfer of the genetic material that controls the degradation of a specific substance. The environmental variables being considered include temperature, moisture, nutrient levels, and the presence of oxygen (since most of the bacteria responsible for the breakdown of large organic molecules in soil are aerobic).

One of the simplest ways to increase the genetic potential at a site—that is, the occurrence of the gene groups, called operons, essential to the degradation process—is to increase bacterial growth in general. This approach, the one most commonly used in bioremediation, is based on the process of **clonal expansion,** by which bacteria divide to create new, identical organisms. When the bacteria with the relevant operons divide, the total number of these operons at the site increases. Some of the operons may occur on the main DNA strand (chromosome) of a bacterial cell. Others may occur on much smaller, generally circular DNA rings, called plasmids.

Genetic information is also transferred naturally between bacteria, under the influence of various environmental factors. One process, **transformation,** involves a cell's taking up free DNA segments, perhaps left over from other bacteria that have died. Another process, **conjugation,** occurs when genetic material moves from one bacterium to another while they are physically joined.

There are also processes whereby the number of operon copies within a single cell increases. During **transposition,** an operon is copied and moves onto one or more plasmids or onto the chromosome. During **plasmid multiplication,** plasmids divide independently of the cell, creating copies of themselves in the bacterium.

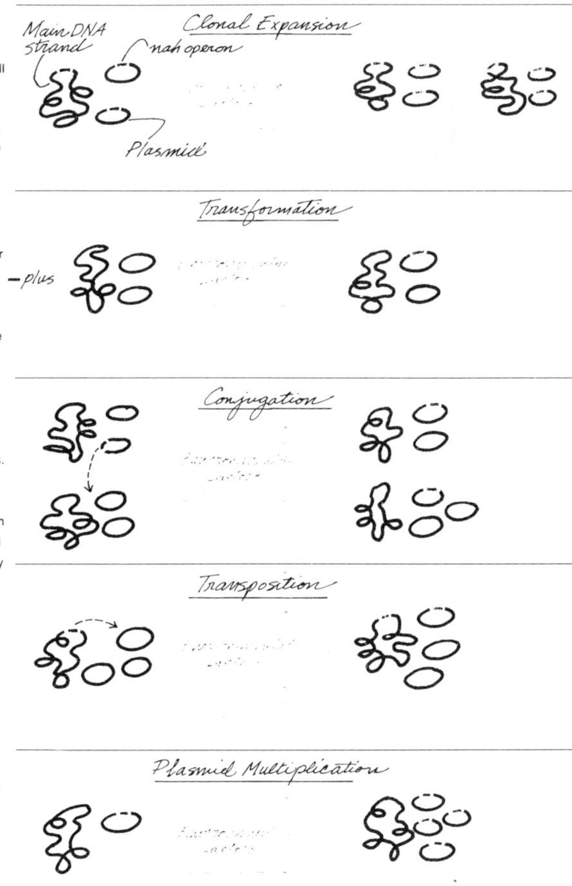

for helping to solve some intractable toxic waste problems, all those involved in the current work stress its exploratory nature and significant cost and regulatory uncertainties. Differences of opinion become particularly evident if one asks when utilities are going to be able to make practical use of the knowledge being developed by genetic ecology research.

"Right now," insists Professor Olson. "Some of our diagnostic techniques could already be used in conjunction with current bioremediation efforts." At sites where efforts are already being made to destroy toxic wastes by stimulating bacterial growth in general, she says, genetic assessment could easily be used to monitor the effectiveness of such work in terms of actual increases in microbial ability to degrade particular toxins.

On the other hand, John Maulbetsch, senior scientist in exploratory research, cautions that "it could be 20 years before genetic ecology becomes an established science. We're dealing with very complex systems and very sophisticated diagnostic tools." Nevertheless, Maulbetsch says, his program is providing funds for the genetic ecology work focusing on mercury because "if they succeed, we'll have not only a new cleanup technique but also a better understanding of how nature deals with contaminants."

Sy Alpert, EPRI senior fellow, is also skeptical. "Practical application of genetic ecology is highly dependent on the results of scientific work that is just getting started," he says. "I believe that the eventual costs and benefits are impossible to ascertain at this time."

Even if the research is successful, two barriers could slow its application, according to John Huckabee, program manager for ecological studies. "Initially," he says, "utilities may be reluctant to be first to try something this new, although the industry representatives we've talked to have been very receptive. Also, I'm concerned that the public may confuse this work with genetic engineering. We've seen tremendous fear that genetically engineered organisms would 'run away' in nature. Those fears shouldn't apply here, since we're not making anything new but rather helping organisms already there to do something better. But I believe the public needs to be educated well about the difference."

Others are more sanguine about public acceptance. "I hope this is not going to be a major issue, since we're not really adding new organisms to a site," says Chin-I Lin, a senior chemist at Pacific Gas and Electric Company. "This work is very, very important for companies, like mine, that have several town gas sites. You can't just hire somebody to clean up these sites. If the genetic ecology work is successful, it could solve many of our problems."

Such expressions of support from the industry are surprisingly frequent, says Ralph Perhac, department director for environmental science at EPRI, who sees at least three reasons for their enthusiasm. "First, I believe they're glad to see EPRI willing to take on high-risk new work that they couldn't afford to do themselves. Second, the utilities are beginning to recognize the shocking cost that would be involved in trying to clean up town gas sites using conventional means. And finally, there is still some time available to do the research. The industry is taking the lead on addressing the problem of town gas sites without waiting to have new regulations held over their heads."

On the issue of when utilities will be able to use some of the new techniques, Perhac takes a middle position: "I believe we're about five years away from first use, if all goes well. The real holdup may not be the complexity of the science, however, but rather a shortage of talented people. I suspect that many molecular geneticists are now working in large commercial ventures aimed at applying genetic engineering to problems of medicine and agriculture. Funding for genetic ecology is very small by comparison, and EPRI is emerging as a leading sponsor of this fledging science."

FURTHER READING

Robert Goldstein and Betty Olson. "Applying Genetic Ecology to Environmental Management," *Environmental Science and Technology,* Vol. 22, No. 4 (1988), pp. 370-372.

This article was written by John Douglas, science writer. Technical background material was provided by Robert Goldstein and Donald Porcella, Environment Division, and Betty Olson, University of California at Irvine. Additional information was provided by Ralph Perhac, Sy Alpert, John Huckabee, John Maulbetsch, and Conrad Kulik.

Subject Index

OFFICERS OF THE PENNSYLVANIA ACADEMY OF SCIENCE